Encyclopedia of Cellulose

Encyclopedia of Cellulose

Edited by **Dwight Cowan**

NY RESEARCH PRESS

New York

Published by NY Research Press,
23 West, 55th Street, Suite 816,
New York, NY 10019, USA
www.nyresearchpress.com

Encyclopedia of Cellulose
Edited by Dwight Cowan

International Standard Book Number: 978-1-63238-127-9 (Hardback)

Printed in China.

Contents

Preface

The world is advancing at a fast pace like never before. Therefore, the need is to keep up with the latest developments. This book was an idea that came to fruition when the specialists in the area realized the need to coordinate together and document essential themes in the subject. That's when I was requested to be the editor. Editing this book has been an honour as it brings together diverse authors researching on different streams of the field. The book collates essential materials contributed by veterans in the area which can be utilized by students and researchers alike.

The aim of this book is to provide state-of-the-art information about cellulose to the readers. Depletion of fossil fuels and environmental awareness are the key reasons behind people turning towards biomaterials as an alternative source of energy, chemicals and materials. Among these, cellulose is emerging as an essential resource in the domain of bio-economy. Cellulose science is evolving at a rapid pace. It is available sufficiently in the nature and has distinct properties which make it ideal for biomass conversion. This book encompasses the basics of cellulose, its properties and intensive knowledge about cellulose science.

Each chapter is a sole-standing publication that reflects each author´s interpretation. Thus, the book displays a multi-facetted picture of our current understanding of application, resources and aspects of the field. I would like to thank the contributors of this book and my family for their endless support.

Editor

Chapter 1

Advanced-Microscopy Techniques for the Characterization of Cellulose Structure and Cellulose-Cellulase Interactions*

Jose M. Moran-Mirabal

Additional information is available at the end of the chapter

1. Introduction

Plants are composed of a variety of different cells with distinct physical characteristics, which are reflected in the different amounts of proteins, lipids, and structural polymers (e.g. polysaccharides and polyphenolic compounds) that are contained within them. One common characteristic to all plant cells is the presence of a rigid cell wall (with thicknesses ranging from 0.1 to 10 μm) that not only conveys to the cell its shape, mechanical strength and pathogen resistance, but also participates in cell-to-cell adhesion, a key interaction that provides plants the robustness that allows them to grow under variable environments.(1,2) The structure of plant cell walls is typically described in terms of three layers: the middle lamella, primary cell wall and secondary cell wall, which have been characterized according to their composition, role during organ growth, and their ability to resist tensile or compressive forces (1,3). Plant cell walls are constructed from a combination of a variety of polysaccharides that can be generally grouped into cellulose, hemicelluloses, and pectic polysaccharides (Figure 1), and whose relative proportions depend on the plant species, specific tissue, and growth stage. Cellulose, the most abundant structural polysaccharide in cell walls (comprising 15-50% of the dry weight of plant biomass),(3) is a linear polymer of glucose units covalently linked through β(1-4) glycosidic bonds. Cellulose glucan chains aggregate through hydrogen bonding and pyranose ring stacking into tightly packed crystalline elementary microfibrils (2-20 nm in diameter), which are the essential scaffold of all plant cell walls.(2,4,5) Hemicelluloses, the second most abundant component of cell walls (10-35% by dry weight),(3) are highly branched hetero-polymers of pentoses and hexoses that can be broadly classified into xylans, mannans, β-glucans and xyloglucans. These

* Fragments of this chapter have been published in: Moran-Mirabal JM. Study of cellulase-cellulose interactions through fluorescence microscopy techniques. Cellulose. 2013; In Press.

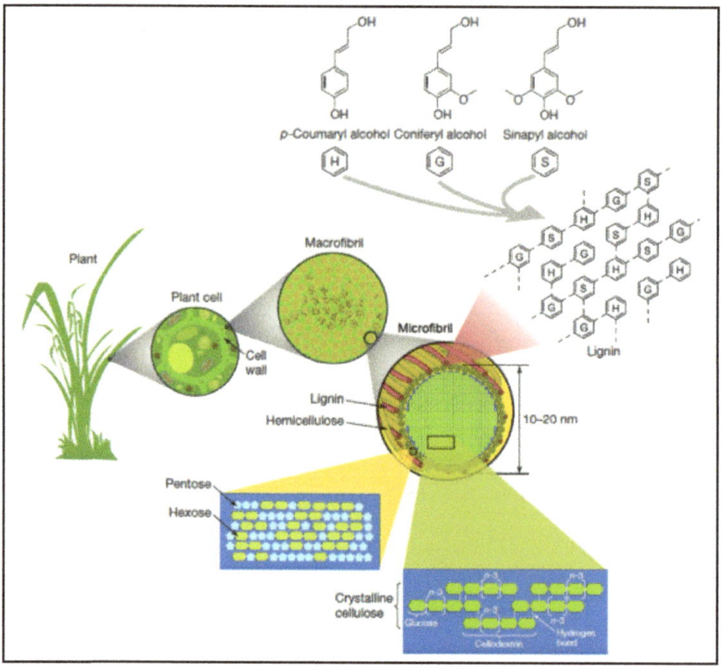

Figure 1. In a simplified representation, the plant cell wall can be visualized as a collection of crystalline cellulose microfibrils sheathed by hemicellulose and lignin. Adapted from (5) with permission from Nature Publishing Group.

branched polysaccharides associate with microfibrils through hydrogen bonds and form, together with pectic polysaccharides a porous matrix (with pores > 10 nm) that sheaths the crystalline cellulose core of the microfibrils.(6–8) In woody biomass the cell wall is further reinforced by lignin (5-30% by dry weight),(3) a three dimensional polymer of phenyl propanoid units that is covalently linked to hemicellulose. The main components that make up lignin polymers are p-coumaryl alcohol, coniferyl alcohol and sinapyl alcohol, the proportional distribution of which varies between plant species and tissue type.(2,5) Lignin acts as nature's glue, forming a protective barrier that limits water and enzyme accessibility to cellulose and gives plants increased resistance to pathogen attack and biomass degradation. The variability in the compositions and relative proportions of structural elements in plant cell walls from different species and tissues results in a high degree of physical and biochemical complexity of plant biomaterials, which creates a challenge for the physicochemical characterization of plant biomass.

Beyond their importance as structural elements in plant morphology and development, the polysaccharides found in the cell wall also represent the major repository of photosynthetically fixed carbon in the biosphere, making them critical for plant, microbial, and animal nutrition and growth, and for the maintenance and balance of Earth's ecosystems.(1) Additionally, structural polysaccharides within the cell wall are widely exploited by humans in the elaboration of a myriad of products such as lumber, food, food additives, paper, fibers, textiles, biocomposites, bioplastics, and biofuels. All these products and their characteristics depend intimately on the composition and structural arrangement of cell walls. In particular, due to their importance in the production of biofuels and bioproducts from plant materials, much recent research has focused on studying the changes effected by thermal, chemical and biochemical processes on the structural arrangement of cellulose, hemicellulose and lignin.

Traditionally, cellulose structure has been studied at high resolution through scanning electron microscopy (7,9–14) or transmission electron microscopy.(9,14–20) In these techniques, cellulose samples must be prepared through drying, coating or staining methods that render the samples compatible with electron microscopy. Despite the exquisite detail that can be gained from these imaging techniques and the wealth of knowledge that has already been obtained through them, the sample preparation procedures can introduce artefacts through modifications to the physical structure of cellulose, such as the collapse of the micro- and nanoporous structure of plant-derived cellulose, the recrystallization and aggregation of elementary microfibrils, or by disrupting the biomolecular interactions occurring at the interface of the insoluble cellulose, such as the dislodgement of proteins weakly adsorbed to the surface of cellulose fibrils. This has resulted in a search for experimental techniques that can reveal the structure of cell wall components with high spatial resolution under environmental conditions that do not modify the structure or the biomolecular interactions. Furthermore, techniques able to visualize the structural and compositional changes that occur due to the biochemical deconstruction of plant material in real time and at temperatures relevant for industrial processing could offer insights into the limitations of biomass conversion processes. This chapter explores a set of advanced-microscopy techniques that allow imaging of cellulosic structures at high resolution under hydrated conditions and how they have been applied to study cellulose and the biomolecular interactions that occur with cell-wall degrading enzymes. The chapter will end with an overview of the possibilities that have opened over the past years due to the development of these novel imaging approaches.

2. Advanced-microscopy techniques

2.1. What can advanced-microscopy offer for the study of cellulose?

The term advanced-microscopy refers in general to the host of techniques that allow us to visualize biological samples with resolution in the submicron to nanometer scale. This level of detail enables the study of biomolecular structures, interactions, and mechanisms of catalysis at the fundamental length scale of proteins, lipids, amino acids, glycans, and other biomolecules. In particular, for the study of cellulosic materials this implies the ability to visualize the arrangement of cellulose from the length scale of the individual glucan chains (sub-nanometer) to that of their aggregates in the form of elementary microfibrils (nanometers) or cell wall macrofibrils (microns). Furthermore, the study of the interactions between cell wall modifying enzymes and cellulosic structures requires the ability to identify individual proteins, their localization, dynamic motions, conformation, and catalytic activities, as well as the changes that they effect on the structural arrangement of cellulose. Ideally, all these measurements would be obtained under experimental conditions that preserve the cellulosic materials in their native state, maintain the full functionality of cell-wall modifying enzymes, and also provide high spatial and temporal resolution to observe any dynamic restructuring and biomolecular interactions in real time. It is clear that a single technique cannot provide all this information simultaneously with the optimal temporal and spatial resolution, and that complementary techniques are needed to achieve a full understanding of the structural and biochemical changes that occur in cellulosic materials as they undergo physical, chemical or biochemical processing.

A number of techniques, including certain types of electron microscopy (EM), scanning probe microscopy (SPM), and optical microscopy can be considered advanced-microscopy techniques because they can yield images with spatial resolution ranging from the micrometer to the sub-nanometer scale. Each one of these types of microscopy has its advantages and limitations in their application to the study of cellulose: EM techniques can yield exquisite detail even at the sub-nanometer scale, but in most cases are limited to dry samples that must ideally be rendered conductive to make them compatible with EM imaging; SPM techniques have the ability to image immobile substrates with sub-nanometer resolution, but must come into contact or be in very close proximity with the sample, cannot effectively image large surface areas, and cannot track multiple biomolecular species simultaneously; and optical microscopy can track multiple biomolecular species with high temporal resolution in large surface areas, but is limited in spatial resolution by the diffraction of light through the optical microscope, as well as by the refraction and scattering of light in thick biological samples. Thus, the choice of experimental advanced-microscopy techniques to image a cellulosic sample depends strongly on the information sought, the complexity of the system, the environmental requirements, and the spatial and temporal resolution desired. In this chapter we focus on a subset of advanced-microscopy techniques which yield images with micron to nanometer scale spatial resolution over periods ranging from seconds to hours, while allowing experimental conditions that provide appropriate temperatures for the catalytic activity of enzymes and keep the samples hydrated in solutions with biologically relevant ionic strength and pH. These capabilities make them ideally suited for the study of cellulose and the interaction between cell wall degrading enzymes and cellulosic substrates.

The discussion throughout the chapter will center on recently developed SPM and fluorescence microscopy techniques, which have been applied to the visualization of the structural arrangement of cellulose and the interactions that occur at the interface between insoluble cellulosic materials and enzymes, with sub-micron to nanometer scale resolution. These techniques have made it possible to study the structural arrangement of cellulose microfibrils and elementary glucan chains, including their intermolecular spacing, cellobiose unit repeat and individual glucose unit distances, both for dry and hydrated cellulosic materials. Furthermore, the use of fast-acquisition SPM and fluorescence techniques has enabled the observation of the binding and molecular motion of individual or multiple cell-wall modifying enzymes and their corresponding binding and catalytic modules even at the single molecule level. These advanced-microscopy techniques have also been recently applied in the study of the depolymerization of cellulose by individual cell wall modifying enzymes and by commercially available cellulolytic enzyme cocktails. The use of advanced-microscopy techniques for the study of cellulose, especially in the complex environment of the cell wall, is still in its infancy and many opportunities exist for the fundamental study of cell wall structure, enzyme-cellulose interactions, and for the study of the changes that occur in cellulose structure and biochemical composition as cell walls are deconstructed through different treatments used in industrial settings. As will be seen in the discussion to follow, the development of new SPM and fluorescence microscopy approaches has made it possible to study cellulose with high spatial and temporal resolution under conditions that are biologically relevant. These studies will surely lead to an improved understanding of the organization of the cell wall and of the way cellulolytic enzymes operate. This in turn can steer the development of deconstruction technologies that more efficiently convert plant biomass into useful raw materials for the manufacturing of renewable and sustainable products.

2.2. Scanning Probe Microscopy

Scanning Probe Microscopy techniques (SPM) measure the interaction between the sample and a small sample probe, which is brought into contact or very close proximity to it. An image of the sample under investigation is then constructed by scanning the tip and collecting pixel by pixel data, which can then be displayed as a three dimensional reconstruction. SPM techniques allow the direct characterization of the surface of biological specimens with very high spatial resolution and with very little sample preparation.(4) Furthermore, SPM techniques allow imaging biological samples under dry or aqueous conditions and with suitable chambers that can provide temperature-controlled environments. Thus, they are ideal for the structural and biochemical characterization of cellulosic materials and for the study of the interactions between cell-wall degrading enzymes and the surface of insoluble cellulose substrates.

Atomic Force Microscopy (AFM) is a high-resolution SPM technique, which has become one of the most widely utilized tools in real-time imaging of biological samples due to its ability to record surface topography and properties at the nanoscale. The basis of AFM operation is quite simple and lies in the interaction between a sharp tip located at the end of a microfabricated cantilever and the sample surface.(21) The tip, with a radius ranging from of a few nanometers (silicon-based tips) to a few angstroms (carbon nanotube tips), senses a distance-dependent potential that is the result of a multitude of interactions with the surface, including van der Waals, chemical, electrostatic, capillary, and magnetic forces, among others.(22) As the tip is brought into close proximity with the surface, these intermolecular forces deflect the cantilever. The deflection is measured by tracking the displacement of a laser spot that is reflected off the back of the cantilever and imaged onto a photodiode array (Figure 2). The tip can be raster-scanned across the sample using a piezoelectric stage, with the recorded cantilever deflection as a function of position yielding a surface map of the biological sample. The resolution of the surface map is limited by the

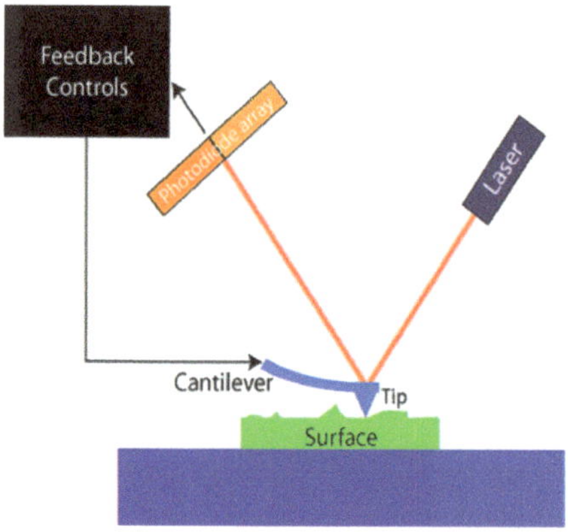

Figure 2. Schematic depiction of a typical atomic force microscope setup. As the tip is brought into close proximity of the surface the cantilever is deflected by attractive or repulsive forces. The displacement is monitored using a laser spot bounced off the back of the cantilever and imaged on a photodiode array. Feedback controls maintain a constant force on the tip.

size of the AFM tip used, and under optimal conditions can reach 0.1 nm in the vertical and 0.5 nm in the lateral directions.(21) In a typical AFM setup, the tip-to-substrate distance is controlled by a feedback mechanism, which adjusts the tip position to maintain a constant force between the tip and the sample. For imaging of soft biological samples, the force applied by the cantilever is usually limited to the range between 50 and 100 pN to minimize sample deformation and avoid the disruption of biomolecular interactions (21).

AFM scanning can be performed in two modes: contact or tapping. In static or contact mode, AFM imaging is performed in contact with the sample, where the overall force is repulsive. The main drawback of contact mode for imaging biological samples is that the forces applied during tip scanning can sweep away weakly bound objects or deform soft samples. On the other hand, tapping mode is achieved by driving the cantilever into oscillation at its resonant frequency and allowing limited interaction between sample and tip. The amplitude, phase and frequency of the oscillations are changed by the tip-sample interactions, leading to a description of the surface characteristics. Nowadays, amplitude modulation is the most widely used non-contact mode. In this mode, changes in the oscillation amplitude and phase provide the feedback for imaging.(23) Amplitude variations allow the mapping of the surface topography, while changes in phase allow the discrimination between different types of materials. Surface topography, rigidity, viscosity, as well as adhesion forces can thus be studied through non-contact AFM.

Until recently, one of the main limitations to AFM was that the acquisition of a single image of a 1-micron square area would take several seconds, which precluded the use of AFM to study molecular motions that occur within the millisecond time regime. Efforts towards realizing high speed AFM were pioneered in the 1990s and by 2001 a complete system for the fast acquisition of high quality images was reported. This system introduced small cantilevers with high resonant frequencies (~600 kHz in water) and a suitable optical detection system, which allowed acquisition times of 80 ms per frame. Later improvements in the z-scanner damping system and the feedback loop increased sample stability and minimized the tip-sample interactions, enabling acquisition times as fast as 30 ms per frame.(24) Nowadays, the most advanced tapping-mode, high-speed AFMs are being utilized to observe the motions of biomolecular motors such as kinesin, myosin V, and dynein, (25,26) as well as the motions of processive cell wall degrading enzymes.(27,28) The ability to monitor surface topography changes along with biomolecular interactions in real time under aqueous environments and under controlled temperature, pH and ionic strength conditions make AFM a very useful tool for the study of cellulose, its structural properties, and the interactions that occur between insoluble cellulosic substrates and cell wall degrading enzymes.

2.3. High-resolution fluorescence microscopy

Fluorescence microscopy has become one of the most utilized techniques for the study of biological systems for its high sensitivity, selectivity and the ability to monitor multiple species under biologically relevant conditions. Fluorescence arises from the spontaneous emission of a photon from a molecule (fluorophore) in which the absorption of an incident photon causes an electron to transition to an excited singlet state and this electron subsequently transitions to the ground state via the emission of a second, lower energy, photon.(29) Fluorophores that emit in the visible wavelength range are usually molecules with multiple aromatic rings, such as the dyes from the cyanine, fluorescein and rhodamine families or fluorescent proteins like those derived from the green, yellow, red, and cyan fluorescent proteins. Commercially available organic fluorophores are further modified with reactive groups that allow the covalent attachment of the organic dyes to thiol, amine, alkyne, and hydroxyl groups, among others. Conversely, fluorescent proteins can be expressed within the target system as proteins fused to the targeted protein or peptides, which allows fluorescent tagging *in vivo*.(30–32) The ability to easily tag proteins, lipids, nucleic acids or glycans with fluorescent moieties allows the selective observation of these biomolecules over the background of the biological system of interest. Furthermore, through the use of high performance optics and very sensitive detectors it is possible to observe the fluorescence emitted from the tagged biomolecules, enabling the study of biomolecular localization, transport, binding, catalysis, and other biomolecular interactions even down to the level of single molecules. A number of microscopy techniques have been developed to exploit the advantages of fluorescence emission to study these biomolecular interactions, including confocal fluorescence microscopy, total internal reflection fluorescence microscopy, fluorescence correlation spectroscopy, Förster resonance energy transfer, single molecule tracking, and super-resolution fluorescence microscopy, among others. Through these techniques researchers aim at maximizing the fluorescence photons collected while minimizing the background noise to enhance the sensitivity and the spatio-temporal resolution of the measurements.

The implementation of fluorescence microscopy can be done in different configurations, all of which aim at achieving the highest fluorescence signal to background noise possible. All

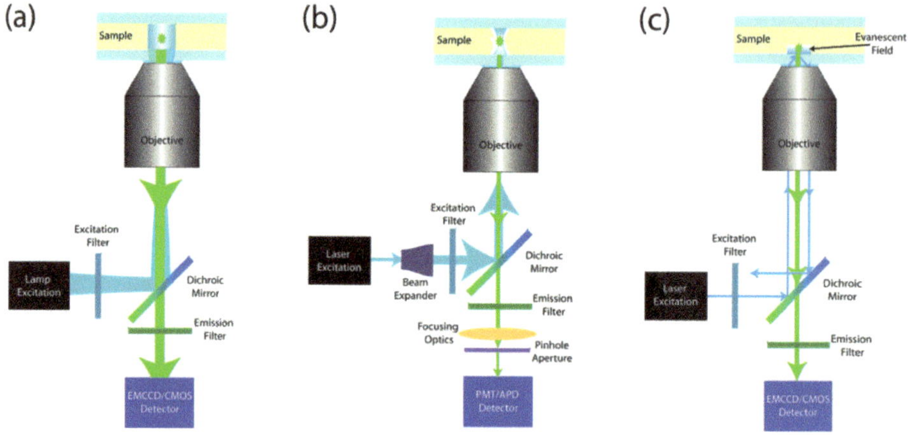

Figure 3. All configurations for fluorescence microscopy imaging share common basic elements: excitation source, excitation and collection optics, and detectors. (a) Epifluorescence microscopy utilizes a broad lamp excitation source for widefield illumination and a high-sensitivity camera for collection. (b) Scanned Confocal and Stimulated Emission-Depletion Super-Resolution microscopy utilize a laser excitation source for point illumination and PMT or APD for collection. (c) Total Internal Reflection and photo-switching based Super-Resolution microscopy utilize a laser excitation source for widefield illumination and a high-sensitivity camera for collection. Adapted from Reference (101).

these experimental configurations share some basic constitutive elements: an excitation source, which can be mono chromatic or wide spectrum and coherent or incoherent; optics for excitation and collection, including excitation and emission filters, dichroic mirrors, and polarizing optics, among others; and one or multiple highly sensitive detectors, such as avalanche photodiodes, photomultiplier tubes, and EMCCD or CMOS cameras (Figure 3). The available microscope configurations can be broadly divided into two main groups: those that detect the emitted fluorescence in wide-field mode (e.g. epifluorescence, TIRF, and photo-switching type Super Resolution microscopy) and those use point detection (e.g. laser scanning confocal microscopy and Stimulated Emission-Depletion Super Resolution microscopy). The selection of the configuration to be used depends on the observables to be measured, the sample and sample requirements, and the spatial and temporal resolution desired. As mentioned before, no single technique can yield all the measurements simultaneously with the maximum spatial and temporal resolution. Thus the selection of the appropriate fluorescence technique is crucial to the successful, reliable and reproducible measurement of sample properties and biomolecular interactions. Below, a few configurations and experimental techniques that have been applied to the study of cellulose or that hold great promise for it are discussed in detail.

Epifluorescence is perhaps the simplest implementation of fluorescence microscopy, but also the most versatile and widely used. The configuration for epifluorescence (Figure 3-a) consists of a broad spectrum light source, such as a mercury arc lap or a metal halide lamp, which emits light in the visible spectrum with enough intensity to excite fluorescence from the specimen; a set of excitation and emission optics, which include an excitation filter that selects the band of wavelengths appropriate for excitation of the fluorophore, a dichroic or dichromatic mirror that selectively transmits or reflects wavelengths of incident light, allowing the separation of excitation and emission photons for the orthogonal observation of fluorescence, and an emission filter with a transmission band for wavelengths of light that separates the reflected and scattered light from fluorescence emitted from the specimen; and a sensitive imaging device, such as an electron multiplication charge-coupled device (EMCCD) or a complementary metal-oxide semiconductor (CMOS) camera, which collects the photons at the image plane of the microscope and translates them into an electronic signal that can be recorded in the form of a picture where the pixel intensity is a reflection of the fluorescence emitted from the sample. The configuration for epifluorescence microscopy relies on the excitation and the collection of the emitted light through a high quality objective with magnifications ranging from 2 to 100×, and numerical apertures as high as 1.49, which allow the efficient collection of photons without chromatic or spherical aberrations. Through this configuration it is possible to image samples containing multiple fluorophores with temporal resolution ranging from milliseconds (EMCCD) to microseconds (CMOS). The use of appropriate calibration standards also allows the use of epifluorescence for the quantitative characterization of concentrations of biomolecules within biological samples. Furthermore, different microscope configurations allow the use of environmental chambers to provide capabilities for temperature control, perfusion of fluids and gases, and integration of electrical measurements, among others. The main limitation to epifluorescence is its poor spatial resolution, which is limited by the diffraction and scattering of light as it travels from the sample to the collection objective and through the optical components of the microscope. Rayleigh's criterion defines the highest achievable resolution in the transverse sample plane:

$$r_{x-y} = \frac{0.61\lambda}{NA} \tag{1}$$

where λ is the wavelength of the emitted light and NA is the numerical aperture of the objective, whereas the axial resolution is typically two to three orders of magnitude larger. As an example, a fluorescent sample emitting at 525 nm, observed with a 60x/1.4 NA objective, would be observed with a maximum resolution in the transverse plane of ~230 nm. Because of its versatility, quantitative epifluorescence has been widely used to visualize a wide range of biological systems. The development of more sensitive cameras, more intense illumination sources and better optics will undoubtedly keep quantitative epifluorescence microscopy as a versatile tool for the study of biological structure and the characterization of biomolecular interactions.

Confocal microscopy improves on the axial resolution of the fluorescence microscope and allows imaging three-dimensional structures through optical sectioning. In the configuration for laser scanning confocal microscopy (Figure 3-b) a collimated laser is used as an excitation source to overfill the back aperture of a high-NA objective, which focuses light into a diffraction-limited spot at the focal plane. The same objective collects the fluorescence emitted from the sample, which is filtered through the dichroic mirror and emission filter, and focuses it onto a small pinhole aperture. This aperture is placed at the conjugate focal plane of the objective and serves to reject out-of-focus light. Typically, the confocal aperture diameter is selected such that it is slightly larger than the transverse dimension of the illumination volume (Eq. 1) multiplied by the objective magnification. (33–35) The spatial filtering provided by the confocal aperture improves the axial resolution of the fluorescence microscope by collecting emitted fluorescence light only from the focal plane. The size of the observation volume depends on the wavelength of the emitted light (λ), the objective's *NA* and the index of refraction of the sample medium (n). Equation 1 defines the transverse dimensions of the observation volume, while the axial dimension is given by

$$r_z = \frac{2n\lambda}{NA^2} \tag{2}$$

From these, the confocal observation volume can be calculated as the volume of an ellipsoid. Following the same example as above, the axial resolution for the observation of a sample in aqueous medium (n = 1.33) with a confocal setup would be ~710 nm, which yields an observation volume of approximately 0.1 fL. After the light passes through the confocal aperture, it is collected by a high-sensitivity photon detector, such as an avalanche photodiode (APD) or a photomultiplier tube (PMT). These detectors convert incident photons into electronic signals that can be recorded. Because the signals recorded from the illuminated focal volume represent a single pixel of information, two-dimensional images can be reconstructed by scanning the illumination across the sample in the x and y planes. Scanning can also be extended to the axial dimension by changing the focal plane, allowing the collection of multiple optical z-sections (z-stacking). This enables the reconstruction of the fluorescence profile arising from complex three-dimensional structures with improved resolution and signal to noise ratio. The main drawback of scanning confocal microscopy is that the collection of a 2D image can take several seconds, while a 3D stack can take up to several minutes, depending on the number of optical sections acquired. Thus, dynamic events that are in the sub-second time regime cannot be observed through confocal microscopy. Yet, the improved spatial resolution and the ability to collect fluorescence in multiple channels simultaneously have made confocal microscopy a widely used technique in the biological sciences, and a good candidate for the study of cellulosic samples with complex structure.

Total Internal Reflection Fluorescence Microscopy (TIRFM) is another configuration, which is well suited to study molecular phenomena because it can selectively excite fluorescence from a thin sample layer adjacent to a glass surface. In TIRFM, the incidence of the excitation laser beam upon the interface between a high and a low index of refraction material (typically a glass-water interface) at an angle beyond the critical angle results in TIR of the incident light. Although under TIR no light propagates through the interface, an exponentially decaying (evanescent) electromagnetic field penetrates the sample. This evanescent field is capable of exciting fluorescence emission from fluorophores located near the interface. The penetrating depth of this evanescent field depends on the incident angle and is usually in the 50-250 nm range. Due to the nature of the evanescent field, the excitation volume is large in the transverse dimensions but highly confined in the axial dimension, which greatly reduces background fluorescence from out-of-focus planes and results in images with very high signal-to-noise ratio. TIRFM imaging can be done in two common configurations: prism-based (36) and objective-based.(37–39) In prism-based TIRFM, the laser beam is coupled into the sample through a trapezoidal or cylindrical prism using an index-matching medium (e.g. glycerol or oil). TIR is achieved by coupling the illumination at angles beyond the critical angle and the microscope objective is used to collect the photons emitted by the sample, which are filtered by the dichroic mirror and the emission filter. Thus, the fluorescent sample is imaged from the other side of the specimen using a high-sensitivity camera, which captures the emitted fluorescence. An advantage of this configuration is the low cost of the components, although the incidence angles are restricted to a few discrete values defined by the prism characteristics. Objective-based TIRFM, on the other hand, couples the laser beam via the same objective used for imaging (Figure 3-c). Through the use of a moving lens, the position of the laser beam can be translated axially from the center to the edge of the back aperture of the objective, causing the incident light to hit the glass/sample interface at an angle. This has become the most popular configuration because it allows a continuous range of incidence angles, with a maximum angle determined by the objective's *NA*. By using a high *NA* objective, supercritical illumination angles can be achieved that are suitable for TIRFM imaging of dense biological samples such as cells or cellulose. The fluorescence emitted from the sample can then be imaged through the same objective onto a high-sensitivity camera (e.g. EMCCD or CMOS). Key advantages of through-the-objective TIRF are that the use of high-magnification, high-*NA* objectives allows improved lateral resolution, that the photons have a shorter path to the collection optics, that the top of the sample can be open and easily accessible, and that switching between epifluorescence and TIRF can be achieved easily by changing the position of the incident beam. In addition, this configuration allows illumination with multiple laser wavelengths either simultaneously or in rapid succession for imaging of multiple fluorophores. Several excellent reviews of the implementation of TIRF and the comparison between the experimental configurations have been published.(36–41)

A number of fluorescence microscopy techniques have been developed over the past decades to take full advantage of the capabilities of the microscope configurations described above in order to measure specific observables with the highest spatial and temporal resolution. These techniques can be implemented using different microscope configurations, depending on the sample characteristics and the sensitivity desired. One such technique is fluorescence recovery after photobleaching (FRAP), which can be implemented in epifluorescence, confocal, or TIRFM setups. First described by Axelrod in 1976,(42) FRAP is a method for measuring the two-dimensional lateral mobility of fluorescent particles or molecules. In this method, a small spot of a fluorescent sample is permanently photobleached through exposure to a high intensity illumination source, such as an arc lamp or a laser beam. Subsequently, imaging of the sample with the same illumination source under attenuated intensity monitors the recovery of the fluorescence due to molecular transport. With this technique it is possible to identify the type of transport process present in the sample (e.g. diffusion, restricted diffusion, flow), measure the lateral diffusion coefficient of the fluorescent species, and identify the fraction of the fluorescent molecules that are mobile within the sample. More recently, models have also been developed to fit FRAP data from systems where binding occurs at the same time scale as the transport phenomenon of interest.(43,44) This has enabled the study of complex biological environments, such as the cytoplasm, where the molecular association to immobile structures hinders the diffusion of the molecular species of interest.(45–49) As will be discussed below, FRAP has already been used to measure the mobility of cellulases on cellulosic substrates.

Förster Resonance Energy Transfer (FRET) is a molecular process that decreases the photon output from fluorophores, a fact that has been tuned into a useful technique to measure molecular distances arising from the close association of two biomolecules or from the mobility of fragments within a single biomolecule. FRET occurs when the emission spectrum of a fluorophore, the donor, overlaps with the absorption spectrum of a second molecule, the acceptor.(50) The basis of FRET lies in the non-radiative transfer of energy from the donor to the acceptor vial long-range dipole-dipole coupling.(51) The energy transfer between the donor and the acceptor depends both on the extent of the spectral overlap and, due to dipole coupling, the distance between them. The FRET efficiency of a donor-acceptor pair is defined as:

$$E = \frac{R_0^6}{R_0^6 + r^6} \tag{3}$$

where r is the distance between donor and acceptor pair and R_0 is the characteristic Förster radius (FRET efficiency of 50%) for the pair. In the practical implementation of FRET where the donor and acceptor are both fluorescent molecules, the sample is illuminated with an excitation source matched to the donor's absorption band. If the FRET pair is in close proximity, energy transfer will occur between the donor and the acceptor, resulting in the excitation and subsequent emission of fluorescence from the acceptor fluorophore. The fluorescence emitted is recorded in both the donor and acceptor channels and the FRET efficiency (E) can be calculated. Because E depends strongly on the separation between donor and acceptor, the ratiometric measurement of fluorescence emission from the pair can be effectively used as a spectroscopic molecular ruler.(50) Typical values of R_0 for fluorescent FRET pairs lie in the 20-100 Å range and optimal sensitivity of donor quenching is achieved at distances where $0.7R_0 < r < 1.5R_0$.(52) Experimentally, FRET can be done in two modalities: through widefield imaging of molecules immobilized on a solid surface or through point probing of molecules in solution. FRET has been used in a wide range of applications, even at the single molecule level, and a number of excellent reviews exist on the theory and applications of this technique.(51,53–56)

Over the past decades, the development of highly sensitive cameras and bright, stable fluorescent dyes has allowed life scientists to image single molecules (SMs) through fluorescence microscopy. The ability to visualize and track individual fluorescent molecules over time permits a more direct characterization of their dynamic behaviour, including molecular association, molecular transport, conformational changes, and catalytic activities.

Furthermore, the study of SMs can evidence heterogeneous behaviour, or help identify rare molecular subpopulations that would otherwise be averaged and masked by ensemble techniques. The downside of SM experimentation is that in order to get statistically significant measurements, large data sets are required to construct histograms that yield discrete values of the experimental observables. Yet, the information obtained through SM experimentation has proven very valuable and the use of SM techniques has increased exponentially over the last decade. In fluorescence microscopy, SM experimentation is performed in two modes: point detection and widefield imaging. In point detection mode, a laser illumination beam is focused into a diffraction-limited excitation volume, which limits the background fluorescence from out-of-focus planes and gives a very high signal-to-noise ratio. As fluorescent molecules are transported through the excitation volume, they emit a fluorescent signature that is collected and recorded by a highly sensitive detector, such as a PMT or an APD. This fluorescence signature, in the form of bursts of recorded photons, can be analyzed through the photon counting histogram (PCH)(57–60) or fluorescence correlation/crosscorrelation spectroscopy (FCS/FCCS) techniques.(61–67) In this way, the collection of fluorescence photon bursts allows the study of dynamic molecular behaviour, such as diffusion, transport, association and conformational changes, with high temporal resolution ranging from the nanosecond to the second time scale. The strength of point detection SM experimentation is the high temporal resolution afforded by the detectors used, with the caveats of having to measure one molecule at a time and that molecular trajectories can only be followed within the focused excitation volume. On the other hand, widefield SM techniques use configurations such as those for epifluorescence or TIRF microscopy to illuminate a sample containing a sparse collection of fluorescent molecules, and SM images are collected using high sensitivity EMCCD or CMOS cameras. In this modality, the field of view can cover areas of hundreds to thousands of square microns where hundreds of molecules can be visualized at a time. This allows the simultaneous collection of SM data from many individual molecules, reducing the number of experiments required to obtain the required measurements for statistical significance. The main limitations of widefield SM techniques are that the molecules can only be followed for a period of time determined by photobleaching and that the temporal resolution is limited by time it takes the camera to acquire the image (typically in the millisecond range). In the analysis of widefield SM images, each molecule within the image is identified and localized through algorithms that fit the SM emission profile to a theoretical point spread function, such as a two-dimensional Gaussian or a Bessel function.(68–70) This results in the ability to pinpoint the location of individual molecules with precisions that are well below the diffraction limit. Experimentally, the number of photons collected for each molecule and the background determine the accuracy of localization(71,72) and values as low as one nanometer have been reported in the literature.(73) Furthermore, the collection of sequential images and their analysis with SM tracking (SMT) algorithms allows tracking the active or passive transport of individual molecules in real time. Thus, through widefield SM techniques it is not only possible to pinpoint the location of individual molecules and their association kinetics, but it is also possible to reconstruct their molecular trajectories. The simultaneous acquisition of SM images from two spectrally distinct channels also permits the extension of widefield SM experimentation to other techniques, such as SM colocalization and SMFRET. SM fluorescence techniques have gained widespread acceptance and are being used for the study of numerous biological systems. The interested reader is referred to the wealth of SM fluorescence reviews for additional information on the different implementations, applications, and the limitations for these advanced-microscopy techniques.(74–86)

Another advanced-microscopy technique that is suitable for the study of cellulose structure and its temporal evolution as a result of physical, chemical or biochemical treatments is super-resolution fluorescence microscopy (SRFM). The term SRFM has been applied to a host of techniques which all aim at obtaining sub-diffraction resolution images from macromolecular assemblies. These assemblies typically contain fluorescently labeled molecules that can be selectively excited to emit photons, and their emission profile can be localized within the field of view with high precision in a manner analogous to the SM localization and tracking approaches described above. SRFM includes methodologies that take advantage of nonlinear optical effects to condition or reduce the size of the excitation point spread function (Stimulated Emission-Depletion or STED, Saturated Structured Illumination microscopy, SSIM)(87–91), or that reconstruct super-resolution images from the emission profiles of individual fluorescent molecules that are selectively (Photo-Activated Localization Microscopy or PALM, and related techniques)(92,93) or randomly (Stochastic Optical Reconstruction Microscopy, STORM, blink microscopy or dSTORM)(94–98) switched to emit fluorescence. While all these approaches obtain similar resolution (< 50 nm), they have different attributes and equipment requirements that make them more or less amenable to implementation, and a number of reviews exist that compare them.(88,99,100) For example, photoswitching techniques allow a facile implementation, without the need for expensive instrumentation, and shift the focus of the SRFM technique development from hardware to the dyes utilized for imaging. Recently, Tinnefeld and collaborators have reported the ability to perform SRFM with conventional fluorescent dyes, provided that suitable imaging wavelength intensities and aqueous medium can be utilized.(95–97) This imaging medium can be utilized with protein or nucleic acid systems, and could be well suited for imaging of cellulose-protein interactions.(101) Despite the difference in operational mechanisms, the principle behind SRFM techniques (Figure 4) is that in a sample that contains a large number of fluorescent molecules, under appropriate experimental conditions, only a few of them are conditioned to emit fluorescence at any given time. The fact that only a sparse collection of molecules is imaged enables their localization with an accuracy that is only limited by the number of photons collected for each emitter. Thus, by continuously illuminating the specimen, and imaging sparse collections of fluorescent molecules in each frame, their position can be determined with great precision, and a final image of the macromolecular structure can be reconstructed. Using these SRFM imaging approaches, the structural arrangement of cellulose microfibrils within a host of substrates could be reconstructed from the aggregated information of fluorescent cellulose-binding proteins. This information could be obtained in real time and with a spatial resolution that can reach the fundamental length scale of proteins (5-20 nm) and elementary cellulose microfibrils (widths of 5-30 nm).

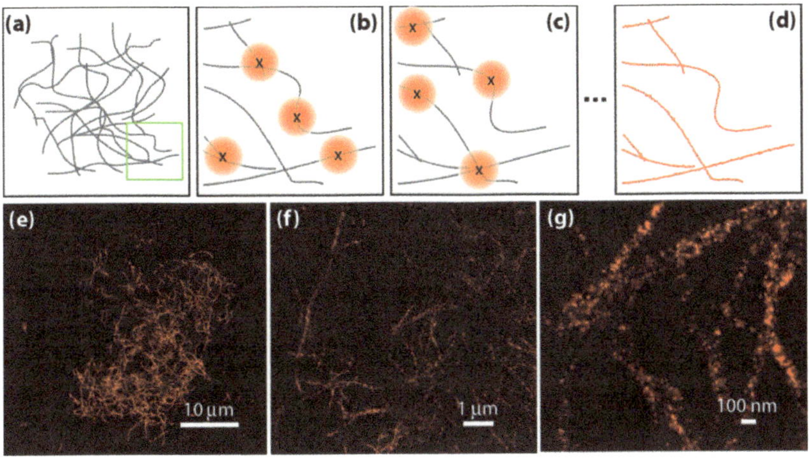

Figure 4. Principle behind photoswitching SRFM. A macromolecular structure (a) is labeled with fluorescent molecules, which are randomly switched between emissive and dark states, causing sparse collections of the fluorescent molecules to be imaged in each frame (b-c). The molecules can be localized with high accuracy, and a super-resolution image (d) is reconstructed using the locations of all molecules imaged. (e-g) Sample SRFM reconstruction images showing sub-diffraction sized fibrils. Adapted from Reference (101)

3. Experimental implementations of advanced-microscopy techniques

3.1. Elucidation of cellulose structure and biochemical composition

Cellulose, in the form of elementary microfibrils, constitutes the core structural component of plant cell walls. As such, the fundamental crystalline packing of glucan chains and microfibril structure determine many of the biological, chemical and physical characteristics of plant derived cellulosic materials. Despite the abundance of plant biomass and its importance to plant biology and the food, pulp and paper, textile, bioenergy and green materials industries, the structure of cellulose in plant cell walls is not known in detail.(102) This stems primarily from the reduced availability of techniques that can access the nanoscale structure of cellulose *in vitro* and *in vivo* under hydrated conditions, which preserve the natural arrangement of the glucan chains and their aggregated fibril forms. Over the past couple of decades, there has been increased interest in the use of advanced-microscopy techniques to study the structure of cellulosic materials both in their natural or treated forms. In particular, AFM has been used extensively to gain insight into both the arrangement of microfibrils within the plant cell wall environment as well as into the crystalline packing of the individual glucan chains that compose these microfibrils.

As early as 1996 Kirby and collaborators used AFM to image the hydrated cell wall structure of apple, carrot, water chestnut and potato parenchyma.(103) In this seminal study, it was demonstrated for the first time that AFM could be used to effectively image and study the structural arrangement of cellulose within partially hydrated plant cell walls (Figure 5).

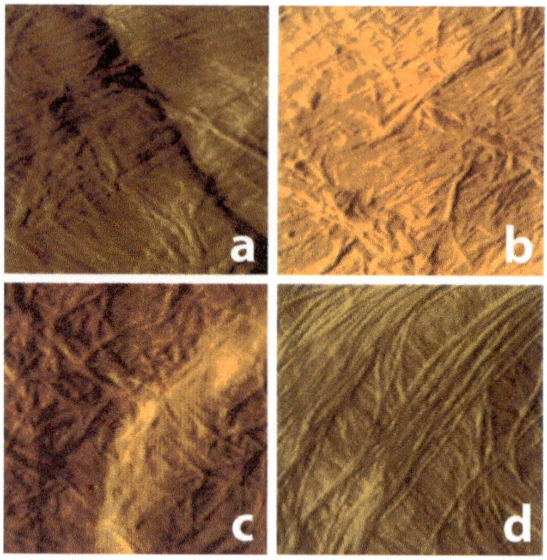

Figure 5. AFM error signal images of (a) apple, (b) carrot, (c) potato, and (d) water chestnut parenchyma cell walls show the polylaminate arrangement of cellulose microfibrils. All images are approximately 1x1 μm. Reproduced form reference (103) with permission.

The topographical and error AFM images obtained from the cell wall of different plant species showed layers of aligned fibrous structures, which on the basis of size and shape were taken to be cellulose microfibrils. These microfibrils could be observed both in freshly prepared samples and in samples that had been stored frozen. The observation of the laminate arrangement of microfibrils with different orientations confirmed the previously held assumption that cell walls are polylaminate structures. It also marked the beginning of the use of AFM as a technique to study the structure of cellulosic materials. Just a year later Pesacreta *et al.* used tapping-mode AFM to investigate the structure of cotton fibre cell walls both in air and under water.(104) They reported widths for dry primary cell wall microfibrils in the 25-40 nm range, while they were able to observe striations within the cotton fibre with widths in the 5-7 nm range under completely hydrated conditions. These striations were considered to be the first observations of elementary cellulose fibrils. Later, Thimm et al. conducted AFM imaging of live celery parenchyma cell walls and studied the effect of dehydration on microfibril size.(105) Under complete hydration, microfibrils as small as 6-8 nm were observed. However, they observed that dehydration increased the observed microfibril size, and postulated an increase in fibre diameter through the coalescence of smaller fibrils and through the coating of the fibres with other polysaccharide components. A study of partially hydrated cell walls from the angiosperms Arabidopsis thaliana (thale cress) and Allium cepa (onion) by Davies and Harris also showed that intact microfibrils had widths of 5.8 and 4.4 nm respectively.(106) Furthermore, they showed that treatments to remove pectic polysaccharides reduced the size of the Arabidopsis microfibrils to 3.2 nm. From these measurements they concluded that each of the observed intact microfibrils consisted of a single crystalline core or crystallite. Recent AFM experimentation by Ding *et al.* with corn parenchyma have given much higher resolution images of the cell wall structure.(107) In this study, in addition to elementary microfibrils with 3x5 nm rectangular cross-sections, they identified cell wall depressions or "pits" with diameters in the micron and depths in hundreds of nanometer range, and macrofibrils that consist of bundles of smaller fibrils. From these observations, they have proposed a model for cellulose synthesis where in growing tissue elementary glucan chains synthesized from CesA protein rosettes aggregate to form macrofibrils, which at later stages disperse into the smaller bundles of elementary chains that constitute the core cellulose microfibrils coated by pectic polysaccharides. Furthermore, form their observations of the cross-sectional dimensions of cellulose microfibrils and computational models they proposed a model where each microfibril consists of 36 glucan chains that assemble through hydrogen bonding and van der Waals forces into the crystalline core and sub-crystalline shell of elementary microfibrils.(107) This model conflicts with recently obtained wide angle X-ray scattering and small angle neutron scattering data from spruce wood cell walls that suggests that elementary microfibrils consist of 24 glucan chains.(102) While there is still controversy about the overall dimensions of the structural components of the cell wall, it is without a doubt that AFM has yielded key insights and will continue to provide useful information of the arrangement of cellulose within plant cell walls. In particular, future implementations where AFM tips are functionalized to detect specific polysaccharides or proteins could provide a richer nanoscale visualization of the cell wall, its biochemical attributes and its biological function.

AFM imaging has also been used to study the fine structure of individual glucan chains and their crystalline arrangement within elementary microfibrils. These studies have been primarily performed with cellulose crystallites extracted from *Valonia ventricosa*, a green alga that is accepted as the standard source of cellulose I because its large crystallites consist of highly ordered 1200-1400 individual glucan chains.(108) The pioneering work of Hanley and collaborators used the Fourier transform of AFM images to show for the first time the periodicity corresponding to the unit cell of glucan chains in *Valonia* crystals.(109) Later on, Baker et al. imaged *Valonia* crystals in water and propanol and reported the direct visualization of the molecular periodicity within glucan chains and the intermolecular periodicity of their crystalline packing.(110) The imaged crystals appeared to be 5-12 nm high and 20-100 nm wide; the discrepancy of the latter dimension with previously reported sizes was attributed to tip-broadening effects. Upon close inspection, the surfaces of these crystals, which at low magnifications seemed smooth, showed corrugations that ran across the crystal. Three spacings were identified in the ultra-high resolution deflection AFM images and their corresponding Fourier Transform plots: the glucose interval with a periodicity of 0.52 nm, the cellobiose repeat with a periodicity of 1.04 nm and the intermolecular spacing between glucan chains of ~0.6 nm.(109,110) Furthermore, the triclinic arrangement of the glucan chains in the I_α cellulose allomorph predominant in algal cells was observed as a 0.26 nm shift of the cellobiose repeat unit along the chain axis. These observations were further compared with computational models of the triclinic arrangement of the glucan chains and the measured periodicities were found to be in good agreement with the predicted spacings.(111,112) While these studies together provided direct

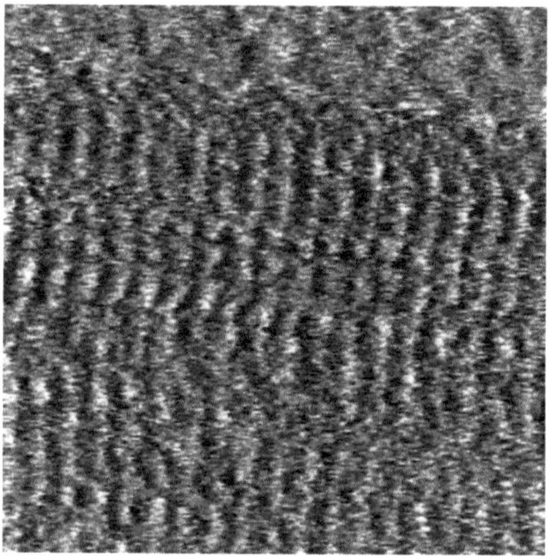

Figure 6. The AFM deflection mode image of a *Valonia* crystal surface shows spacings between the corrugations of 0.55 nm, in good agreement with the spacing between glucose units obtained through crystallography. Glucan chains in the crystal run in the left to right direction. Reproduced from reference (110) with permission.

information of the intra and intermolecular spacing of glucan chains in cellulose I_α crystals from an algal source, no work to date has been performed to observe the spacings in cellulose derived from tunicates or plants, which have cellulose crystallites composed predominantly of the cellulose I_β allomorph. Furthermore, higher resolution probes and more sensitive instruments should allow the acquisition of even higher resolution images, which might help identify the regions of plant cell wall material containing coexisting phases of both allomorphs.

Another area of research that has benefited from advanced-microscopy techniques has been the investigation of how the cellulose structure changes as a result of different mechanical or chemical treatments. This is particularly important for the characterization of pulping processes, the pretreatment of plant biomass, and for the development of novel green materials with enhanced physical and chemical properties based on nanocrystalline cellulose. Pulping is the mechanical or chemical process used to separate cellulose from plant biomass and its outcome, a lignocellulosic fibrous material called pulp, is the primary raw material for papermaking. A number of studies have utilized AFM to investigate the effect of chemical and mechanical treatments used in the pulping process on the resulting cellulose fibres. Koljonen and collaborators used AFM to study the structure of spruce that had been mechanically treated to produce unbleached pressure groundwood pulp (PGWP), thermomechanical pulp (TMP) and chemothermomechanical pulp (CTMP).(113) They found that the surfaces of all pulps exhibited heterogeneous morphology, with the surfaces of PWGP and TMP showing fibrilar and granular structures, while CTMP showed a surface almost completely covered by granular material. This granular structure was interpreted as lignified material that could be removed after peroxide bleaching. Similarly, Gustafsson and collaborators studied the effect of cooking and bleaching on Kraft pulp produced from spruce biomass.(114) It was observed that spruce pulp that had been cooked for short times (15-30 minutes) exhibited highly granular surfaces with granules, containing lignin and extractives, of 40-300 nm size. Increasing the cooking time reduced the amount and size of the granules, but did not eliminate them completely. Further acetone extraction and delignification steps effectively removed the granular material, exposing the underlying fibrilar cell wall construction corresponding to structural polysaccharides. The irregularity observed in fibril size was attributed to the presence of both crystalline cellulose and amorphous hemicellulose.

More recent studies have shown that image analysis techniques can be used to extract more quantitative information from AFM images taken from cellulose samples that have undergone different treatments. Fahlen and Salmen have shown that image analysis techniques can evidence the change in size of cellulose aggregates in spruce wood pulp produced through kraft cooking in presence of different amounts of alkali.(115) Schmied and collaborators recently presented the use of a watershed algorithm to distinguish between cellulose fibre and precipitated lignin aggregates in AFM images taken from softwood Kraft pulps.(116) Through this analysis they were able to investigate the kinetics of lignin precipitation and the structure of these aggregates during the processes of pulping and bleaching. In addition to pulping, the effects of other processing techniques on cellulose structure have been given much attention recently, especially in the context of the pretreatment of biomass for the production of biofuels and bioproducts. For example, Eronen and collaborators characterized the changes induced in pure cellulose samples by the process of mercerization, where alkali induces a change in the crystalline cellulose packing and transforms cellulose I into the cellulose II allomorph.(117) They observed that fibrils imaged in mercerized samples appeared swollen and granular, but that the overall surface roughness of the cellulose samples decreased. These observations were correlated to confocal Raman imaging. AFM has also been recently combined with electron microscopy, x-ray photoelectron spectroscopy and attenuated total reflectance Fourier transform infra-red spectroscopy to study the effects of hydrothermal pretreatment on wheat straw biomass (118) and the effects of active oxygen and solid alkali pretreatment on corn stock biomass. (119)

Similar AFM characterization has been used to determine the physical characteristics of cellulose nanocrystals produced through acid hydrolysis from different plant materials. For example, it has been shown that never-dry materials or materials that are dried under high relative humidity produce longer nanocrystals than those produced from dried materials, an effect ascribed to the tension induced by the drying process and the underlying supramolecular changes in the crystalline cellulose packing.(120) AFM characterization has also helped demonstrate that cellulose II nanocrystals can be produced through the dissolution of cellulose I in sulphuric acid followed by recrystallization under conditions where the acid concentration and treatment duration are carefully controlled.(121) The intense focus on the use of nanocrystalline cellulose for the production of green materials and materials with improved physical and chemical characteristics has spurred an increase in the research devoted to the nanoscale characteristics and structure of these crystalline structures. It is anticipated that advanced-microscopy techniques, such as AFM will play a fundamental role in understanding the nanoscale structure properties of these materials and in helping guide the development of new biocomposites.

3.2. Elucidation of cell wall structure through protein adsorption

Advanced microscopy techniques have been used to image the adsorption of proteins to cellulosic substrates, both with the intent of characterizing the inherent cell wall structure as well as studying the effects that different physical, chemical and biological pretreatments have on cellulosic substrates. In particular, fluorescence microscopy techniques have been widely used for these studies because they can selectively image fluorescently tagged molecules (e.g. antibodies, cell wall binding proteins) with very high sensitivity over the background presented by the cellulosic material. Furthermore, the availability of fluorescence techniques that can perform optical sectioning has enabled the precise localization of the fluorescent molecules in space and the reconstruction of three-dimensional profiles of complex structures contained within cellulosic samples. The application of fluorescence techniques to image protein adsorption has even been extended down to the level of individual molecules, where specific molecular orientations can report on the crystalline arrangement of cellulose fibrils, and with the advent of super-resolution fluorescence microscopy it is foreseen that more detailed structural information from cellulosic substrates will be obtained through fluorescence with resolutions comparable to those achieved by electron microscopy.

Early microscopy studies of the structure of polysaccharide components within cell walls were conducted with whole xyloglucan (XG) binding proteins, galactan-binding proteins or antibodies specifically raised against these cell wall components. One of the first studies of this kind, published in 1984 by Hayashi and Maclachlan, utilized a fluorescently labeled fucose-binding lectin to visualize the spatial distribution of XG in a cellulose/XG macromolecular complex extracted from the elongating regions of etiolated pea.(122) The images obtained through fluorescence microscopy were compared to those obtained through radioisotope imaging and electron microscopy and it was determined that XG was present both on and between the cellulose fibrils of the complex. More recently, Brunecky and collaborators used the immunostaining technique with an XG-directed primary antibody and a fluorescent secondary antibody to study the changes in the distribution of xylan caused by dilute acid pretreatment through confocal laser scanning microscopy.(123) Using this technique they observed that the XG distribution, which was homogeneous in untreated cell walls, was drastically reduced in the center of the cell walls and increased in the middle lamella and the lumen of the cell walls of pretreated materials. Furthermore they observed a decrease in the overall fluorescence signal, which was consistent with the decrease in XG content of the bulk sample. A similar immunostaining and imaging approach has been reported for the visualization of the distribution of XG and galactan in developing (124) and mature tension wood poplar cells.(125) In these reports, the images obtained through fluorescence imaging were also compared to electron microscopy immunostained samples to gain higher resolution of the localization of these hemicellulose components within the cell walls. Despite the ability to visualize the distribution of hemicellulosic components within the cell wall at the micron level, the use of whole proteins and sequential staining with antibodies precludes the study of cell wall structure through fluorescence with high resolution. Because large proteins, such as antibodies, cannot efficiently penetrate interstitial spaces and pores with nanometer dimensions within the cell wall structure, smaller probes are needed to achieve better labeling and higher resolution of hemicellulosic components.

The search for smaller, more specific probes to image and characterize the structural arrangement of polysaccharides in plant cell walls has lead researchers to explore the use of glycoside hydrolase fragments as highly specific molecular imaging probes. Glycoside hydrolases, plant cell wall degrading enzymes produced by a range of bacteria and fungi, have complex molecular architectures consisting of one or more catalytic domains (CDs) and one or more carbohydrate binding modules (CBMs).(126) It is thought that the primary function of CBMs is to promote the attachment of the parent enzyme to its substrate, thus increasing the local concentration of the enzyme and enhancing its catalytic activity. CBMs have evolved in nature to bind to different polysaccharides encountered in the cell wall, as well as starches and other polysaccharides such as chitin that are encountered in a wide range of biological systems. Currently there are 64 families (or types) of CBMs catalogued as functional or putative carbohydrate binding motifs based on their amino acid and structural similarity, as identified in the carbohydrate active enzyme database (CAZy; http://afmb.cnrs-mrs.fr/cazy/).(127) The compact structure of CBMs and their specificity towards the targeted substrate of their parent enzyme have made CBMs prime candidates for the study of the structural arrangement of polysaccharides within the cell wall.

The first studies using CBMs as molecular probes for visualization utilized fluorescent antibodies raised against the CBMs or against motifs appended to them through protein engineering. Linder et al. utilized a recombinant construct where two cellulose-binding modules from *Trichoderma reseii* were linked together to bind specifically to cellulose and established a quick protocol to differentiate between Acanthamoeba, which produces cellulose, and other protozoan parasites that only produce chitin.(128) They visualized the bound CBMs through epifluorescence using an antibody sandwich assay consisting of a primary mouse anti-cellulase antibody and a secondary fluorescent anti-mouse antibody. In 2004, McCartney and collaborators conducted a study that systematically characterized the binding specificity of CBMs from families 2a (bind primarily to crystalline cellulose), 6 (bind to xylan), and 29 (interact with β-(1-4)-linked glucans, mannan, and glucomannan).(129) Using an antibody sandwich assay they were able to visualize the binding of CBMs within thin sections of maize cell walls through epifluorescence microscopy. In this study, they showed for the first time the ability to use recombinant his-tagged CBMs as flexible molecular probes to localize their polysaccharide targets and study the complex cell wall structure in plants. A couple of years later, the same group interrogated the binding specificity of six xylan-binding CBMs (from families 2b, 4, 6, 15, 22 and 35) for their target polysaccharides within the cell walls of a variety of species, such as wheat, flax, tobacco,

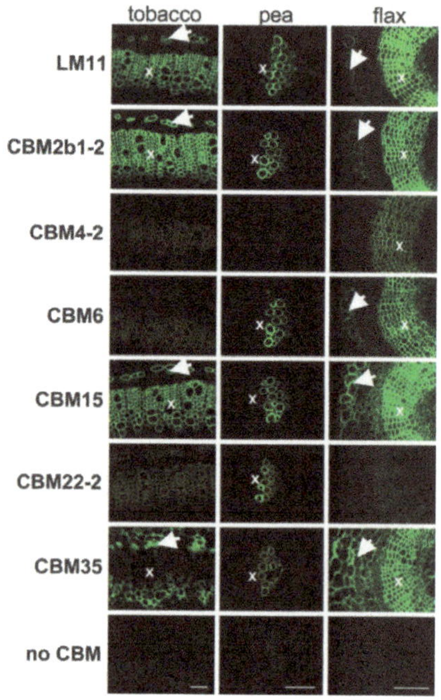

Figure 7. Epifluorescence microscopy of CBMs and monoclonal antibody LM11 binding to secondary cell walls of vascular tissues in thin sections of tobacco, pea, and flax stems. Scale bar 100 μm. Reproduced from reference (130) with permission.

pea, and maize through epifluorescence microscopy (Figure 7).(130) In this study they found that the CBMs tested showed significant differences in the recognition of plant material, and stained the xyloglucans in the primary and secondary cell walls to different extent. They postulated that the context in which the xyloglucans are encountered in the plant is a determinant factor for the ability of CBMs to recognize their targets, and that the variability in recognition in plants must be related to the ability of the different enzymes to hydrolyze xyloglucans in different species. Recently, similar immunofluorescence labeling of cellulose directed CBMs and epifluorescence imaging approaches have been used to study the structural arrangement of the crystalline cellulose contained within the G-layer of tension wood derived from poplar and birch (131) and to study the changes in crystallinity of cellulose II in lyocell fibers derived from eucalyptus wood pulp as they were subjected to NaOH treatment.(132) Altogether these studies have shown that CBMs are suitable probes to study the structure of cellulosic materials derived from plant cell walls through fluorescence microscopy.

Over the past decade, a number of different labeling approaches have been pursued to directly tag CBMs with fluorescent moieties in order to avoid the need for multiple antibody incubation steps. Daniel and collaborators first reported the direct labeling of *T. reseii* Cel7A CBMs with organic dyes (FITC/TRITC) for the visualization of crystalline cellulose within the

G-layer of tension wood.(131) Similarly, Filonova and collaborators directly tagged synthetic xylan-binding CBMs to map the carbohydrate distribution within wood (birch and pine) sections and Kraft pulp fibers (birch).(133) The use of CBMs that were directly labeled with fluorescent molecules effectively reduced the number of sample preparation steps and allowed the acquisition of higher resolution images through epifluorescence microscopy. Yet, a concern of direct labeling of these molecular probes with organic dyes is that because of the randomness of the tagging process, reduced binding could result from labeling of residues that are involved in binding. A second approach that has been tested to directly label CBMs has been tagging them with quantum dots (QDs), nanometer-sized colloidal semiconductor particles with tunable fluorescent emission properties, which are directed towards specific amino acid sequences appended to the CBM. In 2006 Zhang reported the first images of QD-CBM assemblies bound to crystalline cellulose and showed that these constructs could be used in combination with near field scanning optical microscopy to obtain fluorescence images with enhanced spatial resolution.(134) More recently, Xu et al. have reported the labeling of two CBMs from families 2 and 3a with a range of QDs with different spectral properties and their imaging using epifluorescence microscopy.(135) This study used the labeled CBMs to visualize Valonia cellulose crystals with diameters of 20 nm and lengths in the 15 to 20 μm range, and showed that the QD-CBM assemblies preferentially bound to the planar face of the cellulose crystal in the form of linear arrays. Furthermore, Xu and collaborators observed that by using different QD sizes they could control the spacing between CBMs adsorbed on the cellulose surface, which indicated that the steric hindrance between neighboring QDs was the limiting factor for the density of bound CBMs. They also suggested that due to the bright emission of these QD-CBM constructs, they could find use in SM studies to interrogate the catalytic and non-catalytic motion of CBMs on cellulose surfaces.

In 2006, Ding et al. further introduced the use of CBM-fluorescent protein (FP) fusion constructs as molecular probes for the characterization of the complex structure of cell walls and other biomaterials.(136) In this pioneering study, CBMs from families 3, 6, and 20 labeled with QDs or expressed as constructs that incorporated the fusion of either a green or red fluorescent protein (GFP or RFP) were used to label native and phosphoric acid treated Valonia cellulose crystals, starch granules, cellulose/tamarind xyloglucan composites, and maize primary cell walls (Figure 8). FP-CBM constructs were found to be highly specific and bound to their intended targets on complex biomaterials containing mixtures of cellulose, starch and xyloglucan. Since the publication of this work, labeling of CBMs with FPs has become a popular solution for the direct tagging of cell wall binding fragments for a wide variety of studies. CBM-FPs have been used to study the effect of different pretreatments on the structure of plant cell walls. For example, in a report by Porter et al. cellulose and starch directed CBM-GFP/RFP fusion proteins were used to visualize the effect of the wet milling process on corn fibers through confocal microscopy.(137) They observed that the binding of the starch directed CBM-FP correlated well with the remaining starch content in the treated corn fibers and that the cellulose directed CBM-FP construct could not bind significantly to any of the wet milled materials, but would bind well to samples where the starch and most of the hemicellulose had been removed. This evidenced the need to remove as much of the starch and hemicellulosic materials as possible if cellulolytic enzymes are to efficiently

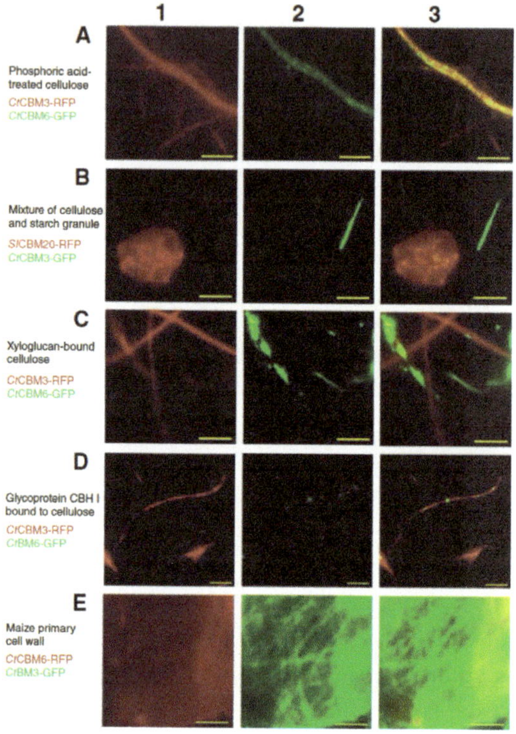

Figure 8. Fluorescence images showing the recognition and binding of two different CBMs to various materials. Column 1 shows fluorescence from RFP labeled samples, column 2 shows the fluorescence from GFP labeled samples, and column 3 shows an overlay of both channels. Each row of images depicts a different material incubated with two distinct CBMs from families 3 (binds to crystalline cellulose), 6 (bind to amorphous cellulose and xyloglucan), and 20 (bind to starch). All scale bars represent 5 microns. Reproduced from reference (137) with permission.

depolymerize crystalline cellulose fibrils within the plant cell wall. Kawakubo et al. also used constructs of family 3 and 28 CBMs fused to cyan fluorescent protein (CFP) as markers for the visualization of crystalline and amorphous cellulose on the complex surfaces of wood tissues pretreated with NaOH, NaOH–Na_2S (kraft pulping), hydrothermolysis, ball-milling, and organosolvolysis.(138) Their assay with CBM–CFPs was proposed as a facile way to measure the extent of exposure of fibrous structures, which are the putative initiation sites of hydrolysis and saccharification in chemically delignified wood pulps. Other applications of CBM-GFP constructs in recent years have been in the study of the distribution of crystalline cellulose in the cell walls within tracheary elements of *Zinnia elegans*,(139) and in the visualization of cell wall regeneration in algal protoplasts.(140)

Up to date, most of the studies conducted to elucidate the structural characteristics of plant cell walls through CBMs that specifically bind to structural polysaccharides have been performed using either epifluorescence or confocal microscopy. Yet, the development of new approaches to perform targeted labeling of CBMs has opened the door for the use of additional forms of advanced fluorescence microscopy techniques. This has already been explored by Dagel and collaborators, who have used a single molecule fluorescence technique termed Defocused Orientation and Position Imaging (DOPI) to visualize the orientation of single CBM-GFP constructs as they bind to Valonia crystals.(141) In this study they used CBMs from families 1, 2 and 3, and showed that binding of these modules was specific for the hydrophobic facet of the cellulose crystals and that they display a preferred orientation with respect to the fiber axis.

Another advanced-microscopy technique that holds promise for the study of plant cell wall structure with molecular resolution is Super Resolution Fluorescence Microscopy (SRFM). In particular, the reconstruction of molecular resolution maps of cellulosic materials is possible through the direct labeling of the targeted polysaccharides (Figure 9) or through binding of molecular reporters such as highly specific CBMs and whole proteins (Figure 9).(101,142) Advanced fluorescence microscopy techniques like these are sure to play an important role in the molecular characterization of the structure and composition not only plant materials, but also of biocomposites and biomaterials.

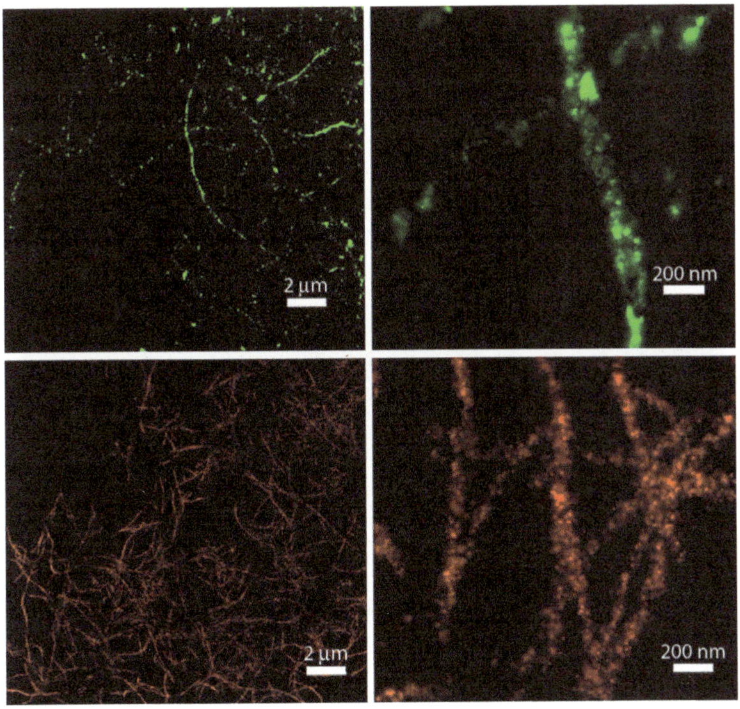

Figure 9. Super-Resolution Fluorescence Microscopy images of cellulose microfibrils (top, direct imaging of cellulose through DTAF labels) and bound cellulases (bottom, T. fusca Cel9A) reveal the nanoscale structure of crystalline cellulose. Adapted from Reference (101).

3.3. CBM and cellulase binding kinetics

The use of highly sensitive cameras in conjunction with stable illumination sources and efficient fluorescent emitters has enabled the quantitative assessment of fluorescence intensity measurements. This has been used to acquire time resolved fluorescence microscopy images for the study of CBM and cellulase binding kinetics. Pinto et al. conducted the first studies using time-lapsed epifluorescence microscopy to study CBM binding kinetics in 2006.(143) In this study they demonstrated a method for the gram-scale production of CBMs derived through mild proteolysis from *T. reseii* CBHI cellulases contained within a commercial cellulolytic cocktail. The purified CBMs were subsequently labeled with an organic dye (FITC) and were used to study CBM binding kinetics on a variety of cellulosic substrates. Image analysis of the fluorescence intensity of the samples showed that CBMs bound preferentially to amorphous cellulose, that the CBMs displayed saturation adsorption kinetics, and that CBM binding was reversible, but that the rate of desorption was slow. The same group a year later published a fluorescence microscopy image analysis method to quantitatively determine the concentration of adsorbed CBMs on cellulosic surfaces.(144) In this method, the fluorescence intensity of images obtained from solutions with known concentrations of fluorophore, which were confined in a chamber with known volume, were used as calibration to extract the concentration of fluorescent CBMs bound to thin cellulose films. Their image analysis routines allowed the quantification of CBM adsorption to homogeneous cellulose surfaces and the reconstruction of binding isotherms that matched those obtained from bulk measurements. This methodology has recently been extended to the quantitation of CBM binding onto cellulosic fibers through epifluorescence and laser scanning confocal fluorescence microscopy.(145) These studies have demonstrated the ability to use fluorescence microscopy and time-lapsed image acquisition to measure the binding kinetics of cellulose binding modules.

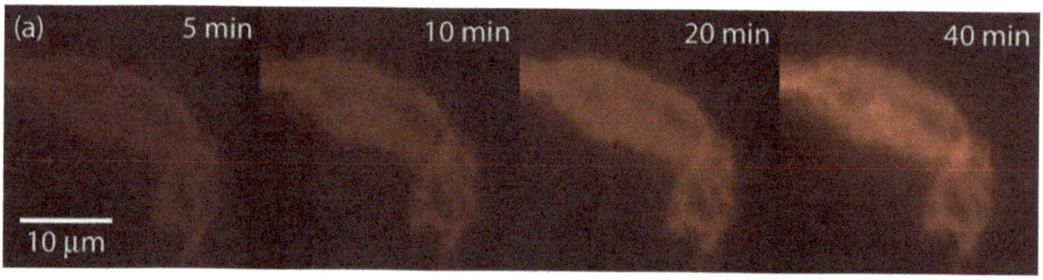

Figure 10. Epifluorescence imaging of real time binding of Cel6B onto immobilized cellulose microfibrils. Representative images of points in time-lapsed experiments in red channel fluorescence show AF-647 cellulase binding onto surface immobilized cellulose. Reproduced from reference (146) with permission.

Quantitative fluorescence microscopy has also been demonstrated for the study of binding kinetics of whole cellulases. In 2008, Moran-Mirabal et al. reported the fluorescent labeling of *Thermobifida fusca* Cel5A (classical endoglucanase), Cel6B (exoglucanase), and Cel9A (processive endoglucanase) cellulases and their use to study binding kinetics on fluorescently labeled bacterial microcrystalline cellulose (BMCC) that had been immobilized on a micropatterned surface.(146) The images obtained through epifluorescence microscopy showed uniform binding of all three cellulases on the surface of the cellulose fibrils, which pointed to binding mediated primarily by the CBM. The time lapse image sequences also evidenced that binding kinetics obtained from individual cellulose fibrils, where cellulases can bind readily, closely followed a saturation binding model, while binding kinetics obtained from cellulose mats and particles, where the porous structure hinders cellulase accessibility to internal cellulose surfaces, significantly deviated form it. This pointed to a strong influence of the physical structure of cellulosic substrates on the ability of cellulases to bind to and ultimately hydrolyze cellulose. A similar approach was taken by Zhu et al. to study the binding kinetics of *T. fusca* cellulases on thermally pretreated wood particles.(147) Confocal laser scanning fluorescence microscopy was used in this study to capture three-dimensional images of the autofluorescence arising from the lignocellulosic particles and fluorescence from cellulase binding onto the cellulose. Fluorescence images were recorded in three spectrally different channels and a deconvolution algorithm was introduced to separate the autofluorescence from the bound cellulase signal. In this way it was observed that cellulases bind preferentially to certain areas of the pretreated particle, arguably those containing easily accessible amorphous cellulose. The time-lapse fluorescence intensity data extracted through image analysis provided a quantitative measure of bound enzymes over time, and were fitted to a transient enzyme-binding model. However, the transient binding model could only be fitted to the initial binding phase and the authors concluded that a more complete model was needed to further explain the binding patterns observed in the complex structure of pretreated lignocellulosic particles.

In recent years advanced imaging techniques have started to be used to study cellulase-cellulose interactions with higher spatial and temporal resolution. The group of Ragauskas has pioneered the use of Förster Resonance Energy Transfer (FRET) for the study of cellulase-cellulose interactions(148) and cellulase-cellulase colocalization.(149) Their first study demonstrated the ability to investigate the temperature dependence of cellulase binding to cellulose by monitoring the fluorescence emission change of donor (or acceptor) as the bound cellulase came into close proximity of the fluorescent labels grafted onto the cellulose substrate.(148) This study demonstrated the use of FRET in a homogeneous liquid environment, with carboxymethyl cellulose as a soluble substrate. In a subsequent study, Wang et al. demonstrated the ability to use FRET together with acceptor photobleaching and spectral unmixing to determine the extent of colocalization of cellulase on cellulose fibrils. In this study, cellulase labeled with two spectrally distinct dyes was incubated with cellulose fibrils and epifluorescence images were acquired before and after bleaching of the donor dye.(149) Although this study was intended as a proof of concept, the authors demonstrated that by changing the concentration of the donor dye on the cellulose fibrils they could observe changes in the acceptor fluorescence, indicating that the cellulases labeled with the different dyes were localized next to each other within a radius corresponding to the molecular size of the enzyme. Another advanced imaging technique that has been successfully applied to study the binding kinetics of cellulases is fluorescence recovery after photobleaching (FRAP). Moran-Mirabal et al. pioneered the use of FRAP with a confocal laser scanning microscopy setup to study the binding reversibility of *T. fusca* cellulases to fibrils and mats composed of bacterial microcrystalline cellulose.(150) This study not only demonstrated that cellulase binding to cellulose is reversible, but also showed that a certain fraction of all adsorbed cellulases is irreversibly bound to the surface. Furthermore, it was the first implementation of fluorescence microscopy imaging of cellulases at temperatures where significant catalysis occurs. The effect of temperature was significant in the binding kinetics observed, especially for processive enzymes, where higher temperatures increased the rate of desorption of the reversibly bound enzymes, while at the same time increasing the irreversibly bound fraction. The implementation of advanced microscopy techniques for the study of binding kinetics is still in its initial stages, and we can foresee an increase in the use of non-traditional techniques to gain a deeper understanding into the mechanisms at play in cellulase-cellulose interactions.

3.4. The molecular transport in cellulosic substrates

The study of diffusion and transport processes on cellulosic substrates is arguably the area that has benefited the most from the employment of advanced microscopy techniques. Scanning probe microscopy as well as fluorescence techniques have been used since the end of the 90's to study the molecular diffusion of glycoside hydrolases on cellulosic surfaces, the processive motion of cellulases as they catalyze the breakage of glycosidic bonds, and the transport of biomolecules through the porous structure of plant derived materials. Jervis and collaborators conducted the first study of the diffusion behaviour of bacterial cellulases and their isolated CBMs on microcrystalline cellulose using FRAP.(151) In this study, they bound fluorescently labeled cellulases and CBMs to a cellulose mat and washed the excess unbound cellulases before bleaching. Areas of the cellulose mats were bleached with a high intensity laser pulse and the recovery of fluorescence was monitored over time. Because it was assumed that the cellulases were irreversibly bound to the cellulose, all recovery was attributed to diffusion along the surface of the cellulose fibrils. Jervis et al. reported a cellulase mobile fraction of 70%, and diffusion coefficients ranging from 2×10^{-11} to 1.2×10^{-10} cm^2/s for cellulases and isolated CBMs. They further concluded that the surface diffusion of cellulases was not a rate limiting process for the hydrolysis of cellulose. A similar FRAP experimental approach was used by Cuyvers et al. to probe the surface diffusion of different *Bacillus subtilis* xylanase mutants on wheat and oat spelt xylans.(152) They observed that the wild-type enzyme was fully mobile and showed complete recovery in bleached areas, whereas an inactive mutant displayed binding but no recovery over the xylan surface. Thus, they concluded that catalytic activity was essential for the mobility of the enzyme on the xylan surface. They also observed that higher binding affinity resulted in slower or hindered recovery. Recently, Moran-Mirabal et al. have used FRAP and single molecule tracking (SMT) techniques to study the surface diffusion of *T. fusca* exo- and endocellulases on bacterial micro-crystalline cellulose at a range of different temperatures.(153) In this study, they used carefully controlled environmental conditions to conduct FRAP experiments and found that *T. fusca* cellulases exhibit limited surface diffusion even at temperatures where significant catalysis occurs. It was shown that when the unbinding and rebinding contribution to fluorescence recovery is removed, a significant fraction of cellulases is immobile and recovery is significantly diminished. This suggested that surface diffusion is limited for *T. fusca* cellulases. In order to confirm these observations, SMT experiments were conducted where the motion of individual enzymes was tracked over time lengths in the order of minutes. They observed that most of the cellulases remained immobile, and that those that exhibited significant mobility displayed heterogeneous behaviours (Figure 11). The authors suggested that these could be explained by the differentiation of the observed displacements into hopping and sliding motions. This study for the first time showed that the dynamics of cellulase motion on cellulose surface are highly heterogeneous and cannot be explained by surface diffusion alone. Furthermore, this study showed that long-range non-catalytic displacements could be tracked using single molecule fluorescence techniques.

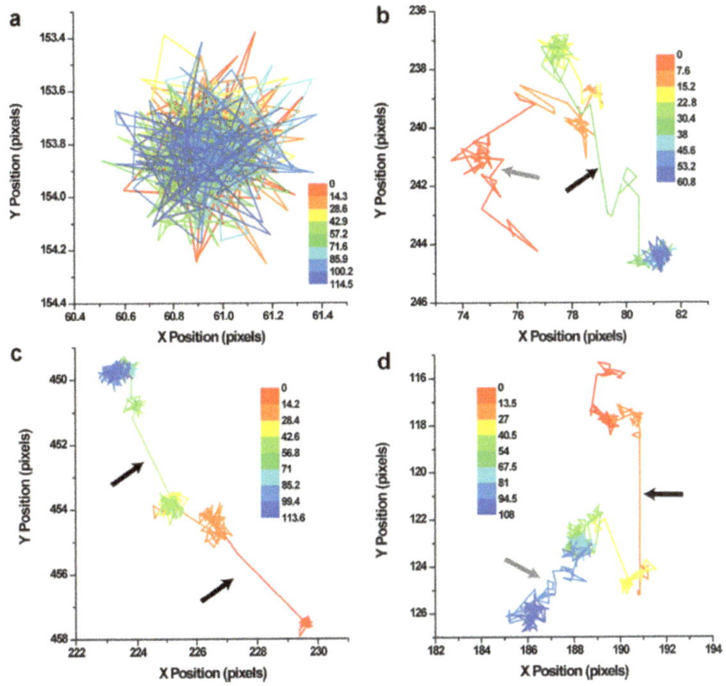

Figure 11. Single molecule tracks show that *T. fusca* cellulases exhibit complex motion patterns that can be related to molecular states. Sample tracks for: (a) immobile Cel5A cellulase, (b) mobile Cel5A cellulase, (c) mobile Cel6B cellulase, and (d) mobile Cel9A cellulase. All tracks displayed were recorded at 45°C. Color of tracks represents time in seconds (0.4 s/frame). Black arrows highlight displacements that correspond to the putative hopping motion, while gray arrow highlight displacements consistent with the putative sliding motion. Reproduced from reference (153) with permission.

The development of High-Speed-AFM (HS-AFM) over the last decade has recently enabled the direct visualization of the displacements resulting from the processive catalytic action of *T. reseii* Cel7A cellobiohydrolase on crystalline cellulose fibrils. These studies, pioneered by Igarashi et al. have already given very valuable insight into the processive action of cellulases as they hydrolyse glycosidic bonds. In a seminal study published in 2009, Igarashi et al. for the first time visualized the processive displacement of individual Cel7A molecules on Cladophora cellulose crystals and reported an average processive speed of 3.5 nm/s and maximum displacements for a single enzyme on the order of 50 nm.(27) They further demonstrated that the same sliding motion could be observed for isolated catalytic domains, but not for inactive mutants. This confirmed that the observed motion arose solely from catalysis and not from non-catalytic surface diffusion. In addition, they observed that the mutation of a tryptophan residue at the entrance of the catalytic cleft significantly reduced the enzyme's mobility, which demonstrated that this residue that promotes the loading of the cellulose fibril into the cellulase catalytic cleft is also key component to the processive motion of cellulases. In a subsequent study, Igarashi et al. presented results of *T. reseii* Cel7A processivity measurements obtained using an improved version of the HS-AFM apparatus.(28) The enhanced temporal and spatial resolution afforded by the improved HS-AFM allowed them to visualize two different Cel7A populations, one that represented enzymes exhibiting processive motions with an average speed of 7.1 nm/s and another that represented enzymes halted on the cellulose surface (with speeds of -0.3 nm/s). They also observed that mobile enzymes exhibited stop-and-go motions that sometimes led to the accumulation of halted enzymes in single file ("traffic jams") on the cellulose fibril surface. The observed traffic jams could be overcome if a significant number of enzymes accumulated, after which the enzymes could proceed with their processive motion. This study further investigated the effect of the cellulose allomorph type on cellulase motion and found that cellulose III allomorphs were able to accommodate higher density and more fluid processive motion of the cellulases. Finally, they investigated the effect of the co-incubation of Cel7A cellulases with Cel6A exocellulase, and found that there was a synergistic effect where the addition of Cel6A reduced the possibility of "traffic jams" and opened up more molecular cellulose lanes for Cel7A cellulases to diffuse in. Together these two studies have shown for the first time that it is possible to visualize the processive motion of cellulases, and have opened the door for subsequent studies of the molecular motion of other glycoside hydrolases.

The visualization of the diffusion of biomolecules through cellulosic substrates can provide significant insight into the transport limitations that enzymes encounter during the process of biomass saccharification. Yet, there are few examples of studies that use advanced imaging techniques to probe the diffusion of biomolecules and polymers into the porous structure of cellulosic materials. This is partly due to the fact that imaging the three dimensional structure of complex lignocellulosic materials through optical techniques is challenging. On one hand, the presence of lignin introduces large background signals in visible channels arising from autofluorescence. On the other, optical sectioning techniques can only probe a small fraction of thick samples because these substrates tend to have high refractive indices and can scatter much of the incident light and the emitted fluorescence. Despite these difficulties, Horvath et al. have recently reported a comprehensive study of the diffusion of cationic polymers into anionic cellulosic fibers using confocal laser scanning microscopy.(154) Through the use of polymer chains with different lengths and different charge densities they were able to show that the charge density of the polyelectrolyte had a larger effect on the diffusion times than did the molecule length. They also observed that the use of high ionic strength solutions could completely screen the interaction between the polyelectrolytes and the cellulosic fibers, leading to faster penetration into the porous structure of the substrates. This study demonstrated the ability to visualize diffusion processes occurring within the complex porous structure of cellulose, and has set the stage for future studies on the diffusive behaviour of glycoside hydrolases during the process of saccharification. It is anticipated that a number of novel advanced microscopy techniques, such as second harmonic generation and multiphoton fluorescence microscopy will find applications in the study of diffusion and transport processes with real time capabilities within cellulosic materials.

3.5. Cellulose depolymerization by glycoside hydrolases

The structural polysaccharides present in plant biomass are an attractive source of fermentable sugars that can be employed in the production of biofuels, chemicals and bioproducts. Yet, it has been recognized that the greatest bottleneck in conversion technology for the release of soluble sugars is the inherent recalcitrance to decomposition of the cellulose within the plant cell wall.(155) Two specific aspects that contribute to the recalcitrance of lignocellulose are: (i) the tight bonding between glucan chains in the highly crystalline cellulose microfibrils and (ii) the structural arrangement of cell wall components, which gives rise to pores and crevices with dimensions from micrometers to nanometers that can hinder the transport of catalysts from solution to the insoluble cellulose. Thus, understanding the mechanisms by which hydrolytic and non-hydrolytic proteins disrupt the structure of recalcitrant cellulosic substrates can be an important step towards designing more efficient and economical biomass conversion strategies. This has resulted in an increased interest in methods for visualizing the process of cellulose depolymerization in real time and with high spatial resolution. Such techniques could yield invaluable information to help elucidate some of the molecular mechanistic details of biomass processing. In recent years, a number of studies have used advanced microscopy techniques to study the depolymerization of cellulose by glycoside hydrolases and obtain qualitative and quantitative information about the catalytic processes occurring at the interface between the insoluble cellulose material and the different enzymes. In 2010 Quirk et al. reported the first study on the direct visualization of the process of biochemical cellulose depolymerization using AFM.(156) This study used *Acetobacter xylinum* crystalline cellulose fibers as the cellulosic substrate and *C. fimi* CenA endoglucanase as the depolymerizing glycoside hydrolase. The depolymerization was observed at room temperature in real time over nine hours and AFM images of the fibrils as they were degraded were obtained in both topography and amplitude mode. Quirk et al. observed that over the experimental time frame cellulose fibrils were degraded preferentially at the ends of fibrils, as well as at the sites where the cellulose fibres exhibited defects and kinks. The visual result of the depolymerization was progressive fraying and shortening of the fibre ends. In addition, they were able to observe significant swelling of the fibers at sites where no visible degradation occurred, which was interpreted as the disruption of the crystalline packing of the cellulose by CenA cellulases. The same year, Santa-Maria and Jeoh reported a detailed study of the depolymerization of crystalline cellulose by a purified *T. reseii* cellobiohydrolase (Cel7A), which combined AFM and confocal fluorescence microscopy imaging.(157) In this study, the authors conducted AFM and confocal imaging of bacterial and algal cellulose that had been degraded in batch preparations, as well real time imaging of cellulose that was degraded in situ. They observed that during the initial degradation the cellulose microfibrils experienced untwisting, which could be interpreted as the relaxation of the inherent internal stress of chiral cellulose crystals. This study also revealed that during degradation the cellulose microfibrils shortened and thinned out, with 0.3-0.6 nm grooves becoming evident along the length of the microfibrils at large extents of hydrolysis (80%). The appearance of such channels was attributed to the processive action of Cel7A cellobiohydrolase, where the hydrolysis of a single glucan chain leaves a void behind. This report was the first to visualize the tension release as a result of cellulose hydrolysis.

Fluorescence microscopy techniques have also been recently applied to study the hydrolysis of cellulosic materials. The capability of fluorescence techniques to monitor individual species within a complex mixture of proteins has made these techniques applicable to the study of cellulose degradation by mixtures of cellulolytic enzymes. Thygesen and collaborators have used polarized light microscopy together with confocal microscopy to study the degradation of filter paper and pretreated wheat straw cellulosic fibres by commercial cellulolytic cocktails.(158) In this study, the authors focused on characterizing the changes occurring at dislocations present in the cellulosic fibres and used a fluorescently labeled cellulase (endoglucanase GH45 from *Humicola insolens*) as a reporter for cellulase binding and hydrolysis. They observed that the dislocations observed in the cellulose fibres were not completely amorphous, as they exhibited the characteristic birefringence of crystalline cellulose. Such dislocations were shown to be the primary site of binding for the fluorescent endoglucanases. It was hypothesized that the dislocations are sites with higher accessibility, which leads to the preferential binding and more rapid hydrolysis. The hypothesis that dislocations are the sites for the initiation of fibre hydrolysis was further supported by the observation that as hydrolysis progresses, the fibres remaining in the sample were shorter and the number of dislocations observed was reduced. This study showed that dislocations play an important role during the initial stages of hydrolysis of lignocellulose. More recently, Luterbacher et al. also used confocal fluorescence microscopy to monitor the degradation of bacterial microcrystalline cellulose using fluorescent *T. reseii* Cel7A cellobiohydrolase as a reporter doped within a commercial cellulolytic cocktail.(159) In this study the authors completely hydrolyzed the cellulosic substrate and used quantitative fluorescence intensity measurements to determine the degree of hydrolysis achieved. They correlated the observed fluorescence intensity with remaining cellulose and were able to fit a theoretical model to the observed fluorescence. The kinetic parameters obtained from these fits were able to correctly predict the total sugars released from the cellulose and compared favourably with values obtained through degradation experiments conducted in bulk. In addition, the kinetic parameters obtained pointed to the availability of cellulase binding sites as a key limiting factor for the degradation of crystalline cellulose. This study demonstrated for the first time that quantitative fluorescence techniques could be used to monitor degradation of cellulosic substrates and extract relevant degradation kinetics.

4. Concluding remarks and future outlook

Advanced microscopy techniques have enabled the acquisition of images of cellulosic substrates with higher spatial and temporal resolution. Both scanning probe microscopy and fluorescence microscopy techniques have already been applied to image the inherent structure of cellulosic materials, such as in the distribution of structural polysaccharides within the cell wall or the arrangement of glucan chains within crystalline cellulose fibrils; the interaction of glycoside hydrolases with cellulosic substrates, such as in the binding, diffusion and hydrolytic activity of cellulases and their cellulose binding domains; and the structural changes that cellulosic materials undergo as they are subjected to different chemical, physical and biochemical treatments. However, new imaging modalities, especially those that exploit the ability to image and track single molecules, are just starting to be applied to the study of cellulose. It is anticipated that such techniques, as they become more widespread, will be of increasing importance to scientists interested in extracting information at the nanometer length scale where the accuracy of localization and the concomitant resolution approximate the molecular size of the basic units of structural polysaccharides and the size of individual glycoside hydrolases. In this way, the application of Fast Scan Atomic Force Microscopy or Single Molecule Fluorescence Microscopy could evidence real time behaviour of the different hydrolases as they interact with cellulose microfibrils. Furthermore, single molecule techniques could probe the enzyme activity and determine the fraction of glycoside hydrolases adsorbed to cellulose chains that are in a catalytic active state. Single molecule measurements could also reveal heterogeneities in the binding of cellulases to different structural motifs within the cell wall. All these measurements could provide further insight into the mechanistic action of the enzymes that depolymerize lignocellulosic materials and could help understand the inactivation of enzymes that occurs during the saccharification process. The ability of fluorescence techniques to monitor multiple species simultaneously could also provide insight into the synergistic action of glycoside hydrolases that are incorporated into cellulolytic cocktails. Other advanced microscopy techniques, such as Super-Resolution Fluorescence Microscopy, will undoubtedly contribute to understanding the changes that lignocellulosic materials undergo during different pretreatments by providing nanoscale resolution images of one or more of the structural polysaccharides within the cell walls of plant biomass. These advances taken together as a whole will further our understanding of how the cell wall is constructed and how to more efficiently harness the richness of plant biomass to produce fuels, plastics and other chemicals in a more sustainable, environmentally friendly, and economically viable fashion. The study of cellulose through advanced microscopy techniques is in its nascent stages and many exciting opportunities exist to develop and apply experimental methods that can offer a glimpse on the nanoscale behaviour of cellulosic materials with high temporal resolution.

Author details

Jose M. Moran-Mirabal
Department of Chemistry and Chemical Biology, McMaster University, Hamilton, ON, Canada

5. References

[1] Lee KJ., Marcus SE, Knox JP. Cell wall biology: perspectives from cell wall imaging. Molecular plant. 2011;4(2):212–9.

[2] Chundawat SP., Beckham GT, Himmel ME, Dale BE. Deconstruction of lignocellulosic biomass to fuels and chemicals. Annual Review of Chemical and Biomolecular Engineering. 2011;2:121–45.

[3] Pauly M, Keegstra K. Cell-wall carbohydrates and their modification as a resource for biofuels. The Plant Journal. 2008;54(4):559–68.

[4] Yarbrough JM, Himmel ME, Ding SY. Plant cell wall characterization using scanning probe microscopy techniques. Biotechnology for biofuels. 2009;2(1):17.

[5] Rubin EM. Genomics of cellulosic biofuels. Nature. 2008 Aug;454(7206):841–5.

[6] MCCANN MC, WELLS B, ROBERTS K. Direct visualization of cross-links in the primary plant cell wall. Journal of Cell Science. 1990 Jun 1;96(2):323 –334.

[7] Carpita N, Sabularse D, Montezinos D, Delmer DP. Determination of the Pore Size of Cell Walls of Living Plant Cells. Science. 1979 Sep 14;205(4411):1144–7.

[8] Grethlein HE. The Effect of Pore Size Distribution on the Rate of Enzymatic Hydrolysis of Cellulosic Substrates. Nat Biotechnol. 1985 Feb;3(2):155–60.

[9] Fengel D, Stoll M. Crystals of cellulose grown from TFA solution. Wood Science and Technology. 1989;23(1):85–94.

[10] Riley R, Gardner JO, Merten U. Cellulose acetate membranes: Electron microscopy of structure. Science. 1964;143(3608):801–3.

[11] Cronshaw J, Myers A, Preston RD. A chemical and physical investigation of the cell walls of some marine algae. Biochimica et Biophysica Acta. 1958 Jan;27:89–103.

[12] Carpita NC, Defernez M, Findlay K, Wells B, Shoue DA, Catchpole G, et al. Cell wall architecture of the elongating maize coleoptile. Plant physiology. 2001;127(2):551–65.

[13] Henriksson M, Berglund LA, Isaksson P, Lindström T, Nishino T. Cellulose Nanopaper Structures of High Toughness. Biomacromolecules. 2008 Jun;9(6):1579–85.

[14] Kvien I, Tanem BS, Oksman K. Characterization of Cellulose Whiskers and Their Nanocomposites by Atomic Force and Electron Microscopy. Biomacromolecules. 2005 Nov;6(6):3160–5.

[15] O'SULLIVAN AC. Cellulose: the structure slowly unravels. Cellulose. 1997;4(3):173–207.

[16] Turner SR, Somerville CR. Collapsed xylem phenotype of Arabidopsis identifies mutants deficient in cellulose deposition in the secondary cell wall. The Plant Cell Online. 1997;9(5):689–701.

[17] Revol JF, Bradford H, Giasson J, Marchessault RH, Gray DG. Helicoidal self-ordering of cellulose microfibrils in aqueous suspension. International journal of biological macromolecules. 1992;14(3):170–2.
[18] Heyn ANJ. The microcrystalline structure of cellulose in cell walls of cotton, ramie, and ute fibers as revealed by negative staining of sections. The Journal of cell biology. 1966;29(2):181–97.
[19] White AR, Brown RM. Enzymatic hydrolysis of cellulose: Visual characterization of the process. Proceedings of the National Academy of Sciences. 1981;78(2):1047.
[20] Hult EL, Iversen T, Sugiyama J. Characterization of the supermolecular structure of cellulose in wood pulp fibres. Cellulose. 2003;10(2):103–10.
[21] Fotiadis D, Scheuring S, Müller SA, Engel A, Müller DJ. Imaging and manipulation of biological structures with the AFM. Micron. 2002;33(4):385–97.
[22] Israelachvili JN. Intermolecular and surface forces. Burlington, MA: Academic Press; 2011.
[23] Deniz AA, Mukhopadhyay S, Lemke EA. Single-molecule biophysics: at the interface of biology, physics and chemistry. Journal of The Royal Society Interface. 2008;5(18):15–45.
[24] Ando T, Uchihashi T, Kodera N, Yamamoto D, Miyagi A, Taniguchi M, et al. High-speed AFM and nano-visualization of biomolecular processes. Pflügers Archiv - European Journal of Physiology. 2007 Dec 20;456(1):211–25.
[25] Ando T, Kodera N, Naito Y, Kinoshita T, Furuta K, Toyoshima YY. A High-speed Atomic Force Microscope for Studying Biological Macromolecules in Action. ChemPhysChem. 2003 Nov 14;4(11):1196–202.
[26] Ando T, Uchihashi T, Kodera N, Miyagi A, Nakakita R, Yamashita H, et al. High-Speed Atomic Force Microscopy for Studying the Dynamic Behavior of Protein Molecules at Work. Japanese Journal of Applied Physics. 2006 Mar 27;45(3B):1897–903.
[27] Igarashi K, Koivula A, Wada M, Kimura S, Penttila M, Samejima M. High Speed Atomic Force Microscopy Visualizes Processive Movement of Trichoderma reesei Cellobiohydrolase I on Crystalline Cellulose. Journal of Biological Chemistry. 2009 Oct 26;284(52):36186–90.
[28] Igarashi K, Uchihashi T, Koivula A, Wada M, Kimura S, Okamoto T, et al. Traffic Jams Reduce Hydrolytic Efficiency of Cellulase on Cellulose Surface. Science. 2011 Sep 1;333(6047):1279–82.
[29] Lakowicz JR. Principles of Fluorescence Spectroscopy. New York, USA: Kluwer Academic/Plenum; 1999.
[30] Shaner NC, Steinbach PA, Tsien RY. A guide to choosing fluorescent proteins. Nature Methods. 2005 Dec;2(12):905–9.
[31] Giepmans BNG. The Fluorescent Toolbox for Assessing Protein Location and Function. Science. 2006 Apr 14;312(5771):217–24.
[32] Tsien RY. THE GREEN FLUORESCENT PROTEIN. Annual Review of Biochemistry. 1998 Jun;67(1):509–44.

[33] Kino GS, Corle TR. Confocal Scanning Optical Microscopy. Physics Today. 1989;42(9):55.

[34] Xiao GQ, Corle TR, Kino GS. Real-time confocal scanning optical microscope. Applied Physics Letters. 1988;53(8):716.

[35] Corle TR, Chou C-H, Kino GS. Depth response of confocal optical microscopes. Optics Letters. 1986 Dec 1;11(12):770.

[36] Axelrod D, Burghardt TP, Thompson NL. Total Internal Reflection Fluorescence. Annual Review of Biophysics and Bioengineering. 1984 Jun;13(1):247–68.

[37] Ambrose WP, Goodwin PM, Nolan JP. Single-molecule detection with total internal reflection excitation: Comparing signal-to-background and total signals in different geometries. Cytometry. 1999 Jul 1;36(3):224–31.

[38] Axelrod D. Selective imaging of surface fluorescence with very high aperture microscope objectives. Journal of Biomedical Optics. 2001;6(1):6.

[39] Paige MF, Bjerneld EJ, Moerner WE. A Comparison of Through-the-Objective Total Internal Reflection Microscopy and Epifluorescence Microscopy for Single-Molecule Fluorescence Imaging. Single Molecules. 2001 Oct;2(3):191–201.

[40] Axelrod D. Cell-substrate contacts illuminated by total internal reflection fluorescence. The Journal of Cell Biology. 1981 Apr 1;89(1):141–5.

[41] Axelrod D. Total Internal Reflection Fluorescence Microscopy in Cell Biology. Traffic. 2001 Nov;2(11):764–74.

[42] Axelrod D, Koppel D, Schlessinger J, Elson E, Webb W. Mobility measurement by analysis of fluorescence photobleaching recovery kinetics. Biophysical Journal. 1976 Sep;16(9):1055–69.

[43] Sprague BL, Pego RL, Stavreva DA, McNally JG. Analysis of binding reactions by fluorescence recovery after photobleaching. Biophysical Journal. 2004;86(6):3473–95.

[44] Sprague B, McNally J. FRAP analysis of binding: proper and fitting. Trends in Cell Biology. 2005 Feb;15(2):84–91.

[45] Sund SE, Axelrod D. Actin dynamics at the living cell submembrane imaged by total internal reflection fluorescence photobleaching. Biophysical Journal. 2000;79(3):1655–69.

[46] Cole NB, Smith CL, Sciaky N, Terasaki M, Edidin M, Lippincott-Schwartz J. Diffusional Mobility of Golgi Proteins in Membranes of Living Cells. Science. 1996 Aug;273(5276):797–801.

[47] Tirnauer JS, Grego S, Salmon ED, Mitchison TJ. EB1-microtubule interactions in Xenopus egg extracts: role of EB1 in microtubule stabilization and mechanisms of targeting to microtubules. Molecular biology of the cell. 2002;13(10):3614.

[48] Stenoien DL, Patel K, Mancini MG, Dutertre M, Smith CL, O'Malley BW, et al. FRAP reveals that mobility of oestrogen receptor-α is ligand-and proteasome-dependent. Nature cell biology. 2000;3(1):15–23.

[49] Mochizuki N, Yamashita S, Kurokawa K, Ohba Y, Nagai T, Miyawaki A, et al. Spatio-temporal images of growth-factor-induced activation of Ras and Rap1. Nature. 2001;411(6841):1065–8.

[50] Stryer L. Energy Transfer: A Spectroscopic Ruler. Proceedings of the National Academy of Sciences. 1967 Aug 1;58(2):719–26.
[51] Jares-Erijman EA, Jovin TM. FRET imaging. Nature Biotechnology. 2003 Nov;21(11):1387–95.
[52] Kapanidis AN, Weiss S. Fluorescent probes and bioconjugation chemistries for single-molecule fluorescence analysis of biomolecules. The Journal of Chemical Physics. 2002;117(24):10953.
[53] Clegg RM. Fluorescence resonance energy transfer. Current Opinion in Biotechnology. 1995 Jan;6(1):103–10.
[54] Ha T. Single-Molecule Fluorescence Resonance Energy Transfer. Methods. 2001 Sep;25(1):78–86.
[55] Selvin PR, others. The renaissance of fluorescence resonance energy transfer. nature structural biology. 2000;7(9):730–4.
[56] Sekar RB. Fluorescence resonance energy transfer (FRET) microscopy imaging of live cell protein localizations. The Journal of Cell Biology. 2003 Mar 3;160(5):629–33.
[57] Chen Y, Müller JD, So PTC, Gratton E. The Photon Counting Histogram in Fluorescence Fluctuation Spectroscopy. Biophysical Journal. 1999 Jul;77(1):553–67.
[58] Puchalla J, Krantz K, Austin R, Rye H. Burst analysis spectroscopy: A versatile single-particle approach for studying distributions of protein aggregates and fluorescent assemblies. Proceedings of the National Academy of Sciences. 2008 Sep 5;105(38):14400–5.
[59] Stavis SM, Corgié SC, Cipriany BR, Craighead HG, Walker LP. Single molecule analysis of bacterial polymerase chain reaction products in submicrometer fluidic channels. Biomicrofluidics. 2007;1(3):034105.
[60] Stavis SM, Edel JB, Samiee KT, Craighead HG. Single molecule studies of quantum dot conjugates in a submicrometer fluidic channel. Lab on a Chip. 2005;5(3):337.
[61] Bacia K, Kim SA, Schwille P. Fluorescence cross-correlation spectroscopy in living cells. Nature Methods. 2006 Feb;3(2):83–9.
[62] Bacia K, Schwille P. A dynamic view of cellular processes by in vivo fluorescence auto- and cross-correlation spectroscopy. Methods. 2003 Jan;29(1):74–85.
[63] Haustein E. Ultrasensitive investigations of biological systems by fluorescence correlation spectroscopy. Methods. 2003 Feb;29(2):153–66.
[64] Haustein E, Schwille P. Single-molecule spectroscopic methods. Current Opinion in Structural Biology. 2004 Oct;14(5):531–40.
[65] Rigler R, Földes-Papp Z, Meyer-Almes F-J, Sammet C, Völcker M, Schnetz A. Fluorescence cross-correlation: A new concept for polymerase chain reaction. Journal of Biotechnology. 1998 Aug;63(2):97–109.
[66] Schwille P, Meyer-Almes FJ, Rigler R. Dual-color fluorescence cross-correlation spectroscopy for multicomponent diffusional analysis in solution. Biophysical Journal. 1997 Apr;72(4):1878–86.

[67] Webb WW. Fluorescence Correlation Spectroscopy: Inception, Biophysical Experimentations, and Prospectus. Applied Optics. 2001;40(24):3969.
[68] Carter BC, Shubeita GT, Gross SP. Tracking single particles: a user-friendly quantitative evaluation. Physical Biology. 2005 Mar;2(1):60–72.
[69] Cheezum MK, Walker WF, Guilford WH. Quantitative Comparison of Algorithms for Tracking Single Fluorescent Particles. Biophysical Journal. 2001 Oct;81(4):2378–88.
[70] Ghosh RN, Webb WW. Automated detection and tracking of individual and clustered cell surface low density lipoprotein receptor molecules. Biophysical Journal. 1994 May;66(5):1301–18.
[71] Basché T, Ambrose WP, Moerner WE. Optical spectra and kinetics of single impurity molecules in a polymer: spectral diffusion and persistent spectral hole burning. Journal of the Optical Society of America B. 1992 May 1;9(5):829.
[72] Jaqaman K, Loerke D, Mettlen M, Kuwata H, Grinstein S, Schmid SL, et al. Robust single-particle tracking in live-cell time-lapse sequences. Nature Methods. 2008 Jul 20;5(8):695–702.
[73] Yildiz A. Myosin V Walks Hand-Over-Hand: Single Fluorophore Imaging with 1.5-nm Localization. Science. 2003 Jun 27;300(5628):2061–5.
[74] Lang E, Baier J, Kohler J. Epifluorescence, confocal and total internal reflection microscopy for single-molecule experiments: a quantitative comparison. J Microsc. 2006 May;222(2):118–23.
[75] Greenleaf WJ, Woodside MT, Block SM. High-Resolution, Single-Molecule Measurements of Biomolecular Motion. Annual Review of Biophysics and Biomolecular Structure. 2007 Jun;36(1):171–90.
[76] Ha T. Single-Molecule Fluorescence Resonance Energy Transfer. Methods. 2001 Sep;25(1):78–86.
[77] Hess ST, Huang S, Heikal AA, Webb WW. Biological and Chemical Applications of Fluorescence Correlation Spectroscopy: A Review †. Biochemistry. 2002 Jan;41(3):697–705.
[78] Joo C, Balci H, Ishitsuka Y, Buranachai C, Ha T. Advances in Single-Molecule Fluorescence Methods for Molecular Biology. Annual Review of Biochemistry. 2008 Jun;77(1):51–76.
[79] Michalet X, Weiss S, Jäger M. Single-Molecule Fluorescence Studies of Protein Folding and Conformational Dynamics. Chemical Reviews. 2006 May;106(5):1785–813.
[80] Moerner WE. Single-molecule Chemistry and Biology Special Feature: New directions in single-molecule imaging and analysis. Proceedings of the National Academy of Sciences. 2007 May 11;104(31):12596–602.
[81] Moerner WE, Fromm DP. Methods of single-molecule fluorescence spectroscopy and microscopy. Review of Scientific Instruments. 2003;74(8):3597.
[82] Nie S, Zare RN. OPTICAL DETECTION OF SINGLE MOLECULES. Annual Review of Biophysics and Biomolecular Structure. 1997 Jun;26(1):567–96.

[83] Tinnefeld P, Sauer M. Branching Out of Single-Molecule Fluorescence Spectroscopy: Challenges for Chemistry and Influence on Biology. Angewandte Chemie International Edition. 2005 Apr 29;44(18):2642–71.
[84] Weiss S. Fluorescence Spectroscopy of Single Biomolecules. Science. 1999 Mar 12;283(5408):1676–83.
[85] Weiss S, others. Measuring conformational dynamics of biomolecules by single molecule fluorescence spectroscopy. Nature structural biology. 2000;7(9):724–9.
[86] Xie XS, Trautman JK. OPTICAL STUDIES OF SINGLE MOLECULES AT ROOM TEMPERATURE. Annual Review of Physical Chemistry. 1998 Oct;49(1):441–80.
[87] Schermelleh L, Carlton PM, Haase S, Shao L, Winoto L, Kner P, et al. Subdiffraction Multicolor Imaging of the Nuclear Periphery with 3D Structured Illumination Microscopy. Science. 2008 Jun;320(5881):1332–6.
[88] Hell SW. Toward fluorescence nanoscopy. Nat Biotechnol. 2003 Nov;21(11):1347–55.
[89] Gustafsson MG. Nonlinear structured-illumination microscopy: Wide-field fluorescence imaging with theoretically unlimited resolution. Proceedings of the National Academy of Sciences. 2005;102(37):13081.
[90] Nagerl UV, Willig KI, Hein B, Hell SW, Bonhoeffer T. Live-cell imaging of dendritic spines by STED microscopy. Proceedings of the National Academy of Sciences. 2008 Nov;105(48):18982–7.
[91] Gustafsson MG. Extended resolution fluorescence microscopy. Current opinion in structural biology. 1999;9(5):627–8.
[92] Shtengel G, Galbraith JA, Galbraith CG, Lippincott-Schwartz J, Gillette JM, Manley S, et al. Interferometric fluorescent super-resolution microscopy resolves 3D cellular ultrastructure. Proceedings of the National Academy of Sciences. 2009 Feb;106(9):3125–30.
[93] Manley S, Gillette JM, Patterson GH, Shroff H, Hess HF, Betzig E, et al. High-density mapping of single-molecule trajectories with photoactivated localization microscopy. Nat Meth. 2008 Jan;5(2):155–7.
[94] Bates M, Huang B, Zhuang X. Super-resolution microscopy by nanoscale localization of photo-switchable fluorescent probes. Current Opinion in Chemical Biology. 2008 Oct;12(5):505–14.
[95] Cordes T, Strackharn M, Stahl SW, Summerer W, Steinhauer C, Forthmann C, et al. Resolving Single-Molecule Assembled Patterns with Superresolution Blink-Microscopy. Nano Lett. 2010 Feb;10(2):645–51.
[96] Heilemann M, van de Linde S, Schüttpelz M, Kasper R, Seefeldt B, Mukherjee A, et al. Subdiffraction-Resolution Fluorescence Imaging with Conventional Fluorescent Probes. Angew. Chem. Int. Ed. 2008 Aug;47(33):6172–6.
[97] Vogelsang J, Cordes T, Forthmann C, Steinhauer C, Tinnefeld P. Controlling the fluorescence of ordinary oxazine dyes for single-molecule switching and superresolution microscopy. Proceedings of the National Academy of Sciences. 2009;106(20):8107.

[98] Huang B, Wang W, Bates M, Zhuang X. Three-Dimensional Super-Resolution Imaging by Stochastic Optical Reconstruction Microscopy. Science. 2008 Feb;319(5864):810–3.

[99] Heintzmann R, Ficz G. Breaking the resolution limit in light microscopy. Briefings in Functional Genomics and Proteomics. 2006 May;5(4):289–301.

[100] Lippincott-Schwartz J, Manley S. Putting super-resolution fluorescence microscopy to work. Nat Meth. 2009 Jan;6(1):21–3.

[101] Moran-Mirabal JM. Study of cellulase-cellulose interactions through fluorescence microscopy techniques. Cellulose. 2013; In Press.

[102] Fernandes AN, Thomas LH, Altaner CM, Callow P, Forsyth VT, Apperley DC, et al. Nanostructure of cellulose microfibrils in spruce wood. Proceedings of the National Academy of Sciences. 2011;108(47):E1195–E1203.

[103] Kirby AR, Gunning AP, Waldron KW, Morris VJ, Ng A. Visualization of plant cell walls by atomic force microscopy. Biophysical Journal. 1996 Mar;70(3):1138–43.

[104] Pesacreta TC, Carlson LC, Triplett BA. Atomic force microscopy of cotton fiber cell wall surfaces in air and water: quantitative and qualitative aspects. Planta. 1997;202(4):435–42.

[105] Thimm JC, Burritt DJ, Ducker WA, Melton LD. Celery (Apium graveolens L.) parenchyma cell walls examined by atomic force microscopy: effect of dehydration on cellulose microfibrils. Planta. 2000;212(1):25–32.

[106] Davies LM, Harris PJ. Atomic force microscopy of microfibrils in primary cell walls. Planta. 2003;217(2):283–9.

[107] Ding S-Y, Himmel ME. The Maize Primary Cell Wall Microfibril: A New Model Derived from Direct Visualization. J. Agric. Food Chem. 2006 Feb;54(3):597–606.

[108] Sugiyama J, Harada H, Fujiyoshi Y, Uyeda N. Lattice images from ultrathin sections of cellulose microfibrils in the cell wall of Valonia macrophysa Kütz. Planta. 1985;166(2):161–8.

[109] Hanley SJ, Giasson J, Revol J-F, Gray DG. Atomic force microscopy of cellulose microfibrils: comparison with transmission electron microscopy. Polymer. 1992;33(21):4639–42.

[110] Baker AA, Helbert W, Sugiyama J, Miles MJ. High-Resolution Atomic Force Microscopy of NativeValoniaCellulose I Microcrystals. Journal of Structural Biology. 1997 Jul;119(2):129–38.

[111] Baker AA, Helbert W, Sugiyama J, Miles MJ. Surface structure of native cellulose microcrystals by AFM. Applied Physics A: Materials Science & Processing. 1998;66:559–63.

[112] Baker AA, Helbert W, Sugiyama J, Miles MJ. New Insight into Cellulose Structure by Atomic Force Microscopy Shows the I [alpha] Crystal Phase at Near-Atomic Resolution. Biophysical Journal. 2000;79(2):1139–45.

[113] Koljonen K, Österberg M, Johansson LS, Stenius P. Surface chemistry and morphology of different mechanical pulps determined by ESCA and AFM. Colloids and Surfaces A: Physicochemical and Engineering Aspects. 2003;228(1-3):143–58.

[114] Gustafsson J, Ciovica L, Peltonen J. The ultrastructure of spruce kraft pulps studied by atomic force microscopy (AFM) and X-ray photoelectron spectroscopy (XPS). Polymer. 2003;44(3):661–70.

[115] Fahlen J, Salmen L. Cross-sectional structure of the secondary wall of wood fibers as affected by processing. Journal of materials science. 2003;38(1):119–26.

[116] Schmied FJ, Teichert C, Kappel L, Hirn U, Schennach R. Analysis of precipitated lignin on kraft pulp fibers using atomic force microscopy. Cellulose. 2012 Feb 18;19(3):1013–21.

[117] Eronen P, Österberg M, Jääskeläinen A-S. Effect of alkaline treatment on cellulose supramolecular structure studied with combined confocal Raman spectroscopy and atomic force microscopy. Cellulose. 2008 Oct 11;16(2):167–78.

[118] Kristensen JB, Thygesen LG, Felby C, Jørgensen H, Elder T, others. Cell-wall structural changes in wheat straw pretreated for bioethanol production. Biotechnol Biofuels. 2008;1(5):1–9.

[119] Pang C, Xie T, Lin L, Zhuang J, Liu Y, Shi J, et al. Changes of the surface structure of corn stalk in the cooking process with active oxygen and MgO-based solid alkali as a pretreatment of its biomass conversion. Bioresource Technology. 2012 Jan;103(1):432–9.

[120] Kontturi E, Vuorinen T. Indirect evidence of supramolecular changes within cellulose microfibrils of chemical pulp fibers upon drying. Cellulose. 2008 Jun 12;16(1):65–74.

[121] Sèbe G, Ham-Pichavant F, Ibarboure E, Koffi ALC, Tingaut P. Supramolecular structure characterization of cellulose II nanowhiskers produced by acid hydrolysis of cellulose I substrates. Biomacromolecules [Internet]. 2012 [cited 2012 Jul 6]; Available from: http://pubs.acs.org/doi/abs/10.1021/bm201777j

[122] Hayashi T, Maclachlan G. Pea xyloglucan and cellulose: I. Macromolecular organization. Plant Physiol. 1984;75(3):596.

[123] Brunecky R, Vinzant TB, Porter SE, Donohoe BS, Johnson DK, Himmel ME. Redistribution of xylan in maize cell walls during dilute acid pretreatment. Biotechnology and Bioengineering. 2009 Apr 15;102(6):1537–43.

[124] Sandquist D, Filonova L, von Schantz L, Ohlin M, Daniel G. Microdistribution of xyloglucan in differentiating poplar cells. Bio. Res. 2010;5(2):796–807.

[125] Arend M. Immunolocalization of (1, 4)-β-galactan in tension wood fibers of poplar. Tree Physiology. 2008;28(8):1263–7.

[126] Boraston AB, Bolam DN, Gilbert HJ, Davies GJ. Carbohydrate-binding modules: fine-tuning polysaccharide recognition. Biochemical Journal. 2004 Sep 15;382(3):769.

[127] Cantarel BL, Coutinho PM, Rancurel C, Bernard T, Lombard V, Henrissat B. The Carbohydrate-Active EnZymes database (CAZy): an expert resource for Glycogenomics. Nucleic Acids Research. 2009 Jan 1;37(Database):D233–D238.

[128] Linder M, Winiecka-Krusnell J, Linder E. Use of Recombinant Cellulose-Binding Domains of Trichoderma reesei Cellulase as a Selective Immunocytochemical Marker for Cellulose in Protozoa. Applied and Environmental Microbiology. 2002 May 1;68(5):2503–8.

[129] McCartney L, Gilbert HJ, Bolam DN, Boraston AB, Knox JP. Glycoside hydrolase carbohydrate-binding modules as molecular probes for the analysis of plant cell wall polymers. Analytical Biochemistry. 2004 Mar;326(1):49–54.

[130] McCartney L, Blake AW, Flint J, Bolam DN, Boraston AB, Gilbert HJ, et al. Differential recognition of plant cell walls by microbial xylan-specific carbohydrate-binding modules. Proceedings of the National Academy of Sciences of the United States of America. 2006;103(12):4765–70.

[131] Daniel G, Filonova L, Kallas ÅM, Teeri TT. Morphological and chemical characterisation of the G-layer in tension wood fibres of Populus tremula and Betula verrucosa: Labelling with cellulose-binding module CBM1 Hj Cel7A and fluorescence and FE-SEM microscopy. Holzforschung. 2006 Jan 1;60(6):618–24.

[132] Široký J, Benians TAS, Russell SJ, Bechtold T, Paul Knox J, Blackburn RS. Analysis of crystallinity changes in cellulose II polymers using carbohydrate-binding modules. Carbohydrate Polymers. 2012 Jun;89(1):213–21.

[133] Filonova L, Gunnarsson L, Daniel G, Ohlin M. Synthetic xylan-binding modules for mapping of pulp fibres and wood sections. BMC Plant Biology. 2007;7(1):54.

[134] Zhang P. Investigation of Novel Quantum Dots/Proteins/Cellulose Bioconjugate Using NSOM and Fluorescence. Journal of Fluorescence. 2006 Feb 14;16(3):349–53.

[135] Xu Q, Tucker MP, Arenkiel P, Ai X, Rumbles G, Sugiyama J, et al. Labeling the planar face of crystalline cellulose using quantum dots directed by type-I carbohydrate-binding modules. Cellulose. 2008 Jun 11;16(1):19–26.

[136] Ding S-Y, Xu Q, Ali M, Baker J, Bayer E, Barak Y, et al. Versatile derivatives of carbohydrate-binding modules for imaging of complex carbohydrates approaching the molecular level of resolution. BioTechniques. 2006 Oct;41(4):435–43.

[137] Porter SE, Donohoe BS, Beery KE, Xu Q, Ding S-Y, Vinzant TB, et al. Microscopic analysis of corn fiber using starch- and cellulose-specific molecular probes. Biotechnology and Bioengineering. 2007 Sep 1;98(1):123–31.

[138] Kawakubo T, Karita S, Araki Y, Watanabe S, Oyadomari M, Takada R, et al. Analysis of exposed cellulose surfaces in pretreated wood biomass using carbohydrate-binding module (CBM)â cyan fluorescent protein (CFP). Biotechnology and Bioengineering. 2010 Feb 15;105(3):499–508.

[139] Lacayo CI, Malkin AJ, Holman H-YN, Chen L, Ding S-Y, Hwang MS, et al. Imaging Cell Wall Architecture in Single Zinnia elegans Tracheary Elements. PLANT PHYSIOLOGY. 2010 Jun 30;154(1):121–33.

[140] Umemoto Y, Araki T. Cell Wall Regeneration in Bangia atropurpurea (Rhodophyta) Protoplasts Observed Using a Mannan-Specific Carbohydrate-Binding Module. Marine Biotechnology. 2009 May 23;12(1):24–31.

[141] Dagel DJ, Liu Y-S, Zhong L, Luo Y, Himmel ME, Xu Q, et al. In Situ Imaging of Single Carbohydrate-Binding Modules on Cellulose Microfibrils. J. Phys. Chem. B. 2011 Feb;115(4):635–41.

[142] Fox JM, Jess P, Jambusaria RB, Moo GM, Liphardt J, Clark DS, et al. A single-molecule analysis reveals morphological targets for cellulase synergy. Nature Chemical Biology [Internet]. 2013 Apr 7 [cited 2013 Apr 23]; Available from: http://www.nature.com/doifinder/10.1038/nchembio.1227

[143] Pinto R, Carvalho J, Mota M, Gama M. Large-scale production of cellulose-binding domains. Adsorption studies using CBD-FITC conjugates. Cellulose. 2006 Apr 12;13(5):557–69.

[144] Pinto R, Amaral AL, Carvalho J, Ferreira EC, Mota M, Gama M. Development of a Method Using Image Analysis for the Measurement of Cellulose-Binding Domains Adsorbed onto Cellulose Fibers. Biotechnology Progress. 2007 Dec 7;23(6):1492–7.

[145] Pinto R, Amaral AL, Ferreira EC, Mota M, Vilanova M, Ruel K, et al. Quantification of the CBD-FITC conjugates surface coating on cellulose fibres. BMC Biotechnology. 2008;8(1):1.

[146] Moran-Mirabal JM, Santhanam N, Corgie SC, Craighead HG, Walker LP. Immobilization of cellulose fibrils on solid substrates for cellulase-binding studies through quantitative fluorescence microscopy. Biotechnology and Bioengineering. 2008 Dec 15;101(6):1129–41.

[147] Zhu P, Moran-Mirabal JM, Luterbacher JS, Walker LP, Craighead HG. Observing Thermobifida fusca cellulase binding to pretreated wood particles using time-lapse confocal laser scanning microscopy. Cellulose. 2011 Feb 6;18(3):749–58.

[148] Wang L, Wang Y, Ragauskas AJ. A novel FRET approach for in situ investigation of cellulase–cellulose interaction. Analytical and Bioanalytical Chemistry. 2010 Aug 8;398(3):1257–62.

[149] Wang L, Wang Y, Ragauskas AJ. Determination of cellulase colocalization on cellulose fiber with quantitative FRET measured by acceptor photobleaching and spectrally unmixing fluorescence microscopy. The Analyst. 2012;137(6):1319.

[150] Moran-Mirabal JM, Bolewski JC, Walker LP. Reversibility and binding kinetics of Thermobifida fusca cellulases studied through fluorescence recovery after photobleaching microscopy. Biophysical Chemistry. 2011 Apr;155(1):20–8.

[151] Jervis EJ, Haynes CA, Kilburn DG. Surface diffusion of cellulases and their isolated binding domains on cellulose. Journal of Biological Chemistry. 1997;272(38):24016.

[152] Cuyvers S, Hendrix J, Dornez E, Engelborghs Y, Delcour JA, Courtin CM. Both Substrate Hydrolysis and Secondary Substrate Binding Determine Xylanase Mobility as Assessed by FRAP. The Journal of Physical Chemistry B. 2011 Apr 28;115(16):4810–7.

[153] Moran-Mirabal JM, Bolewski JC, Walker LP. Thermobifida fusca cellulases exhibit limited surface diffusion on bacterial micro-crystalline cellulose. Biotechnology and Bioengineering. 2013 Jan;110(1):47–56.

[154] Horvath AT, Horvath AE, Lindström T, Wågberg L. Diffusion of Cationic Polyelectrolytes into Cellulosic Fibers. Langmuir. 2008 Oct 7;24(19):10797–806.

[155] Himmel ME, Ding S-Y, Johnson DK, Adney WS, Nimlos MR, Brady JW, et al. Biomass Recalcitrance: Engineering Plants and Enzymes for Biofuels Production. Science. 2007 Feb;315(5813):804–7.

[156] Quirk A, Lipkowski J, Vandenende C, Cockburn D, Clarke AJ, Dutcher JR, et al. Direct Visualization of the Enzymatic Digestion of a Single Fiber of Native Cellulose in an Aqueous Environment by Atomic Force Microscopy. Langmuir. 2010 Apr 6;26(7):5007–13.

[157] Santa-Maria M, Jeoh T. Molecular-Scale Investigations of Cellulose Microstructure during Enzymatic Hydrolysis. Biomacromolecules. 2010 Aug 9;11(8):2000–7.

[158] Thygesen LG, Hidayat BJ, Johansen KS, Felby C. Role of supramolecular cellulose structures in enzymatic hydrolysis of plant cell walls. Journal of Industrial Microbiology & Biotechnology. 2010 Sep 18;38(8):975–83.

[159] Luterbacher JS, Walker LP, Moran-Mirabal JM. Observing and modeling BMCC degradation by commercial cellulase cocktails with fluorescently labeled Trichoderma reseii Cel7A through confocal microscopy. Biotechnology and Bioengineering. 2013 Jan;110(1):108–17.

Chapter 2

Cellulosic Fibers: Role of Matrix Polysaccharides in Structure and Function

Polina Mikshina, Tatyana Chernova,
Svetlana Chemikosova, Nadezhda Ibragimova,
Natalia Mokshina and Tatyana Gorshkova

Additional information is available at the end of the chapter

1. Introduction

Cellulose, being the major cell wall component and the most abundant organic matter, produced by living organisms, is not uniformly distributed within plant tissues. There are numerous cells, like the parenchyma ones, which even at maturity have thin cell wall. Thick cell walls are characteristic for the tissues with mechanical function. Among those, there are cell walls, which contain several major components, and those, the predominant component of which is cellulose. The most pure natural cellulose is considered to be present in cotton seed hairs (sometimes erroneously called "cotton fibers") – over 90% of cell wall [1]. Very close to this value is a special group of plant fibers – cellulosic or gelatinous fibers, the proportion of cellulose in which amounts for 85-90% [2,3]. The cell wall thickness in such fibers may reach 15 μm, as compared to 0.2 μm in cells with thin cell wall. So, the very significant portion of total plant cellulose may be concentrated within the gelatinous fibers, making them the important source for production of biofuels and bio-based products. An additional attractiveness of cellulosic fibers for such applications comes from the fact that gelatinous cell wall layers are devoid of lignin – the major hurdle in using plant biomass [1].

Cell wall of cellulosic fiber is of very special design, which provides unusual properties. Such fibers serve as a kind of plant "muscles" [4,5]. The revealing of mechanisms of the formation and function of cellulosic fibers is important for understanding the determinants of general plant architecture and can be useful in construction of new bio-based materials.

2. Definition of plant cellulosic fibers

In terms of plant biology, a fiber is an individual cell of mechanical tissue (sclerenchyma), characterized by the extreme cell length and well developed secondary cell wall, architecture of which is the major determinant of fiber properties. Plant fibers are one of the most economically important raw material, used both for traditional and innovative technologies [6,7]. Due to the massive cell wall, fibers comprise a significant proportion of plant biomass, thus being the valuable source of bio-based products and biofuels. For plant itself fibers are very important for the general architecture and mechanical properties of certain organs.

The functional roles of fibers within the plant and their numerous commercial applications are largely based on the characteristics of their well-developed cell wall of considerable thickness. Fibers of different origin are not uniform in their structure and cell wall composition. The thick cell walls of fully differentiated fibers can be categorized into two broad types – the xylan and the gelatinous ones [2] (Figure 1).

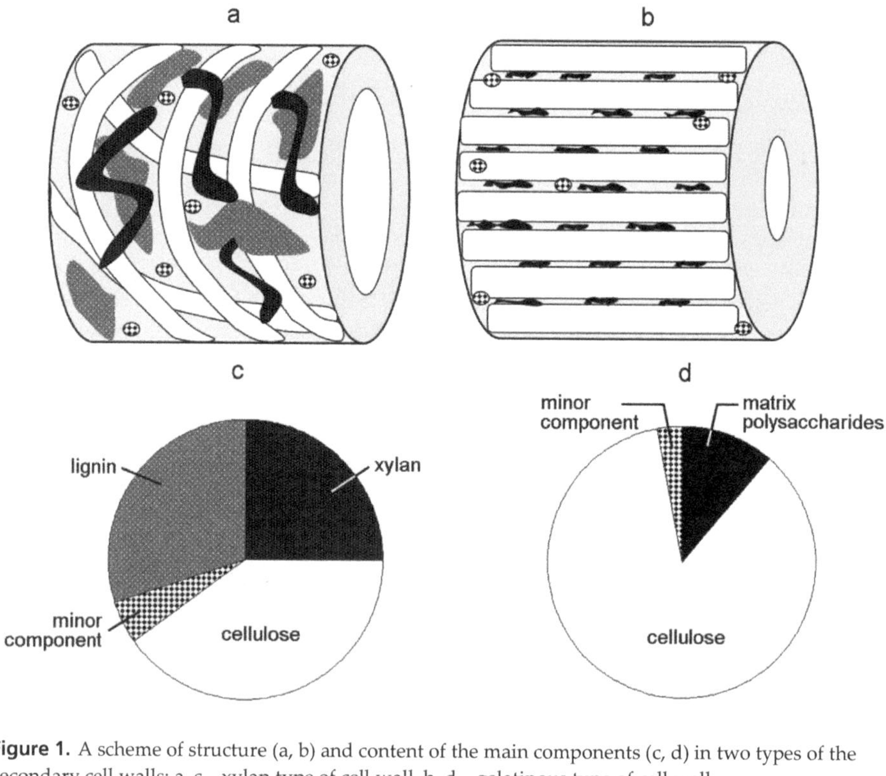

Figure 1. A scheme of structure (a, b) and content of the main components (c, d) in two types of the secondary cell walls: a, c – xylan type of cell wall, b, d – gelatinous type of cell wall.

A secondary cell wall of the xylan type is the most common one in various types of cells with secondary cell walls in land plants. The xylan type secondary cell walls are characterized by helical orientation of cellulose microfibrils, predominance of xylan in non-cellulose matrix, and high degree of lignification. The orientation of cellulose microfibrils may be significantly changed several times through the development of the xylan type secondary cell wall, leading to the formation of distinct layers, designated as S1, S2, and S3 layers (S from "secondary") in the order of deposition. Total thickness of the xylan type secondary cell walls is between 1 and 4 μm. They comprise the bulk of secondary cell walls in various types of wood cells, including vessels and wood parenchyma, and are also present in the bast fibers of some species, for example, jute and kenaf.

The second (gelatinous) type of thick cell wall is present only in fibers. It was firstly described by Th. Hartig at the end of the XIX century as a peculiar layer (G-layer) produced in the reaction wood of dicotyledonous plants (cited after [8]). The reason for such name came from the artefactual swelling of this layer in the cross-section due to the presence of certain components (e.g. alkali) of the solutions used to prepare the sample for microscopy. Fibers, which have developed G-layer of cell wall, get the name "gelatinous fibers". With modern techniques of sample preparation for microscopy, this layer doesn't look like a jelly. Moreover, G-layer was described as having exclusively high content of cellulose (up to 90%) with high degree of crystallinity [9], giving the reason for the alternative name – "cellulosic fibers". However, the justification for a term "gelatinous" was recently provided by description of the gel-like performance of G-layer upon drying (large shrinkage [8,10-12] and high rigidification [13]), and hydrogel type of structure, which has special characteristics of mesoporosity. G-layer has high content of mesopores (pore size between 2 nm and 50 nm); in tension wood fibers the pore surface areas may be more than 30 times higher than that in normal wood as was revealed by nitrogen adsorption technique [8, 14]. Mesoporosity was suggested as a new parameter for G-layer characterization [14].

Gelatinous cell-wall layer is deposited inward to the xylan type secondary wall layers; the degree of S-layer development in fibers with G-layer differs from well pronounced, like in tension wood [15], to barely detectable, as in flax [2]. Though not appropriately recognized, this type of fibers is widespread and is present in various organs of plants from many taxa [2,3]. Among others, phloem fibers of flax, hemp, and ramie, gelatinous fibers of tension wood, some fibers of bamboo and *Equisetum* belong to this group. Arabidopsis was shown to have the potential for gelatinous fiber formation [16], same as some other plant species where this type of fibers was not well known, like alfalfa [17]. Fibers able to form the gelatinous cell wall layer can originate from both primary and secondary meristems and be located within phloem or xylem [18].

Specific characteristics of the gelatinous layer of cell wall include: a) the overwhelming content of cellulose (80-90%); b) high crystallinity of cellulose; c) very low angle of cellulose microfibrils, which are laid almost parallel to the fiber's longitudinal axis throughout the whole layer; d) considerable thickness, which in some species can reach more than 15 μm; e) the absence of xylan, f) the absence of lignin, g) special composition of matrix polysaccharides, presented mostly by galactose-containing pectins with rhamnogalacturonan backbone; h) high water content, as compared to S-layers, i) mesoporosity, j) exclusive presence in fibers, and, as discussed below, k) contractile properties.

The very peculiar characteristic of fibers with the gelatinous cell wall is their contractile properties. Such fibers may serve to move the plant parts in space. For instance, the tension wood fibres contract longitudinally during differentiation and generate longitudinal tensile stresses of up to about 70 MPa [19], providing rightning force in the tilted tree. A high-tensile growth stress generated on the surface of the xylem in the tension wood region often becomes ten times as large as that in the normal wood region [20]. The ability of a plant organ to contract is proportional to the degree of the development of fibers with G-layers [21,22].

Plants do not possess animal-type muscles, which contract due to protein-protein interactions. However, they have a different mechanism, which has the ability to move even very heavy plant parts in space. This mechanism is specifically developed in cellulosic fibers. Their contractile properties are based on tension developing within the specially designed thick cell walls. The efficacy of such fibers is remarkable. Thus, the gelatinous fibers may to a certain extent be named as "cell-wall-based plant muscles", though they do not have the ability to relax and the time-scale of their contraction is very different from that of animal muscles.

The examples of contractile action of cellulosic fibers (Figure 2) include the restoration of young tree vertical position, if disturbed [15,23], deepening of geophyte shoot in the soil by contraction of roots in the course of adaptive reaction [24], shortening of aerial roots when they reach the soil to form the effective support to heavy branches [21]; correction of lateral branch angle and, in case of apical meristem death, upward bending of lateral branch in order to transform into dominant [25-27]. The cellulosic gelatinous fibers are especially developed in the plants exposed to high mechanical stresses due to high ratio between stem height and stem diameter (most of fiber crops, like flax, hemp, ramie, nettle) [2] – this gives the properties of spring to the plant stem, helping to restore vertical position when disturbed, for example by wind. Such plants, similar to plants with developed tension wood [28,29], also exhibit the pronounced negative gravitropism, being able to turn back to vertical position a long stem, if bent, far away from the growing apical stem region. Cellulosic fibers seem to be present in the spines of cacti [30], and are developed in peduncles helping to support heavy fruits, like in the sausage tree, fruits of which may weight up to 10 kg [31]. The gelatinous fibers were demonstrated to be widely involved in the twining of vines and the coiling of tendrils [32,33].

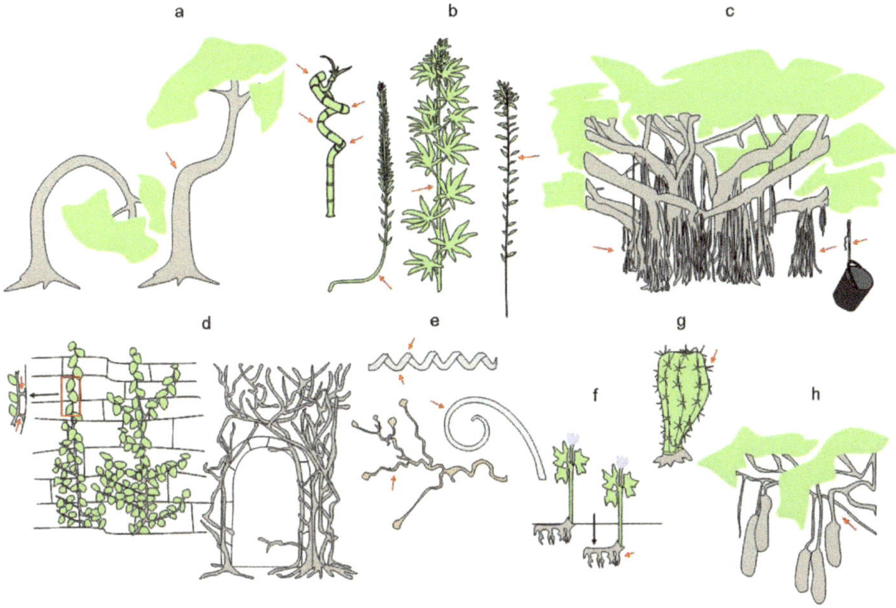

Figure 2. The examples of contractile action of cellulosic fibers: a – formation of tension wood during restoration of young tree vertical position, if disturbed, b – development of gelatinous fibers in the plants with high ratio between stem height and stem diameter, c – shortening of aerial roots after they reach the soil to form the effective support to the heavy branches (when rooted in pot, aerial roots may rise it up), d – development of cellulosic fibers in parts of stem due to which plant attaches to substrate, e – involvement of gelatinous fibers in the twining of vines, the coiling of tendrils and the expansion of climbing plants, f – deepening of geophyte shoot into the soil by contraction of roots in the course of adaptive reaction, g – presence of cellulosic fibers in the spines of cacti, h – development of gelatinous fibers in peduncles to help support heavy fruits.

3. Cellulose microfibrils in gelatinous cell wall

The very specific characteristic of cellulose microfibrils in the gelatinous cell wall is their axial orientation, which is not observed in any other cell wall type of any other but fiber cell type. This is especially remarkable, if one remembers the total thickness of G-layers. The axial orientation of cellulose microfibrils throughout the gelatinous cell wall layer was known long ago [9,34] and was confirmed by several techniques, including microbeam X-ray diffraction [35], wide-angle X-ray scattering [36], and scanning Raman microscopy [24,37]. In accordance to that, cortical microtubules, which are considered to rule the microfibril orientation, are axially oriented during deposition of the gelatinous cell wall layers [38].

In the gelatinous fibers cellulose microfibrils are characterized by a higher degree of crystallinity and a larger size of crystalline regions (crystallites) as compared with most other plant tissues [39, 40]. The diameter of cellulose crystallites was measured in various species by several authors and though the absolute values might differ, the general conclusion was that in G-layers it was larger than in the S-layers [12,35,41-44]. The roentgen structural analysis showed that the diameter of cellulose crystal transverse sections in tension wood G-layer (6.5 nm) is markedly larger than in the neighboring S-layer of the xylan cell wall (about 3 nm), i.e., its section area is approximately fourfold higher [35]. Thus, cellulose microfibrils within the gelatinous layer exist in the form of aggregates. This allowed a supposition that individual cellulose microfibrils, each of which is formed by individual cellulose-synthesizing complex, so-called "rosette", in the gelatinous layer interact laterally [4]. Such lateral interaction is stimulated due to similar (axial) orientation of all microfibrils, the absence of lignin and of considerable amount of matrix polysaccharides, which separate microfibrils, like in S-layer. Despite the high degree of crystallinity of the cellulose, G-layer has a remarkable hygroscopicity and high water content [37,45].

Cellulose microfibrils in the gelatinous layer are under tension. This was proved by the increase in cellulose lattice spacing revealed by synchrotron radiation microdiffraction [46]. The visual demonstration of tension comes from shrinkage of G-layer along the cell longitudinal axis upon the release of tension, as observed by scanning microscopy of tension wood cross-sections [10].

For a long time, the gelatinous cell wall was believed to be composed of cellulose only [9]. Correspondingly, the ideas on tension origin were based on cellulose microfibril properties. Cellulose microfibrils themselves are virtually incontractible. So, the problem was how to get contraction, having incontractible basis of cellulose microfibrils. One of the possible solutions suggests that the contraction of the fibre is not caused by the G-layer directly, but by interaction of the G-layer with the surrounding S-layer. It was proposed that the origin of tension in cell wall of cellulosic fibers lays in the differential parameters of swelling of the S- and G-layers due to different orientation of cellulose microfibrils [36].

The cell walls with helicoidal orientation of microfibrils increase in length upon swelling, while the ones with axial orientation shrink [47]. The stress–strain-curves of cell walls show the influence of the cellulose microfibril orientation on the deformation behavior of plant tissues [48]. Within plant organism, such differences are indeed exploited in some mechanisms, for instance in opening of pine cones [49]. Fibers and sclereids located at the opposite sides of a cone scale have different angle between the long axis of the cell and the direction of cellulose microfibrils (MFA): it is high in sclerids and low in fibers. Correspondingly, the coefficient of hygroscopic expansion of fibers is significantly lower than that of sclerids. Due to that, the increase in relative humidity leads to the increase of the angle between the scale and the frame, leading to cone opening. Similar is the mechanism of wheat awn opening aimed to seed dispersal [48,50].

Similarly, the differences in swelling of S- and G-layers were suggested to explain the formation of tension in fibers of reaction wood [36]. The idea is based on the established fact that the enzymatic removal of the G-layer lead to the longitudinal extension and tangential shrinkage of tissues within the tension wood slice. It was proposed that in the living plant, a lateral swelling of the G-layer forced the surrounding S-layers to shrink in the axial direction.

Such forces may indeed be the part of the tension creation mechanisms in gelatinous fibers. But: 1) such system would be highly dependent on humidity, same as the openings of pine cones and wheat awns; 2) in some species, like flax or ramie, S-layer in the gelatinous fibers is poorly developed [2] and hardly may serve as a mechanical counterpart of very thick G-layer; 3) mesoporosity of the G-layer is not explained; 4) tension is argumented to be developed within the G-layer itself [10,51]. Finally, specific matrix polysaccharides, which were not considered in the above hypothesis, appear at the onset of G-layer formation.

4. Matrix polysaccharides in cellulosic fibers

The presence of a polymer within a certain cell wall layer is not easy to prove. The biochemical analysis in the majority of studies is usually performed without separation into different cell wall layers, which is rather hard to achieve. In most of the experiments on the analysis of the gelatinous fiber composition, primary cell wall was not detached from the secondary one, and the xylan layers were not separated from the gelatinous ones. Moreover, such analysis is often done on the samples, like wood, which contain complex mixture of various cell types (e.g. parenchyma, vessels, fibers) at different stages of the development. A significant amount of data concerning the composition and structure of the gelatinous type cell walls was obtained by the analysis of phloem fiber bundles, which extreme strength permited their mechanical or enzymatic separation from surrounding tissues. Although gelatinous layers predominate in such fiber cell walls, the primary cell wall and S-layer of the secondary wall are also present. That's why, for instance, polygalacturonic acid (PGA) or rhamnogalacturonan I (RG I), described in numerous papers on plant fibers (e.g. [52-55] can not be attributed to a certain cell wall layer. The presence of polymers just in the G-layers must be additionally proved.

To do so, several approaches can be used or, better, combined: a) isolation of the G-layers and the biochemical analysis of constituents; b) cytochemistry, including immunocytochemistry; c) the analysis of deposition dynamics: search for the marker monomer, sugar linkage type or other specific characteristic of a certain polysaccharide, the formation of which goes in parallel to the G-layer deposition; pulse-chase experiments with labeled precursors can be especially effective since they permit to exclude the background of previously synthesized polymers; d) tracing the transcription of the identified genes, involved in the metabolism of certain cell wall polysaccharide, in the course of the G-layer formation; e) detection within G-layer of the enzyme or enzymatic activity, involved in modification of matrix polysaccharide, by various types of staining. The best way to analyze the components of the gelatinous cell walls is isolation of the G-layers, like it was done for poplar tension wood [28,56]. To this end, thin tissue sections (20 μm) are prepared and treated with ultrasound; however, this procedure permits obtaining only small amount of the material.

Among neutral monosaccharides present in polymers of the isolated G-layers, rhamnose, arabinose, galactose, and xylose were found along with glucose [56]. The cell wall matrix polysaccharides, which were identified or suggested to be present in the G-layer include xyloglucan [56], arabinogalactan proteins [57], pectic galactan [58], and, probably some arabinans [30]. Usually these polymers are considered to be the components of the primary cell walls. The obtained data on their presence in the G-layer were recently summarized in the review [5], so here we consider them only briefly.

Arabinogalactan proteins (AGPs) are highly water soluble polymers, which consist of protein backbone and carbohydrate side chains of variable structure, which can comprise over 90% of the molecule. Glycan component of AGPs has various length chains of β-(1→4)-galactan and β-(1→6)-galactan units, often decorated by terminal arabinose residues and connected to each other by (1→3,1→6)-linked branch points, which are indicative of AGPs. Protein backbone may also vary in structure and is encoded by a large gene family, the several representatives of which are always detected among the most up-regulated upon the G-layer induction genes, both in tension wood [59-62] and in fiber crops [63-65]. Carbohydrate constituents of AGPs were detected within the G-layer by immunocytochemical [57,60], cytochemical (by staining with Yariv reagent) [58,66] and biochemical [56] approaches. AGPs, different both in carbohydrate and protein part of the molecules, are present in many, if not all, plant tissues, but their exact function is still unknown.

There is not much information about another possible constituents of the gelatinous fibers – the arabinans. They were reported to be the major cell wall matrix component of cellulosic fibers in cactus spines [30]. It is not clear if the arabinans are attached to the RG I backbones.

The most substantial evidence for the matrix cell wall polysaccharides of the G-layers was collected on xyloglucan. This cross-linking glycan is composed of a backbone, which is built similar to cellulose molecule as β-(1→4)-glucan. The side chain of xylose, which is sometimes additionally substituted by galactose and further – by fucose, are attached to the backbone. Xyloglucan is the major noncellulosic polysaccharide in the isolated G-layers of poplar tension wood; its content was assessed to be 10–15% of the cell wall mass [28,56,67]. The presence of xyloglucan was detected by several methods, including the biochemical analysis of the types of bonds between monosaccharides and immunocytochemistry. Moreover, the presence of xyloglucan endotransglycosylase, an enzyme providing for connection between the regions of two different xyloglucan molecules, was demonstrated in the G-layers of the secondary cell wall. Two main functions were suggested for xyloglucan in the secondary cell walls of tension wood fibers [67]. The first one is binding of the G-layer to the neighboring xylan layer because xyloglucan and xyloglucan endotransglycosylase are localized just at the boundary between these two layers, as was shown immunocytochemically. The second supposed function is the creation of tension – it will be considered in the next chapter.

One more component of the G-layers is pectic galactan, built as a very complex rhamnogalacturonan I with a high degree of branching and a varying structure of side chains, which are mainly composed of β-(1→4)-galactose [52,58,68]. The predominant monosaccharide in the polymer is galactose, which determines the polymer name as a galactan [69]. The side chains may include only one or two galactose residues; long chains of several tens galactose residues, likely branched side chains, which are not cleaved by galactanase; side chains, decorated with a single pentose, most likely arabinose or a galactose residue connected by other than β-(1→4) linkage [68,70]. This type of pectic galactan is fiber- and stage-specific, being present only in fibers, while forming G-layer [2,71].

Pectic galactan may be of the specific three-dimensional organization, the signs of which are revealed upon the treatment with specific glycanases. The hydrolysis of considerable part of galactose side chains of galactan as well as the partial degradation of its backbone do not change the total hydrodynamic volume, which determines the efficiency of elution from gel-filtration column, and the polymer elutes in the same part of profile, as before enzymatic treatments [68,72]. The unusual property of pectic galactan from the gelatinous fibers is the ability to form water-soluble associates, so that the charged backbone is located at the periphery of it, while the neutral side chains form the core zone (Mikshina, Gorshkova, in preparation).

The presence of the galactan within the gelatinous layer was confirmed by the analysis of the dynamics of its formation, which coincides with the G-layer deposition [71], by immunolocalization of the galactan side chains [71,73-75], by presence of tissue- and stage specific β-(1→4)-galactosidase, the substrate of which is the described galactan [75-77]. The gene of this galactosidase is highly upregulated at the onset of the G-layer formation [63, 78] and the activity is detected within fibers, forming gelatinous cell wall [64,75].

The complex galactans built mainly from β-(1→4)-galactose were found in tension wood fibers in 60-ties of the XXth century [79,80], though the linkage with the RG I backbone was not proved at that time. The content of galactose was even suggested as an indicator of the extent of the G-layer development [81]. However, these old data were actually put away for several decades due to the overwhelming notion that the G-layers were pure cellulosic, so that the published in 2008 paper describing the detection of rhamnogalacturonan I by cytochemical approaches in tension wood of several species was entitled "...gelatinous fibers contain more than just cellulose" [57].

5. Matrix polymers as the causative agent for cellulose tension in gelatinous cell wall

Presence of specific matrix polysaccharides within G-layer suggests their importance for function of cellulosic fibers, including tension creation to form contractile properties. Mellerowicz et al. [4] put forward an idea that matrix polysaccharides are entrapped by laterally interacting cellulose microfibrils. The presence of such entrapped polysaccharides between cellulose microfibrils limits their interaction and results in creation of tension, which underlies specific mechanical properties of cellulosic fibers (Figure 3).

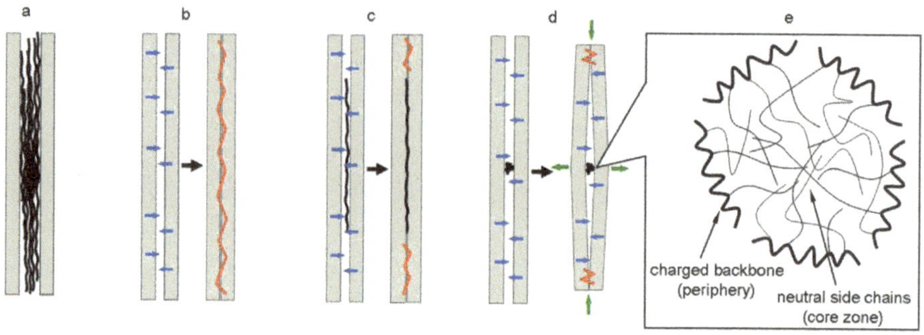

Figure 3. Possible ways of interaction between matrix polysaccharides and cellulose microfibrils in various types of cell walls: a – high content of matrix polysaccharides in xylan secondary cell wall prevents lateral interaction of cellulose microfibrils, b – microfibrils of the G-layer (gelatinous cell wall) with low content of matrix polysaccharides, cellulose microfibrils tend to lateral interactions, giving reason for higher degree of crystallinity and larger size of crystallites, c – theoretically, if matrix polysaccharides has high affinity to cellulose, being entrapped they won't cause much of tension, d – the most effective to provide longitudinal tensile stress in the cellulose microfibrils is compact polysaccharide of considerable size with low affinity to cellulose, e – a model of pectic galactan associates, in which negatively charged RG I backbone is at the periphery, and long galactose side chains form the core zone.

Originally, xyloglucan was proposed as the polymer entrapped by the laterally interacting cellulose microfibrils. Xyloglucan is, indeed, very important for the function of gelatinous fibers in tension wood of some species, like poplar. This is proved by the fact that in transgenic poplars with the expressed gene of fungal xyloglucanase, which decreased the content of xyloglucan, righting of stem basal regions in placed horizontally young plants was completely abolished, while the G-layer formation was not affected [29,82]. The expression of other endoglycanases, which decreased the levels of xylan or arabinogalactan, had no effect on the ability of transgenic plant to exhibit gravitropic reaction, restoring the stem vertical position.

However, the exact function of xyloglucan in tension wood fibers is still a matter of debate. Firstly, xyloglucan was detected in the G-layers only in limited plant species, it was never conclusively reported (though searched) to be present in thick secondary walls of cellulosic fibers in fiber crops, like flax, hemp, etc. Further evidence comes from the analysis of polysaccharides, strongly retained by cellulose microfibrils upon extraction: if entrapped between the interacting laterally cellulose microfibrils, a polymer should not be extracted by the conventional methods and should come out only after degradation of microfibrils by chemical or enzymatic means. However, due to high crystallinity of cellulose in the gelatinous layers, in natural form it is poorly degraded by specific enzymes [66] and thus, has to be first dissolved by corresponding chemicals.

To analyze the polysaccharides, which are especially strongly retained within cell wall, a special protocol was developed [83]. After removal of the extractable polysaccharides by chelators and concentrated alkali, the residual cell wall material was dissolved in solution of lithium chloride in N,N-dimethylacetamide and afterwards cellulose was precipitated by water. Such treatment turned natural cellulose I (with parallel orientation of individual cellulose chains) into cellulose II (with antiparallel orientation of individual cellulose chains) and made it completely degradable by purified cellulase. The matrix polysaccharides, which were present in the fraction, remained in polymeric form, making possible to separate them by gel-filtration for further analysis.

We have compared the composition of matrix polysaccharides, strongly retained by cellulose microfibrils, in fibers with different proportions of the secondary cell walls of xylan and gelatinous types (Figure 4). The polymers from fibers with only xylan type secondary cell wall eluted in the region below 30 kDa. The monosaccharide analysis and antibody binding indicated that the major component of this fraction was xylan. It is known that small proportion of matrix polymers, both in the primary and the secondary cell walls get entrapped by cellulose microfibrils in the process of their crystallization [84,85]. Some polygalacturonic acid was also present, which could be originated from the primary cell wall.

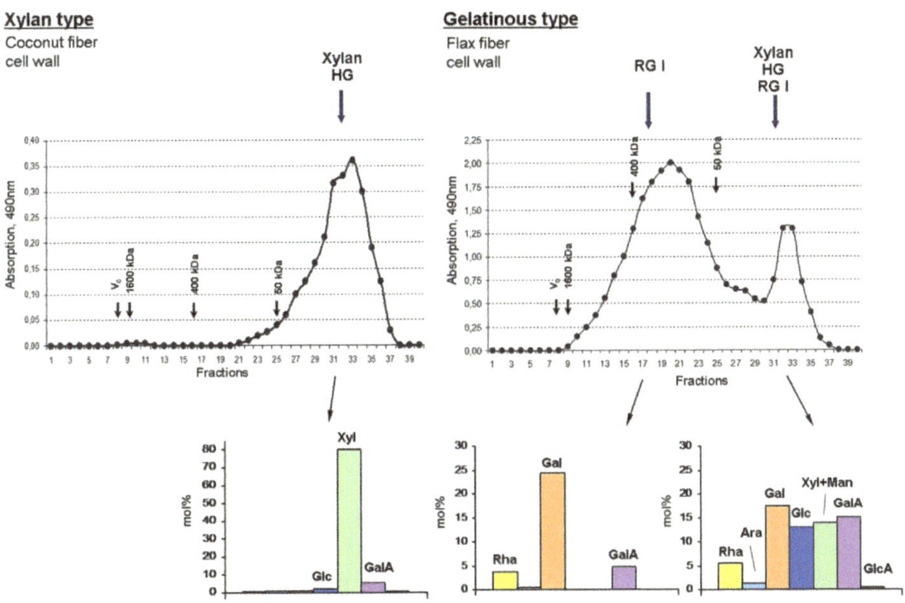

Figure 4. Elution profiles of polysaccharides strongly retained by cellulose microfibrils in the xylan and gelatinous cell walls and the relative monosaccharide content (mol%) of the main fractions of these polysaccharides.

In fibers with the G-layers, the major peak of matrix polysaccharide eluted between 100 and 400 kDa; its predominating component was pectic galactan. This galactan from flax fibers was characterized by various techniques, including ^{1}H and ^{13}C NMR and antibody binging [83]. The ratio between high and low molecular mass peaks on the elution profile depended on the proportion of the S- and G-layers within the fiber cell wall. Antibody to xyloglucan epitopes didn't bind any fraction on the elution profile.

The proportion of pectic polymers, which were strongly retained by cellulose microfibrils, from their total content in cell wall of the gelatinous type, could be much higher than that of xylan in the S-layers. Such selectivity in entrapping of certain polymers can not be explained by their affinity to cellulose as the charged pectic molecules are far less competitive, compared to xylan. The obtained data suggest the alternative mechanism of interaction between cellulose and pectic galactan, which is specifically developed in cell walls of the gelatinous type.

The above data suggest that in cellulosic fibers it is pectic galactan that is entrapped by laterally interacting cellulose microfibrils. This polymer, due to ability to form associates, can perfectly fit the proposed function in tension creation, as illustrated in Figure 3. In the xylan type of secondary cell wall (a), high content of matrix polysaccharides prevents the lateral interaction of cellulose microfibrils. At low content of matrix polysaccharides in G-layer (b), cellulose microfibrils tend to lateral interactions, giving reason for higher degree of crystallinity and larger size of crystallites. Matrix polysaccharides with high affinity to cellulose, if entrapped (c) won't cause much of tension. Most effective would be compact polysaccharide of considerable size with low affinity to cellulose (d). Associates of pectic galactan with RG I backbone may be a good choice of Nature for such purpose. They have compact structure of considerable volume, which has poor ability (due to charged surface) to interact with cellulose (e).

Additional arguments for the important role of pectic galactans in creation of tension come from the analysis of the course of the G-layer formation and of *in muro* modifications of matrix polymers, which was in detail performed on flax cellulosic fibers.

6. Dynamics of the G-layer formation and in muro modifications of cell wall polymer

Formation of a cell wall layer is a complicated event. Partly it is based on the processes of polysaccharides' self-assembly in specific surroundings. Besides, the cell wall formation may involve modification of the interacting polysaccharides. The very illustrative example of the latter is the remodeling of the deposited G-layer in flax cellulosic fibers. In the dynamics of the G-layer formation two stages are clearly visualized at microscopic investigation of the flax fiber cell wall formation [71,86]. Under electron and/or light microscope, one can see that the inner part of the cell wall has a characteristic appearance of the loose structure where the electron dense parallel bands alternate with light regions; the outer part has much more homogenous structure (Figure 5). These two parts of the cell wall are designated as the Gn- and G-layers. During formation of the secondary cell wall, the thickness of outer layer gradually increases, while additional portions of the Gn-layer are added by the protoplast. This indicates that with time the Gn-layer is transformed into the G-layer.

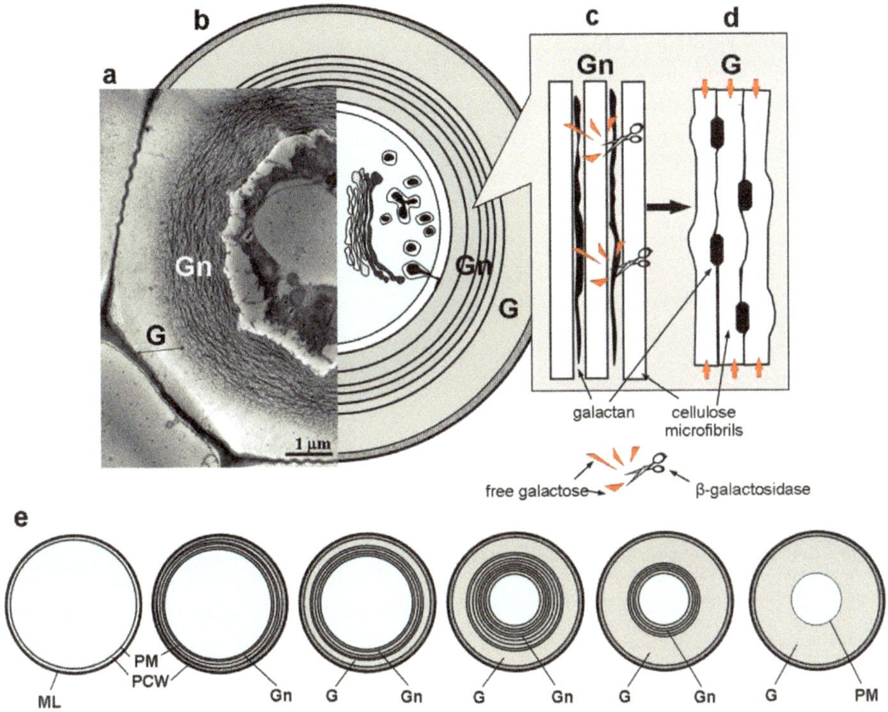

Figure 5. A model of the Gn-layer to the G-layer transformation in gelatinous fibers: a – electron microscopy of developing flax fiber cross-section; two layers (Gn and G) are obvious; b – scheme of developing flax fiber cross-section, showing a tissue-specific galactan delivered by specific Golgi vesicles to the developing Gn-layer; c – the nascent galactan is interspersed between cellulose microfibrils, preventing their association and maintaining the loosely packed morphology characteristic of the Gn-layer of secondary cell wall. During cell wall maturation, high molecular galactan partially digested by β-galactosidase, releasing free galactose; d – reducing of side chain length of galactan by galactosidase allows cellulose microfibrils to interact laterally, entrapping the galactan. Thus densely packed G-layer that is rich in crystalline cellulose is formed. The presence of entrapped galactan during lateral interactions of axially oriented microfibril causes longitudinal tensile stress in cellulose; e – dynamics of gelatinous layers deposition and remodeling in cellulosic fibers (left to right). ML – middle lamellae, PM – plasmalemma, PCW – primary cell wall, Gn – newly deposited gelatinous layer of secondary cell wall, G – mature gelatinous layer of secondary cell wall.

Transition from the Gn- to G-layer is coupled with changes in cellulose crystallinity. It is confirmed by the cytochemical analysis using the enzyme–gold complex, which showed that as distinct from G-layer, Gn-layer poorly bound cellobiohydrolase, the substrate of which is crystalline cellulose [58,87]. Besides, in pulse-chase experiments with $^{14}CO_2$ with intact flax plant, the dynamics of cellulose crystallization in fiber-enriched peels from all other analyzed samples. In roots, stem xylem, and stem apical part, which do not contain fibers with the gelatinous cell wall, the proportion of crystalline cellulose did not change during the entire experiment, while in fibers starting at the same level as in other tissues, it increased twice through the first day of chase and only later attained the plateau, which was at much higher level than in other tissues [88] (Figure 6). This indicated that crystallization of cellulose microfibrils in the G-layer was a biphasic process: the first stage occured right after the individual cellulose chain synthesis, similar to other plant tissues, while the second stage, which gave additional increase in crystallinity, occured *in muro* – within cell wall.

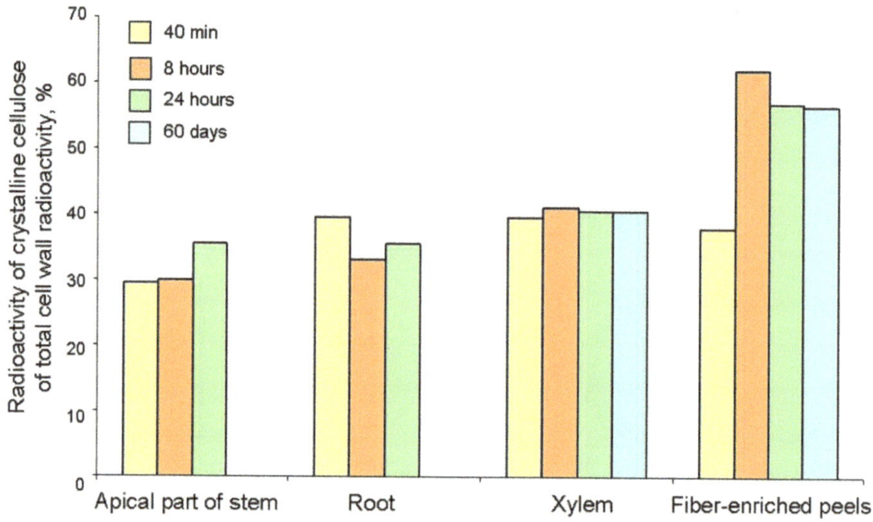

Figure 6. Radioactivity of crystalline cellulose of total cell wall radioactivity in the different part of flax plant after 40 min of photosynthesis with $^{14}CO_2$ (pulse) and during different periods of plant growth in the absence of a radioactive substrate (chase). Modified from data in [88].

The Gn- and G-layers differently bind not only cellobiohydrolase probe, but also the LM5 antibody, which is specific for β-(1→4)-galactan [89]. With LM5 the number of gold particles per area unit in the Gn-layer was fivefold higher than in the G-layer [71]. So, the reverse pattern was observed with binding the probes for cellulose crystallinity and for pectic galactan. Keeping in mind that antibody binding depends not only on the presence of the epitope but also on its availability, we consider it possible to suppose that changes in the degree of cellulose crystallization were related to *in muro* modification of tissue-specific galactan. An additional argument for such suggestion is a disappearance in the G-layer of dark bands, which are produced in the Gn-layer at galactan secretion by the Golgi apparatus and are well distinguished under electron microscope [71,74].

The pectic galactans are subjected to intensive *in muro* modifications. The investigation of galactan metabolism using the pulse-chase approach [2] confirmed that this polymer is synthesized in the Golgi apparatus, secreted outside the plasma membrane, and interacts with cellulose microfibrils. Flax fibers, while forming the secondary cell walls, have a peculiar mechanism of polysaccharide secretion. Golgi-derived vesicles first accumulate in the cytoplasm and only later fuse with the plasma membrane to give their contents to the apoplast [74]. These Golgi derivatives as well as the layers of the secondary cell wall, especially the inner "striated" layer, bind the LM5 antibody, indicating that all these structures contain galactan. It suggests that this peculiar type of galactan secretion permitting for filling large spaces of the periplasm facilitates the contact between galactan and cellulose microfibrils when they are in the process of assembly and may be necessary for preventing lateral interaction of cellulose microfibrils right at their deposition. Such a mechanism of secretion allows to accumulate a sufficient amount of the nascent pectic galactan before it is incorporated into the cell wall. This nascent form of the pectic galactan can be collected from the tissue homogenization buffer and compared to the polymer strongly retained by cellulose microfibrils. The composition and structure of these polysaccharides together with tracing in pulse-chase experiments [68-70,83] permit to consider the entrapped by cellulose microfibrils galactan as a derivative of the nascent galactan. The comparison of these polymers revealed the following differences: the nascent polysaccharide elutes at gel-filtration as having higher molecular mass (in the 700-2000 kDa region) and has higher degree of branching and longer side chains, as compared to cell wall galactan [83].

The detected differences between the nascent and entrapped galactans suggested that they might be the result of *in muro* galactan modification by the enzyme cleaving off a part of the galactan side chains. Indeed, the histochemical staining of stems and hypocotyls with corresponding chromogenic or fluorogenic substrates shows β-galactosidase activity to be localized to developing fibers [64,76]. The gene of β-galactosidase is among the most up-regulated ones upon induction of the G-layer formation [63,78]. The substantial amounts of free galactose, which is the product of β-galactosidase action is present specifically in fibers forming gelatinous cell wall [76].

Shortening of the galactan side chains permits microfibril lateral interaction, due to which an additional portion of galactan is captured by them. The necessity of pectic galactan modification with the participation of β-galactosidase for the remodeling of cell wall supramolecular structure and transformation of the Gn-layer into mature the G-layer was demonstrated [75]. The role of fiber-specific β-galactosidase in providing the particular mechanical properties of gelatinous fibers was confirmed with transgenic flax plants (reduced galactan modifications – less mechanical strength) [75]. Antibodies raised to fiber-specific β-galactosidase of flax, revealed similar protein in the G-layers of cellulosic fibers in other plants (poplar tension wood fibers and both primary and the secondary phloem fibers of hemp), indicating that the process of the G-layer remodeling may be similar in fibers of different origin [77].

Thus, in the last several years the views on matrix polysaccharides of the gelatinous cell walls have changed dramatically: from rejecting their presence – to ascribing the major role to them in the development and function of cellulosic fibers.

7. Conclusions and perspectives for future research

Summary of our ideas on the cell wall design of cellulosic fibers and the origin of their contractile properties include the following statements, based on the considered in the current review literature data and our own results:

- Tension is caused due to lateral interaction of cellulose microfibrils and entrapment of matrix polysaccharides.
- Lateral interaction is possible because of very high cellulose content, absence of xylan and lignin.
- Similar axial orientation of all cellulose microfibrils in thick G-layer helps to cumulate tension of individual microfibrils and to develop it in the necessary direction. The effect is increased due to extreme length of fiber cells.
- The entrapped polysaccharide – complex rhamnogalacturonan I with galactan side chains of specific structure and distribution, which is able to form water-soluble associates.
- Entrapment of such associates leads to increased mesoporosity and to the development of cellulose microfibril tension.
- High hydroscopic capacity of RG I helps to keep water in the G-layer.
- Conditions for lateral interaction of cellulose microfibrils may be provided by *in muro* modification of deposited polysaccharide by fiber-specific galactosidase.
- Additional important factors may be the interaction of the G-layer with the S-layer through the action of xyloglucan-modifying enzyme, the activity of which is mainly detected at the boundary between layers, and/or different deformation behavior of the S- and the G-layers upon swelling due to different orientation of cellulose microfibrils.

Cellulosic fibers are the example of very peculiar cell wall type. Its formation includes significant reprogramming of synthesis and secretion of matrix polysaccharides, reorientation of cellulose microfibrils, active remodeling of the deposited cell wall layers, specific inter- and intra-molecular interactions between cell wall polymers. The study of these processes may give additional clues for general understanding of the plant cell wall formation, which still belongs to the most enigmatic biological processes. Of special interest is the investigation of specific three-dimensional organization of pectic galactans from cellulosic fibers in order to elucidate the largely unknown principles of supramolecular structure of complex polysaccharides. Comparison of the gelatinous cell wall formation in fibers of various organs may help to figure out the biological determinants of plant fiber yield and quality in order to improve the characteristics of plant biomass for effective conversion into biofuels and bio-based products.

Author details

Polina Mikshina, Tatyana Chernova, Svetlana Chemikosova, Nadezhda Ibragimova, Natalia Mokshina and Tatyana Gorshkova
Kazan Institute of Biochemistry and Biophysics, Kazan Scientific Centre, Russian Academy of Sciences, Kazan, Russia

Acknowledgement

This work was partially supported by the Russian Foundation for Basic Research (project no. 11-04-01602), and the Program of State Support of Leading Scientific Schools (project no. 825.2012.4, 12-04-31418).

8. References

[1] Albersheim P, Darvill A, Roberts K, Sederoff R, Staehelin A. Plant Cell Walls. New York: Garland Science; 2011.

[2] Gorshkova TA, Gurjanov OP, Mikshina PV, Ibragimova NN, Mokshina NE, Salnikov VV, Ageeva MV, Amenitskii SI, Chernova TE, Chemikosova SB. Specific Type of Secondary Cell Wall Formed by Plant Fibers. Russian Journal of Plant Physiology 2010;57 328-341.

[3] Gorshkova T, Brutch N, Chabbert B, Deyholos M, Hayashi T, Lev-Yadun S, Mellerowicz EJ, Morvan C, Neutelings G, Pilate G. Plant Fiber Formation: State of the Art, Recent and Expected Progress, and Open Questions. Critical Reviews in Plant Sciences 2012;31(3) 201-228.

[4] Mellerowicz EJ, Immerzeel P, Hayashi T. Xyloglucan: the Molecular Muscle of Trees. Annals of Botany 2008;102 659-665.

[5] Mellerowicz E, Gorshkova TA. Tensional Stress Generation in Gelatinous Fibres: a Review and Possible Mechanism Based on Cell-Wall Structure and Composition. Journal of Experimental Botany 2012;63(2) 551-565.

[6] Bledzki AK, Gassan J. Composites Reinforced with Cellulose Based Fibres. Progress in Polymer Science 1999;24 221-274.

[7] Jacob M, Thomas S. Biofibres and Biocomposites. Carbohydrate Polymers 2008;71 343-364.

[8] Clair B, Gril J, Di Renzo F, Yamamoto H, Quignard F. Characterization of a Gel in the Cell Wall to Elucidate the Paradoxical Shrinkage of Tension Wood. Biomacromolecules 2008;9 494-498.

[9] Norberg PH, Meier H. Physical and Chemical Properties of Gelatinous Layer in Tension Wood Fibers of Aspen (*Populus Tremula* L.). Holzforschung 1966;20 174-178.

[10] Clair B, Thibault B. Shrinkage of the Gelatinous Layer of Poplar and Beech Tension Wood. IAWA Journal 2001;22 121-131.

[11] Yamamoto H, Abe K, Arakawa Y, Okuyama T, Gril J. Role of the Gelatinous Layer (G-Layer) on the Origin of the Physical Properties of the Tension Wood of *Acer Sieboldianum*. Journal of Wood Science 2005;51 222-233.

[12] Yamamoto H, Ruelle J, Arakawa Y, Yoshida M, Clair B, Gril J. Origin of the Characteristic Hygro-Mechanical Properties of the Gelatinous Layer in Tension Wood from Kunugi Oak (*Quercus Acutissima*). Wood Science and Technology 2010;44 149-163.

[13] Clair B, Ruelle J, Thibaut B. Relationship between Growth Stress, Mechanical-Physical Properties and Proportion of Fibre with Gelatinous Layer in Chestnut (*Castanea Sativa* Mill.). Holzforschung 2003;57 189-195.

[14] Chang SS, Clair B, Ruelle J, Beauchene J, Di Renzo F, Quignard F, Zhao GJ, Yamamoto H, Gril J. Mesoporosity as a New Parameter for Understanding Tension Stress Generation in Trees. Journal of Experimental Botany 2009;60 3023-3030.
[15] Mellerowicz EJ, Baucher M, Sundberg B, Boerjan W. Unraveling Cell Wall Formation in the Woody Dicot Stem. Plant Molecular Biology 2001;47 239-274.
[16] Wyatt SE, Sederoff R, Flaishman MA, Lev-Yadun S. *Arabidopsis Thaliana* as a Model for Gelatinous Fiber Formation. Russian Journal of Plant Physiology 2010;57(3) 363-367.
[17] Patten AM, Jourdes M, Brown EE, Laborie M-P, Davin LB, Lewis NG. Reaction Tissue Formation and Stem Tensile Modulus Properties in Wild-Tipe and *P*-Coumarate-3-Hydroxylase Downregulated Lines of Alfalfa, *Medicago Sativa* (Fabaceae). American Journal of Botany 2007;94(6) 912-925.
[18] Chernova TE, Gorshkova TA. Biogenesis of Plant Fibers. Russian Journal of Developmental Biology 2007;38 221-232.
[19] Okuyama T, Yamamoto H, Yoshida M, Hattori Y, Archer RR. Growth Stresses in Tension Wood: Role of Microfibrils and Lignification. Annals of Forest Science 1994;51 291-300.
[20] Yamamato H. Role of the Gelatinous Layer on the Origin of the Physical Properties of the Tension Wood. Journal of Wood Science 2004;50 197-208.
[21] Zimmermann MH, Wardrop AB, Tomlinson PB. Tension Wood in Aerial Roots of *Ficus Benjamina* L. Wood Science Technology 1968;2 95-104.
[22] Fang CH, Clair B, Gril J, Liu SQ. Growth Stresses are Highly Controlled by the Amount of G-Layer in Poplar Tension Wood. IAWA Journal 2008;29 237-246.
[23] Pilate G, Chabbert B, Cathala B, Yoshinaga A, Leple JC, Laurans F, Lapierre C, Ruel K. Lignification and Tension Wood. Comptes Rendus Biologies 2004;327 889-901.
[24] Schreiber N, Gierlinger N, Putz N, Fratzl P, Neinhuis C, Burgert I. G-Fibres in Storage Roots of *Trifolium Pretense* (Fabaceae): Tensile Stress Generators for Contraction. The Plant Journal 2010;61 854-861.
[25] Fisher JB, Stevenson JW. Occurrence of Reaction Wood in Branches of Dicotyledons and Its Role in Tree Architecture. Botanical Gazette 1981;142 82-95.
[26] Hsu YS, Chen SJ, Lee CM, Kuo-Huang LL. Anatomical Characteristics of the Secondary Phloem in Branches of *Zelkova Serrata* Makino. Botanical Bulletin of Academia Sinica 2005;46 143-149.
[27] Hamilton JR, Thomas CK, Carvell KL. Tension Wood Formation Following Release of Upland Oak Advance Reproduction. Journal Wood and Fiber Science 1985;17(3): 380-390.
[28] ://swst.metapress.com/content/q70t6764831872r1/ (accessed 27 June 2007).
[29] Kaku T, Serada S, Baba K, Tanaka F, Hayashi T. Proteomic Analysis of the G-Layer in Poplar Tension Wood. Journal of Wood Science 2009;55 250-257.
[30] Hayashi T, Kaida R, Kaku T, Baba K. Loosening Xyloglucan Prevents Tensile Stress in Tree Stem Bending but Accelerates the Enzymatic Degradation of Cellulose. Russian Journal of Plant Physiology 2010;57 316-320.
[31] Vignon MR, Heux L, Malainine ME, Mahrouz M. Arabinan-Cellulose Composite in *Opuntia Ficus-Indica* Prickly Pear Spines. Carbohydrate Research 2004;339(1) 123-131.

[32] Sivan P, Mishra P, Rao KS. Occurrence of Reaction Xylem in the Peduncle of *Couroupita Guianensis* and *Kigelia Pinnata*. IAWA Journal 2010;31 203-216.
[33] Meloche CG, Knox JP, Vaughn KC. A Cortical Band of Gelatinous Fibers Causes the Coiling of Redvine Tendrils: a Model Based upon Cytochemical and Immunocytochemical Studies. Planta 2007;225 485-498.
[34] Bowling AJ, Vaughn KC. Gelatinous Fibers are Widespread in Coiling Tendrils and Twining Vines. American Journal of Botany 2009;96 719-727.
[35] Dadswell HE, Wardrop AB. The Structure and Properties of Tension Wood. Holzforschung 1955;9 97-103.
[36] Müller M, Burghammer M, Sugiyama J. Direct Investigation of the Structural Properties of Tension Wood Cellulose Microfibrils Using Microbeam X-Ray Fibre Diffraction. Holzforschung 2006;60 474-479.
[37] Goswami L, Dunlop JWC, Jungnikl K, Eder M, Gierlinger N, Coutand C, Jeronimidis G, Fratzl P, Burgert I. Stress Generation in the Tension Wood of Poplar is Based on the Lateral Swelling Power of the G-Layer. The Plant Journal 2008;56 531-538.
[38] Gierlinger N, Schwanninger M. Chemical Imaging of Poplar Wood Cell Walls by Confocal Raman Microscopy. Plant Physiology 2006;140 1246-1254.
[39] Ageeva MV, Petrovska B, Kieft H, Sal'nikov VV, Snegireva AV, van Dam JEG, van Veenendaal WLH, Emons AMC, Gorshkova TA, van Lammeren AAM. Intrusive Growth of Flax Phloem Fibers is of Intercalary Type. Planta 2005;222 565-574.
[40] Viëtor RJ, Newman RH, Ha MA, Apperley DC, Jarvis MC. Conformational Features of Crystal-Surface Cellulose from Higher Plants Plant Journal 2002;30 721-731.
[41] Sturcova A, His I, Apperley DC, Sugiyama J, Jarvis MC. Structural Details of Crystalline Cellulose from Higher Plants. Biomacromoles 2004;5 1333-1339.
[42] Washusen R, Evans R. The Association Between Cellulose Crystallite Width and Tension Wood Occurrence in *Eucalyptus Globulus*. IAWA Journal 2001; 22235-243.
[43] Hillis WE, Evans R, Washusen R. An Unusual Formation of Tension Wood in a Natural Forest *Acacia* sp. Holzforschung 2004;58 241-245.
[44] Ruelle J, Yamamoto H, Thibaut B. Growth Stresses and Cellulose Structural Parameters in Tension and Normal Wood from Three Tropical Rainforest Angiosperms Species. Bioresources 2007;2 235-251.
[45] Yamamoto M, Saito T, Isogai A, Kurita M, Kondo T, Taniguchi T, Kaida R, Baba K, Hayashi T. Enlargement of Individual Cellulose Microfibrils in Transgenic Poplars Over Expressing Xyloglucanase. Journal of Wood Science 2011;57 71-75.
[46] Abe K, Yamamoto H. The Influences of Boiling and Drying Treatments on the Behaviors of Tension Wood with Gelatinous Layers in *Zelkova Serrata*. Journal of Wood Science 2007;53 5-10.
[47] Clair B, Almeras T, Pilate G, Jullien D, Sugiyama J, Riekel C. Maturation Stress Generation in Poplar Tension Wood Studied by Synchrotron Radiation Microdiffraction. Plant Physiology 2010;152 1650-1658.
[48] Burgert I, Eder M, Gierlinger N, Fratzl P. Tensile and Compressive Stresses in Tracheids are Induced by Swelling Based on Geometrical Constraints of the Wood Cell. Planta 2007;226 981-987.

[49] Burgert I, Fratzl P. Plants Control the Properties and Actuation of Their Organs through the Orientation of Cellulose Fibrils in Their Cell Walls. Integrative and Comparative Biology 2009;49 69-79.
[50] Dawson C, Vincent JFV, Rocca AM. How Pine Cones Open. Nature 1997;390 668.
[51] Elbaum R, Gorb S, Fratzl P. Structures in the Cell Wall that Enable Hygroscopic Movement of Wheat Awns. Journal of Structural Biology 2008;164 101-107.
[52] Clair B, Almeras T, Pilate G, Jullien D, Sugiyama J, Riekel C. Maturation Stress Generation in Poplar Tension Wood Studied by Synchrotron Radiation Microdiffraction. Plant Physiology 2011;155 562-570.
[53] Davis EA, Derouet C, Herve du Penhoat C, Morvan C Isolation and N.M.R. Study of Pectins from Flax (*Linum usitatissirnum* L.). Carbohydrate Research 1990;197 205-215.
[54] Mooney C, Stolle-Smits T, Schols H, de Jong E. Analysis of Retted and Non Retted Flax Fibres by Chemical and Enzymatic Means. Journal of Biotechnology 2001;89 205-216.
[55] Morvan C, Andeme-Onzighi C, Girault R, Himmelsbach DS, Driouich A, Akin DE. Building Flax Fibres: More Than One Brick in the Walls. Plant Physiology and Biochemistry 2003;41 935-944.
[56] Cronier D, Monties B, Chabbert B. Structure and Chemical Composition of Bast Fibers Isolated from Developing Hemp Stem. Journal of Agricultural and Food Chemistry 2005;53 8279-8289.
[57] Nishikubo N, Awano T, Banasiak A, Bourquin V, Ibatullin F, Funada R, Brumer H, Teeri TT, Hayashi T, Sundberg B, Mellerowicz EJ. Xyloglucan Endotransglycosylase (XET) Functions in Gelatinous Layers of Tension Wood Fibers in Poplar – A Glimpse into the Mechanism of the Balancing Act of Trees. Plant and Cell Physiology 2007;48 843-855.
[58] Bowling AJ, Vaughn KC. Immunocytochemical Characterization of Tension Wood: Gelatinous Fibers Contain More Than Just Cellulose. American Journal of Botany 2008;95 655-663.
[59] Gorshkova T, Morvan C. Secondary Cell-Wall Assembly in Flax Phloem Fibers: Role of Galactans. Planta 2006;223 149-158.
[60] Dejardin A, Leple J-C, Lesage-Descauses M-C, Costa G, Pilate G. Expressed Sequence Tags from Poplar Wood Tissues – a Comparative Analysis from Multiple Libraries. Plant Biology 2004;6 55-64.
[61] Lafarguette F, Leple J-C, Dejardin A, Laurans F, Costa G, Lesage-Descauses M-C, Pilate G. Poplar Genes Encoding Fasciclin-Like Arabinogalactan Proteins are Highly Expressed in Tension Wood. New Phytologist 2004;164 107-121.
[62] Andersson-Gunneras S, Mellerowicz EJ, Love J, Segerman B, Ohmiya Y, Coutinho PM, Nilsson P, Henrissat B, Moritz T, Sundberg B. Biosynthesis of Cellulose-Enriched Tension Wood in *Populus*: Global Analysis of Transcripts and Metabolites Identifies Biochemical and Developmental Regulators in Secondary Wall Biosynthesis. The Plant Journal 2006;45 144-165.
[63] Qiu D, Wilson IW, Gan S, Washusen R, Moran GF, Southerton SG. Gene Expression in Eucalyptus Branch Wood with Marked Variation in Cellulose Microfibril Orientation and Lacking G-Layers. New Phytology 2008;179 94-103.

[64] Roach MJ, Deyholos MK. Microarray Analysis of Flax (*Linum Usitatissimum* L.) Stems Identifies Transcripts Enriched in Fibre-Bearing Phloem Tissues. Molecular Genetics and Genomics 2007;278 149-165.

[65] Roach MJ, Deyholos MK. Microarray Analysis of Developing Flax Hypocotyls Identifies Novel Transcripts Correlated with Specific Stages of Phloem Fibre Differentiation. Annals of Botany 2008;102 317-330.

[66] Hobson N, Roach MJ, Deyholos MK. Gene Expression in Tension Wood and Bast Fibers. Russian Journal of Plant Physiology 2010;57 339-345.

[67] Girault R, His I, Andeme-Onzighi C, Driouich A, Morvan C. Identification and Partial Characterization of Proteins and Proteoglycans Encrusting the Secondary Cell Walls of Flax Fibres. Planta 2000;211 256-264.

[68] Mellerowicz E, Sundberg B. Wood Cell Walls: Biosynthesis, Developmental Dynamics and Their Implication for Wood Properties. Current Opinion in Plant Biology 2008;11 293-300.

[69] Gurjanov OP, Gorshkova TA, Kabel MA, Schols HA, van Dam JEG. MALDI-TOF MS Evidence for the Linking of Flax Bast Fibre Galactan to Rhamnogalacturonan Backbone. Carbohydrate Polymers 2007;67 86-96.

[70] Gorshkova TA, Wyatt SE, Salnikov VV, Gibeaut DM, Ibragimov MR, Lozovaya VV, Carpita NC. Cell-Wall Polysaccharides of Developing Flax Plants. Plant. Physiology 1996;110(2) 721-729.

[71] Mikshina PV, Gurjanov OP, Mukhitova FK, Petrova AA, Shashkov AS, Gorshkova TA. Structural Details of Pectic Galactan from the Secondary Cell Walls of Flax (*Linum Usitatissimum* L.) Phloem Fibres. Carbohydrate Polymers 2012;87 853-861.

[72] Gorshkova TA, Chemikosova SB, Salnikov VV, Pavlencheva NV, Gurjanov OP, Stoll-Smits T, van Dam JEG. Occurrence of Cell-Specific Galactan is Coinciding with Bast Fibre Developmental Transition in Flax. Industrial Crops and Products 2004;19 217-224.

[73] Gorshkova TA, Mikshina PV, Ibragimova NN, Mokshina NE, Chernova TE, Gurjanov OP, Chemikosova SB. Pectins in Secondary Cell Walls: Modifications during Cell Wall Assembly and Maturation. In: Schols HA, Visser RGF, Voragen AGJ. (eds) Pectins and Pectinases. Wageningen: Academic Publishers; 2009. p. 149-164.

[74] Arend M, Stinzing A, Wind C, Langer K, Latz A, Ache P, Fromm J, Hedrich R. Polar-Localized Poplar K^+ Channel Capable of Controlling Electrical Properties of Wood-Forming Cells. Planta 2005;223 140-148.

[75] Salnikov VV, Ageeva MV, Gorshkova TA. Homofusion of Golgi Secretory Vesicles in Flax Phloem Fibers during Formation of Gelatinous Secondary Cell Wall. Protoplasma 2008;233 269-273.

[76] Roach MJ, Mokshina NY, Snegireva AV, Badhan A, Hobson N, Deyholos MK, Gorshkova TA. Development of Cellulosic Secondary Walls in Flax Fibers Requires β-Galactosidase. Plant Physiology 2011;156 1351-1363.

[77] Mikshina PV, Chemikosova SB, Mokshina NE, Ibragimova NN, Gorshkova TA Free Galactose and Galactosidase Activity in the Course of Flax Fiber Development. Russian Journal of Plant Physiology 2009;56 58-67.

[78] Mokshina NE, Ibragimova NN, Salnikov VV, Amenitskii SI, Gorshkova TA. Galactosidase of Plant Fibers with Gelatinous Cell Wall: Identification and Localization. Russian Journal of Plant Physiology 2012;59(2) 246-254.

[79] Amenitsky SI, Snegireva AV, Gorshkova TA. Genomics of plant fibers. Mutants with modified character of the fiber development. In: Gorshkova T. (ed.) Biogenesis of plant fibers. Moscow: Nauka; 2009. p189-207.

[80] Meier H. Studies on a Galactan from Tension Wood of Beech (*Fagus Silvatica* L.). Acta Chemica Scandinavica 1962;16 2275-2283.

[81] Kuo CM, Timell TE. Isolation and Characterization of a Galactan from Tension Wood of American beech (*Fagus Grandifolia* Ehrl.). Svensk Papperstid 1969;72 703-716.

[82] Ruel K, Barnoud F, Goring DAI. Lamellation in the S2 Layer of Softwood Tracheids as Demonstrated by Scanning Transmission Electron Microscopy. Wood Science and Technology 1978;12(4) 287-291.

[83] Baba K, Park YW, Kaku T, Kaida R, Takeuchi M, Yoshida M, Hosoo Y, Ojio Y, Okuyama T, Taniguchi T, Ohmiya Y, Kondo T, Shani Z, Shoseyov O, Awano T, Serada S, Norioka N, Norioka S, Hayashi T. Xyloglucan for Generating Tensile Stress to Bend Tree Stem. Molecular Plant 2009;2 893-903.

[84] Gurjanov OP, Ibragimova NN, Gnezdilov OI, Gorshkova TA. Polysaccharides, Tightly Bound to Cellulose in Cell Wall of Flax Bast Fibre: Isolation and Identification. Carbohydrate Polymers 2008;72 719-729.

[85] Hayashi T. Xyloglucans in the primary cell wall. Plant Physiology. Plant Molecular Biology1989;40 139-168.

[86] Pauly M, Albersheim P, Darvill A, York WS. Molecular Domains of the Cellulose/Xyloglucan Network in the Cell Walls of Higher Plants. Plant Journal 1999;20 629-639.

[87] Salnikov VV, Ageeva MV, Yumashev VN, Lozovaya VV. The Ultrastructure of Bast Fibers. Russian Journal of Plant Physiology 1993;40 458-464.

[88] Andeme-Onzighi C, Girault R, His I, Morvan C, Driouich A. Immunocytochemical Characterization of Early-Developing Flax Fiber Cell Walls. Protoplasma 2000;213 235-245.

[89] Gorshkova TA, Chemikosova SB, Lozovaya VV, Carpita NC. Turnover of Galactans and Other Cell Wall Polysaccharides during Development of Flax Plants. Plant Physiology 1997;114(2) 723-729.

[90] Jones L, Seymour G, Knox JP. Localization of Pectic Galactan in Tomato Cell Wall Using a Monoclonal Antibody Specific to (1→4)-β-D-Galactan. Plant Physiology 1997;113 1405-1412.

Chapter 3

Structural Characteristics and Thermal Properties of Native Cellulose

Matheus Poletto, Vinícios Pistor and Ademir J. Zattera

Additional information is available at the end of the chapter

1. Introduction

In 1838 the French chemist Anselme Payen discovered and isolated cellulose from green plants [1-2]. After more than 170 years of the discovery of the "sugar of the plant cell wall", consumers, industry and government are increasingly demanding products from renewable and sustainable resources that are biodegradable, non-petroleum based, carbon neutral and at the same time generating low environmental, animal/human health and safety risks [3]. Therefore, cellulose is one of the most abundant material on earth and the most common organic polymer, representing about 1.5 x 10^{12} tons of the annual biomass production [1,3]. Cellulose is considered an almost inexhaustible source of raw material for the increasing demand for environmentally friendly and biocompatible products. Therefore, wood remains one of the most important raw material source for obtaining cellulose, but other sources can be used as well. Natural cellulose-based materials (wood, hemp, cotton, sisal, ramie, etc.) have been used as engineering materials for thousands of years and their use currently continues as demonstrated by the huge number of forest products-based worldwide industries, such as paper, textiles, etc. Such cellulose derivatives produced on an industrial scale are used for coatings, laminates, optical films and sorption media, as well as for property-determining additives in building materials, composites and nanocomposites, pharmaceuticals, foodstuffs and cosmetics [2-3]. As a consequence, several reviews and scientific papers have been published on cellulose research in the last two decades [2-4].

At this time one question can be asked: what makes cellulose such an important material? The fascination for cellulose results from its specific structure. The cellulose macromolecule is made up of repeating glucose units that generate surprising specificity and impressively diverse architectures, reactivities and functions [2]. The reactions and properties of cellulose are determined by the isolation process used, the number of inter- and intramolecular hydrogen bonds, the chain lengths, the chain length distribution, the crystallinity and by the distribution of functional groups within the repeating units and along the polymer chains [2, 5-6]. These important parameters make cellulose a unique material.

2. Structure of cellulose

Cellulose is a natural polymer consisting of ringed glucose molecules. The repeat unit showed in Figure 1 is comprised of two anhydroglucose rings $(C_6H_{10}O_5)_n$, linked together through an oxygen covalently bonded to the C1 of one glucose ring and the C4 of the adjoining ring (1 → 4 linkage) and so called the β 1-4 glucosidic bond [2-3]. The degree of polymerization, *n*, varies between 10 000 and 15 000, where *n* is dependent on the cellulose source material [3,7].

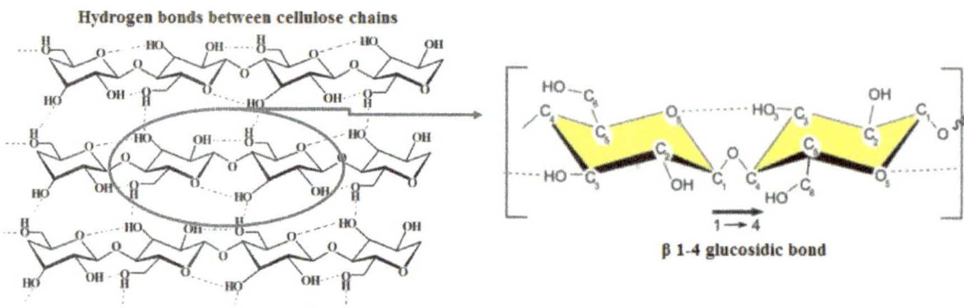

Figure 1. Molecular structure of a cellulose unit, showing the β 1-4 glucosidic bond and the intrachain hydrogen bonding (dotted line) (Adapted from [3]).

As can be seen in Figure 1, each repeating unit contains three hydroxyl groups. These hydroxyl groups and their ability to make hydrogen bonds between cellulose chains govern the physical properties of cellulose [7]. The intrachain hydrogen bonding between hydroxyl groups and oxygens of the adjoining ring molecules stabilizes the linkage and results in the linear configuration of the cellulose chain [3]. During cellulose formation, van der Waals and intermolecular hydrogen bonds between hydroxyl groups and oxygens of adjacent molecules promote aggregation of multiple cellulose chains forming fibrils [2-3]. The intra- and inter-chain hydrogen bonding network makes cellulose a relative stable polymer, and gives the cellulose fibrils high axial stiffness [3]. The high cohesive energy ensuing from these physic-chemical interactions explains why cellulose does not possess a liquid state [8] and these cellulose fibrils are the main reinforcement phase in trees and plants. Within these cellulose fibrils there are regions where the cellulose chains are arranged in a highly ordered crystalline structure and regions that are low order or amorphous regions [3,7].

2.1. Crystal structure

The polymorphy of cellulose and its derivatives has been well documented. These are cellulose I, II, III and IV [1-3]. Cellulose I, or native cellulose, is the form found in nature. Its structure is thermodynamically metastable and can be converted to either cellulose II or III [3,7]. This work focuses on the characterization of the cellulose I structure, which is the crystal structure naturally produced by a variety of organisms.

Cellulose I has two polymorphs, a monoclinic structure Iβ and a triclinic strucuture Iα, which coexist in various proportions depending on the cellulose structure [3,9]. The Iα is a rare form, whereas Iβ is the dominant polymorph for higher plants [10]. The Iα polymorph is metastable and can be converted into Iβ by hydrothermal treatments in alkaline solution [3,9].

The Iα and Iβ polymorph structures are shown in Figure 2. The Iα unit cell contains one cellulose chain, the unit cell parameters being a = 0.672 nm, b = 0.596 nm, c = 1.040 nm, α = 118.08°, β = 114.80°, γ = 80.375° [3]. The Iβ unit cell contains two cellulose chains, and the unit cell parameters are a = 0.778 nm, b = 0.820 nm, c = 1.038 nm, and γ = 96.5° [1,10]. Three lattice planes with approximate d-spacings of 0.39 nm, 0.53 nm and 0.61 nm correspond to the Iα lattice planes (110), (010), and (100) for the triclinic structure, and to the Iβ lattice planes (200), (110) and ($1\bar{1}0$) for the monoclinic structure [3,9]. The main difference between the Iα and Iβ polymorph structures is the relative displacement of cellulose sheets along the (110) lattice plane in the triclinic structure and the (200) lattice plane in the monoclinic structure, called "hydrogen-bonded" planes, in the chain axis direction [2-3]. In Iα there is a relative displacement of c/4 between each subsequent hydrogen-bonded plane, while for Iβ the displacement alternates between ± c/4 through van der Waals interactions [3,10].

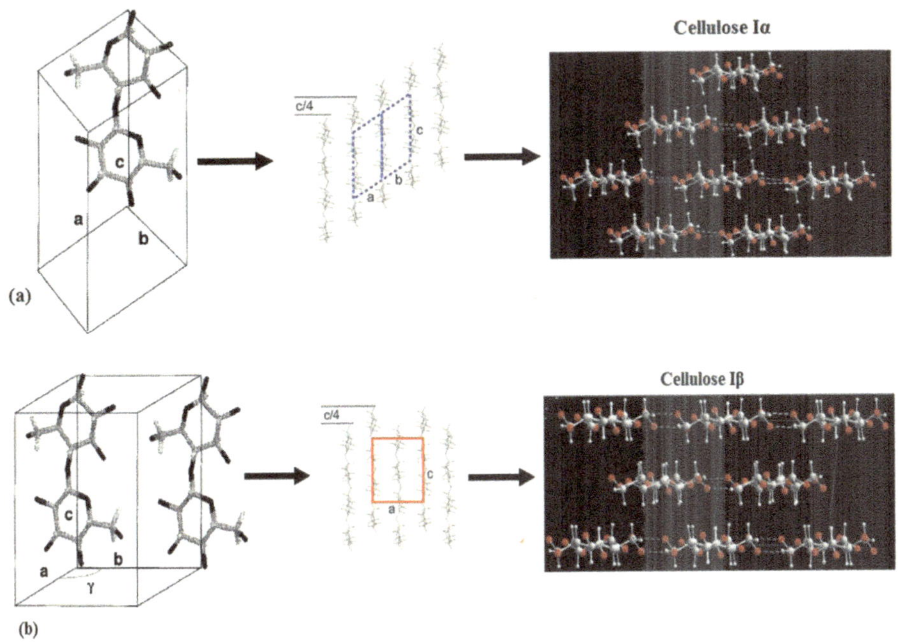

Figure 2. Schematic representation of the unit cells for cellulose structures Iα (a), and Iβ (b) and the displacement of the hydrogen bonding sheets for Iα of + c/4, and for Iβ alternating + c/4 and - c/4. (Adapted from [3, 11-12]).

2.2. Hydrogen bonding

Three hydroxyl groups are available for reaction in each repeating unit of cellulose, the structure of cellulose being largely affected by hydrogen bonds and van der Waals forces. Hydrogen bonding within neighboring cellulose chains may act to determine the straightness of the chain [1] and impart improved mechanical properties and thermal stability to the cellulose fibers. Interchain hydrogen bonds might introduce order or disorder into the system depending on their regularity [1].

So, understanding hydrogen bonding within the Iα and Iβ structures is important as it governs the stability and properties of these polymorphs [3] and of cellulose itself. With the hydroxyl groups being equatorial to the cellulose ring plane, intra- and inter-chain hydrogen bonding is most prevalent within the (110) plane in the triclinic structure and within the (200) plane in the monoclinic structure, hence the name " hydrogen-bonded" plane [3]. On the other hand, intrachain hydrogen bonding is dominated by strong O3-H···O5 bonds [1,3], as shown in Figure 1.

Inter-chain hydrogen bonding within the other planes (010), (100) in the triclinic structure and the planes (110) and $(1\bar{1}0)$ in the monoclinic structure is substantially lower and attractive van der Waals forces are believed to dominate the cohesion forces between cellulose chains [3]. Within these planes the number of weak inter-chain hydrogen bonds in the Iβ structure is believed to be larger than in the Iα polymorph and it has been suggested that it would contribute to the higher stability of the Iβ form, as compared to the Iα form [3,9]. The Iα hydrogen bonds thermally degrade at lower temperatures, contributing to the lower Iα thermal stability [3].

In this way, this study focuses on the characterization of structure and thermal properties of cellulose I, sometimes referred to as native cellulose. This work investigates the relationship between chemical structure, hydrogen bond interactions, crystallite size and crystallinity and the influence of these parameters on the thermal stability and decomposition kinetics of cellulose fibers obtained by two different pulping processes. However, in order to better understand the parameters used in this work for cellulose characterization a brief theoretical background is presented.

3. Theoretical background

3.1. X-ray diffraction parameters

The *d*-spacings were calculated using the Bragg equation [10,13]:

$$n\lambda = 2dsen\theta \tag{1}$$

where n is the order of reflection, λ is the wavelength of the incident X-rays, d is the interplanar spacing of the crystal and θ is the angle of incidence. The crystalline index (Eq.2), proposed by Hermans et al. [13-14] is:

$$Cr.I. = \frac{A_{cryst}}{A_{total}} \tag{2}$$

where Cr.I. is the crystalline index, A_{cryst} is the sum of crystalline band areas, and A_{total} is the total area under the diffractograms.

The second approach used to determine the crystalline index (Eq. 3) was the empirical method proposed by Segal [13,15]:

$$Cr.I. = \frac{I_{200} - I_{am}}{I_{200}} \times 100 \tag{3}$$

where I_{200} is the maximum intensity of the (200) lattice diffraction and I_{am} is the intensity diffraction at 18° 2θ degrees. The apparent crystallite size (L) (Eq. 4) was calculated using the Scherrer equation [14]:

$$L = \frac{K \times \lambda}{H \times \cos\theta} \tag{4}$$

where K is a constant of value 0.94, λ is the X-ray wavelength (0.1542 nm), H is the half-height width of the diffraction band and θ is the Bragg angle corresponding to the (200) plane. The surface chains occupy a layer approximately 0.57 nm thick so the proportion of crystallite interior chains [14,16] is:

$$X = \frac{(L - 2h)^2}{L^2} \tag{5}$$

where L is the apparent crystallite size for the reflection of plane (200), and h=0.57 nm is the layer thickness of the surface chain. In this study, the Z-function developed by [13] for determination of the crystalline monoclinic and triclinic structures of cellulose was used.

By employing discriminant analysis it is possible to categorize cellulose as belonging to the Iα or Iβ predominant form. The Z-value indicates whether cellulose is Iα or Iβ [9]. The function which discriminates between Iα or Iβ [9] is given by:

$$Z = 1693d_1 - 902d_2 - 549 \tag{6}$$

where d_1 is the d-spacing of the Iβ ($1\bar{1}0$) peak and d_2 is the d-spacing of the Iβ (110) peak.

Z>0 indicates that cellulose is rich in the Iα form and Z<0 indicates that Iβ is the predominant form.

3.2. Fourier Transform Infrared (FTIR) spectroscopy

The ratio between the heights of the bands at 1372 cm^{-1} and 2900 cm^{-1} proposed by Nelson and O'Connor as total crystalline index (TCI) [17] was used to evaluate the infrared (IR) crystallinity ratio. The band at 1430 cm^{-1} is associated with the amount of crystalline structure of cellulose, while the band at 898 cm^{-1} is assigned to the amorphous region in cellulose [17]. The ratio between the areas of the bands at 1430 cm^{-1} and 898 cm^{-1} is used as a lateral order index (LOI) [17]. Considering the chain mobility and bond distance, the hydrogen bond intensity (HBI) of cellulose is closely related to the crystal system and the degree of intermolecular regularity, that is, crystallinity [6]. The ratio of the absorbance bands at 3400 and 1320 cm^{-1} was used to study the cellulose samples HBI. The energy of the hydrogen bonds E_H for the OH stretching band was calculated using equation 7 [18]:

$$E_H = \frac{1}{K}\left[\frac{(\nu_0 - \nu)}{\nu_0}\right] \tag{7}$$

where ν_0 is the standard frequency corresponding to free OH groups (3650 cm^{-1}), ν is the frequency of the bonded OH groups, and K is a constant ($1/K = 2.625 \times 10^2$ kJ).

3.3. Thermogravimetric analysis (TGA)

For a reaction occurring during a differential thermal analysis (DTA), the change in the sample heat content and thermal properties is indicated by a deflection or a derivative peak. If the reaction is carried out using different heating rates, the level of activation energy (E_a) associated with this phenomenon, and therefore, the position of the peak of the derivative or the gradient of the deflection of the thermogravimetric curve varies with the heating rate whereas other experimental conditions are kept constant. The cellulose thermal decomposition is complex and may involve many reactions. It is very difficult to obtain precise kinetic parameters, however thermogravimetric analysis (TGA) has been used in recent decades as a quick evaluation for thermal stability. However, for the better understanding of the kinetic parameters determination a theoretical background is presented below.

3.4. Degradation kinetics

Information on the kinetics of degradation can be obtained by different methods. Kinetic studies assume that the isothermal conversion rate, $d\alpha/dt$, is a temperature-dependent linear function while the conversion (α) is independent of the temperature function according to Equation (8) [19-21]:

$$\frac{d\alpha}{dt} = k(T)f(\alpha) \tag{8}$$

where $f(\alpha)$ is a function dependent on the mechanism of decomposition and k is the rate constant. Equation (8) represents the rate of conversion at constant temperature according to the concentration of reactants at a constant rate. In this study, the conversion rate α is defined by [19-21]:

$$\alpha = \frac{m_0 - m_t}{m_0 - m_f} \quad (9)$$

where m_0, m_f and m_t are the initial and final weights of the sample and its weight at time (t), respectively.

The rate constant k is given by the Arrhenius equation [20-23]:

$$k = A\exp\left({E_a}/{RT} \right) \quad (10)$$

where A is the pre-exponential factor (independent of temperature), and E_a is the activation energy, T is the absolute temperature and R is the gas constant. By combining Equations (8) and (10) the relationship described by Equation (11) is obtained:

$$\frac{d\alpha}{dt} = Af(\alpha)\exp\left({E_a}/{RT} \right) \quad (11)$$

Whenever the sample temperature is controlled by a heating rate constant (β=dT/dt), the degree of conversion can be analyzed as a function of temperature. In this case the temperature becomes dependent of time at a heating rate, β, and the rate of reaction can be rewritten as:

$$\frac{d\alpha}{dt} = \frac{d\alpha}{dT}\frac{dT}{dt} = \beta\frac{d\alpha}{dT} \quad (12)$$

By considering the heating rate, again we can rewrite Equation (11) using the relationship shown in Equation (12), as shown in Equation (13):

$$\frac{d\alpha}{dT} = \frac{A}{\beta}e^{-\frac{E_a}{RT}}f(\alpha) \quad (13)$$

Integrating Equation (13) considering the initial temperature (T_0) and the initial conversion (α_0=0), we have:

$$\int_{\alpha_0}^{\alpha_\infty} \frac{d\alpha}{f(\alpha)} = \frac{A}{\beta}\int_{T_0}^{T_\infty} e^{\frac{E_a}{RT}} dT \quad (14)$$

Considering that T_0 is low and assuming $\alpha_0 = 0$ and that no reaction occurs between 0 and T_0:

$$g(\alpha) = \int_0^{\alpha_\infty} \frac{d\alpha}{f(\alpha)} = \frac{A}{\beta}\int_{T_0}^{T_\infty} e^{\frac{E_a}{RT}} dT \quad (15)$$

where $g(\alpha)$ is the integral function of conversion.

The degradation process can follow sigmoidal and deceleratory functions. These functions are shown in Table 1 through different $g(\alpha)$ expressions for the different solid state mechanisms [20, 22-23].

Mechanism	$g(\alpha)$	$f(\alpha)$
A_2, Nucleation and growth (Avrami equation (1))	$[-\ln(1-\alpha)]^{1/2}$	$2(1-\alpha)[-\ln(1-\alpha)]^{1/2}$
A_3, Nucleation and growth (Avrami equation (2))	$[-\ln(1-\alpha)]^{1/3}$	$3(1-\alpha)[-\ln(1-\alpha)]^{2/3}$
A_4, Nucleation and growth (Avrami equation (3))	$[-\ln(1-\alpha)]^{1/4}$	$4(1-\alpha)[-\ln(1-\alpha)]^{3/4}$
R_1, Controlled reaction on the surface (motion in one dimension)	α	1
R_2, Controlled surface reaction (dimensional contraction)	$[1-\ln(1-\alpha)^{1/2}]$	$2(1-\alpha)^{1/2}$
R_3, Controlled reaction on the surface (migration volume)	$[1-\ln(1-\alpha)^{1/3}]$	$3(1-\alpha)^{2/3}$
D_1, Diffusion in one dimension	α^2	$(1/2)\alpha^{-1}$
D_2, Diffusion in two dimensions (Valensi equation)	$(1-\alpha)\ln(1-\alpha)+\alpha$	$-[\ln(1-\alpha)]^{-1}$
D_3, Diffusion in three dimensions (Jander equation)	$[1-(1-\alpha)^{1/3}]^2$	$(3/2)[1-(1-\alpha)^{1/3}]^{-1}(1-\alpha)^{2/3}$
D_4, Diffusion in three dimensions (Ginstling–Brounshtein equation)	$[1-(2/3)\alpha]-(1-\alpha)^{2/3}$	$(3/2)[1-(1-\alpha)^{1/3}]^{-1}$
F_1, Random nucleation with one nucleus of individual particle	$-\ln(1-\alpha)$	$1-\alpha$
F_2, Random nucleation with two nuclei of individual particles	$1/(1-\alpha)$	$(1-\alpha)^2$
F_3, Random nucleation with three nuclei of individual particles	$1/(1-\alpha)^2$	$(1/2)(1-\alpha)^3$

Table 1. Expressions for $g(\alpha)$ and $f(\alpha)$ for the most frequently used mechanisms of degradation processes.

The mechanisms presented in Table 1 are essentially separated into four different groups shown schematically in Figure 3. The nucleation and growth (A_n) and random nucleation (F_n) are the most common types of mechanisms. Nucleation occurs through the breaking of bonds between molecules within the structure followed by rearrangement to release one molecule of product gas and a molecule referred to as the solid core of reaction [24]. The degradation reaction through nucleation is random, however, the speed of the degradation reaction tends to rise due to the fact that the formation of cores increases the concentration of degradation sites propagating along the material structure.

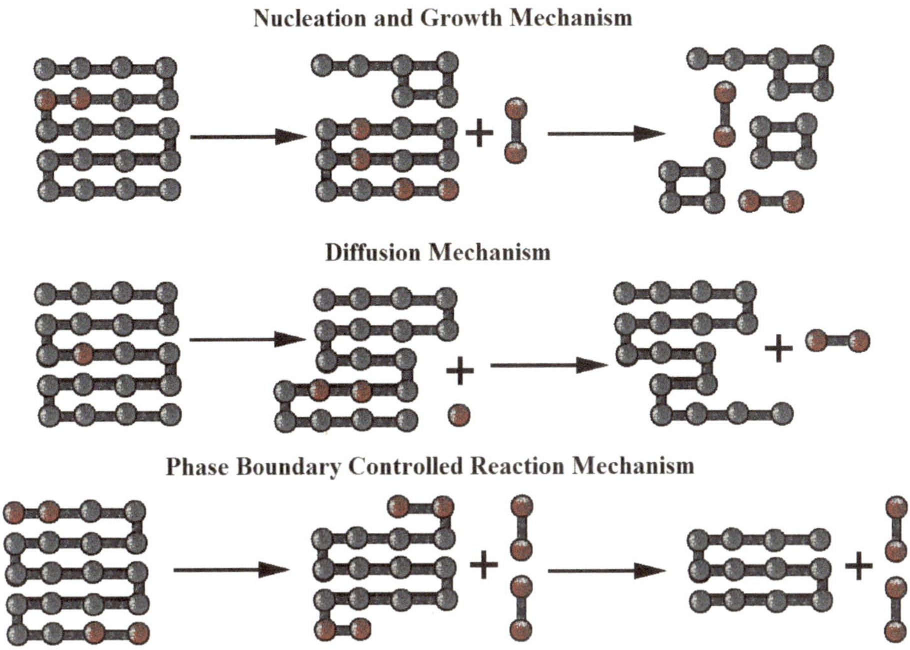

Figure 3. Schematic representation of the degradation processes reported by the mechanisms described in Table 1.

There are also controlled reactions at the interface (R_n). In mechanisms such R_n, degradation occurs from one end to the other one across the structure and this kind of mechanism is associated with the drawback of random breaking of bonds within the material structure. Factors that influence the material to follow the R_n mechanism are: high packing factor, molecular crosslinking and strong intermolecular interactions like hydrogen bonds between chains. Another class, diffusion (D_n), depends on the presence of one or more products formed by reaction or formation of gaseous products able to diffuse across the solid structure. Furthermore, in the case of macromolecules, the diffusion process becomes also dependent on the free volume and therefore the lower crystallinity and molecular packing factor can contribute for the degradation mechanism to occur by diffusion.

3.5. Flynn-Wall-Ozawa (FWO) method

In the FWO method [25-26], Equation (13) is integrated with the Doyle [27] approach and the result of the integration under logarithmic form is illustrated in Equation (16):

$$\log \beta = \log\left(\frac{AE_a}{g(\alpha)R}\right) - 2.315 - \frac{0.457E_a}{RT} \qquad (16)$$

Using the FWO equation, the activation energy (E_a) can be determined for each fraction of conversion (α) required. The isoconversional FWO method assumes that the reaction rate at a given conversion (α) is a function of temperature only. Therefore, through different values of the temperature (T) it is possible to observe a linear relationship by plotting *log* β *vs.* $1/T \times 10^{-3}$. By linear fit the apparent activation energy (E_a) of degradation can be determined by the slope of the straight line obtained, where, *slope=(0.457* E_a/R*)* [25-26].

3.6. Kinetic mechanisms of degradation

The activation energy of a solid state reaction can be determined, no matter the mechanism of degradation, by different methods, isothermal or non-isothermal. After determining the E_a, the mechanism of degradation can be estimated by the method proposed by Criado *et al.* [28] through the Z(α) function described in Equation (17):

$$Z(\alpha)=\frac{\left(\frac{d\alpha}{dt}\right)}{\beta}\pi(x)T \tag{17}$$

where $x = (E_a/RT)$ and $\pi(x)$ is an integral function obtained by approximations and can not be obtained as an algebraic function. However, a relationship between π(x) and the function $P(x)$ can be assessed by the following expression proposed in Equation 18:

$$\pi(x)=xe^{x}P(x) \tag{18}$$

For the P(x) function, Senum and Yang's [29], proposed expressions of rotational 2nd and 4th degree to assess the accuracy of the integral of Arrhenius and ensure a margin of error precisely controlled. These expressions, to the 8th degree, are illustrated in Table 2. Using the expression of the 4th degree one can assume that for $x > 20$ the expression results in rotational errors of less than 10^{-5}% [29].

By combining Equations (8), (17) and (18) one can obtain the relationship shown in Equation (19):

$$Z(\alpha)=f(\alpha)g(\alpha) \tag{19}$$

Equation 19 allows the determination of the thermogravimetric master curves represented by the g(α) and f(α) functions as shown in Table 1. To confront the theoretical curves shown in Table 1, it is possible to superimpose the experimental data determined by Equation (20):

$$Z(\alpha)=\frac{d\alpha}{dT}\frac{E_a}{R}e^{\frac{E_a}{RT}}P(x) \tag{20}$$

So, Equation (19) is used to plot the master Z(α) versus α curves for the different models listed in Table 2, whereas Equation (20) is used to represent the experimental curve. By comparing these two curves, the kind of mechanism involved in the thermal degradation can be identified.

Degree	$P(x)$
1º	$\frac{\exp(-x)}{x}\frac{1}{x+2}$
2º	$\frac{\exp(-x)}{x}\frac{x+4}{x^2+6x+6}$
3º	$\frac{\exp(-x)}{x}\frac{x^2+10x+18}{x^3+12x^2+368x+24}$
4º	$\frac{\exp(-x)}{x}\frac{x^3+18x^2+86x+96}{x^4+20x^3+120x^2+240x+120}$
5º	$\frac{\exp(-x)}{x}\frac{x^4+28x^3+246x^2+756x+600}{x^5+30x^4+300x^3+1200x^2+1800x+720}$
6º	$\frac{\exp(-x)}{x}\frac{x^5+40x^4+552x^3+3168x^2+7092x+4300}{x^6+42x^5+630x^4+4200x^3+12600x^2+15120x+5040}$
7º	$\frac{\exp(-x)}{x}\frac{x^6+54x^5+1070x^4+9720x^3+41112x^2+71856x+35280}{x^7+56x^6+1176x^5+11760x^4+58800x^3+141120x^2+141120x+40320}$
8º	$\frac{\exp(-x)}{x}\frac{x^7+70x^6+1886x^5+24920x^4+170136x^3+577584x^2+844560x+357120}{x^8+72x^7+2024x^6+28560x^5+216720x^4+880320x^3+1794240x^2+1572480x+403200}$

Table 2. Expressions for one to eight degrees rational approximations for the Arrhenius integral.

4. Experimental

4.1. Materials

Bleached sulfite cellulose fibers from *Pinus taeda* (CPT) were supplied by Cambará S.A (Cambará do Sul, Brazil) obtained at the cooking temperature of 140°C and bleaching with hydrogen peroxide. Bleached kraft cellulose fibers from *Eucalyptus grandis* (CEG) were supplied by CMPC S.A. (Guaíba, Brazil), obtained at the cooking temperature of 155°C and bleaching with hydrogen peroxide. The samples were dried at 70°C for 24h in a vacuum oven before the tests. The average fiber particle length for both CTP and CEG is around 150 μm.

4.2. Methods

The X-ray diffractograms were collected using a sample holder mounted on a Shimadzu diffractometer (XRD-6000), with monochromatic Cu Kα radiation (λ = 0.1542 nm), the generator operating at 40 kV and 30 mA. Intensities were measured in the range of $5 < 2\theta <$ 35°C, typically with scan steps of 0.05°C and 2s/step (1.5° min^{-1}). Peak separations were carried out using Gaussian deconvolution.

Fourier transform infrared spectroscopy spectra were obtained using a Nicolet IS10- Thermo Scientific spectrometer. Samples of the finely divided celluloses (5 mg) were dispersed in a KBr matrix (100 mg) followed by compression to form pellets. The analysis was obtained in triplicate using 32 scans, from 4000 cm^{-1} to 400 cm^{-1}, at a resolution of 4 cm^{-1}.

Thermogravimetric analysis (TGA50 – Shimadzu) was carried out under N_2 atmosphere, from 25 up to 600°C. Approximately 10 mg of each sample was used. The analysis was carried out at four different heating rates (5, 10, 20 and 40 °C min^{-1}). The results obtained were used to calculate the kinetic parameters.

5. Results and discussion

5.1. X-ray diffraction

X-ray diffraction is a method used generally to evaluate the degree of crystallinity of several materials. The free hydroxyl groups present in the cellulose macromolecules are likely to be involved in a number of intramolecular and intermolecular hydrogen bonds, which may give rise to various ordered crystalline arrangements [14, 21].

Figure 4 shows the X-ray diffractograms of the cellulose samples studied. In order to examine the intensities of the diffraction bands, establish the crystalline and amorphous areas more exactly and determine the crystallite sizes the diffractograms were deconvoluted using Gaussian profiles. Crystallographic planes are labeled according to the native cellulose structure as described by Wada et al. (2001) [30].

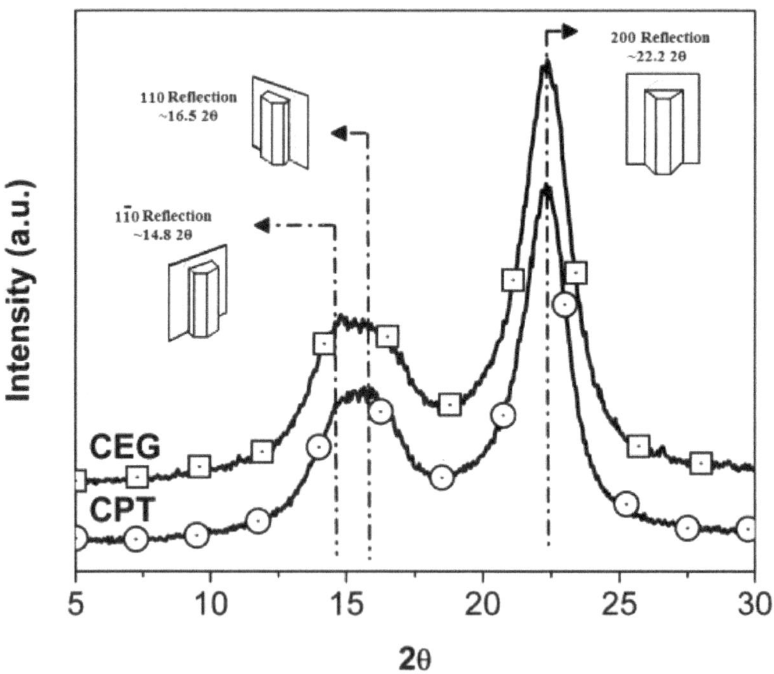

Figure 4. X-ray diffractograms of celluloses studied and corresponding crystal planes, adapted from [31], and most common 2θ values.

Following deconvolution, the two diffractograms show the 14.3-14.6°C 2θ reflection assigned to the ($1\bar{1}0$) crystallographic plane, the 16.00°C 2θ reflection assigned to the (110) crystallographic plane, the 18.30-18.40°C 2θ reflection assigned to the amorphous phase and the 22.20-22.40°C 2θ reflection assigned to the (200) crystallographic plane [14,30]. In Figure 5 a model is shown to represent the cellulose chains and the crystallographic planes described above.

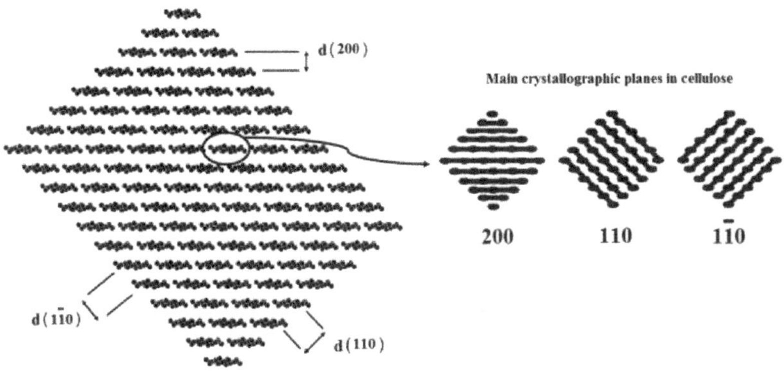

Figure 5. Model to represent cellulose chains (left), showing the d-spacings along the cellulose structure, adapted from [32]. Lines indicate the crystallographic planes in native cellulose (right), each circle on the line representing a chain normal to the paper, adapted from [33].

The band position (2θ values) and d-spacings of the celluloses calculated from X-ray diffractograms profiles are depicted in Table 2. Values of band position and d-spacings were similar.

Samples	($1\bar{1}0$)		(110)		Amorphous	(200)	
	2θ	d (nm)	2θ	d (nm)	2θ	2θ	d (nm)
CEG	14.30	0.618	16.00	0.553	18.30	22.40	0.397
CPT	14.60	0.605	15.95	0.555	18.40	22.20	0.399

Table 3. Band position (2θ) and d-spacings of crystalline and amorphous cellulose regions for the samples studied.

The degree of cellulose crystallinity is one of the most important crystalline structure parameters. The rigidity of cellulose fibers increases and their flexibility decreases with increasing ratios of crystalline to amorphous regions [15]. The crystallinity index calculated according to the Hermans (Eq. 2) and Segal methods (Eq. 3) showed that the CPT crystallinity is higher than that of CEG, as presented in Table 3. These differences are confirmed when the values of the crystallite size along the three crystallographic planes are taken into consideration. Crystallinity indices increased with increasing crystallite sizes because the crystallites surface corresponding to amorphous cellulose regions diminished

[10]. The values of X were used as estimates of the fraction of cellulose chains contained in the interior of the crystallites [33]. The proportion of crystallite interior chains, X, is similar for both samples. As can be seen in Table 3, the Z-values for CEG and CPT indicate that the cellulose samples belong to the Iβ dominant type, because Z<0 indicates that Iβ is the predominant form [9,13].

The crystallinity index (CrI) shows slight differences in crystallinity between the two cellulose samples. However, the d-spacing value for CPT in ($1\bar{1}0$) was around 20% higher than for CEG. The increase in the crystallite size for CPT in ($1\bar{1}0$), might be associated with a reduction in the corresponding amorphous region [10, 33]. If the amorphous domains of cellulose are attacked during the pulping treatment, chain scission and peeling reactions can occur, which reduce the total amount of amorphous regions and therefore increase the CPT crystallite size.

Samples	L ($1\bar{1}0$) (nm)	L (110) (nm)	L (200) (nm)	Cr.I.	C.I.	X	Z-values
CEG	3.783	2.370	3.826	60.4	74.9	0.493	-1.532
CPT	4.731	2.370	3.825	62.6	75.5	0.493	-25.345

Table 4. Parameters obtained from the XRD analysis of the cellulose samples studied.

These results confirm that CPT contains more cellulose chains in a highly organized form than CEG. This can lead to higher hydrogen bond intensity among neighboring cellulose chains resulting in a more packed cellulose structure besides higher crystallinity. On the other hand, the thermal stability of cellulose was found to depend mainly on its crystallinity index, crystallite size and degree of polymerization [10, 21, 33].

5.2. FTIR spectroscopy

FTIR spectroscopy has been used as a simple technique for obtaining rapid information about the chemical structure and crystallinity of cellulose samples [34-37]. Contrary to conventional chemical analysis, this method requires small sample sizes and short analysis time, besides being non-destructive [14].

Because of their complexity, the spectra were separated into two regions, namely: the OH and CH stretching vibrations in the 4000-2700 cm^{-1} region, showed in Fig. 6(a), and the "fingerprint" region which is assigned to different stretching vibrations of different groups in the 1800-800 cm^{-1} region, Figure 6(b). In Fig. 6(a) a strong broad band can be observed in the region of 3700-3000 cm^{-1} which is assigned to different OH stretching modes and another band in the region of 3000-2800 cm^{-1} is ascribed to the stretching of asymmetric and symmetric methyl and methylene cellulose groups [37]. The band at around 3360 cm^{-1} related to OH stretching modes is more prominent for CPT than for CEG. This is probably due to a larger number of hydroxyl groups in CPT which may be associated with an increase in the number of hydrogen bonds formed [14]. Thus, a mixture of intermolecular and intramolecular hydrogen bonds is considered to cause the broadening of the OH band in the IR spectra [14].

Fig. 6(b) shows that in the "fingerprint" region the spectra revealed several bands. The band at 1642 cm^{-1} is associated with adsorbed water in cellulose and probably some hemicelluloses [17, 37-38]. The bands at 1430, 1370, 1335 and 1320 cm^{-1} are attributed to CH_2 symmetric bending, CH bending, in-plane OH bending, CH_2 rocking vibration, respectively [17, 38-39], and the bands at 1162, 1111, 1057, 1033, 898 cm^{-1} are assigned to asymmetric C-O-C bridge stretching, anhydroglucose ring asymmetric stretching, C-O stretching, in-plane C-H deformation and C-H deformation of cellulose, respectively [17, 38-41].

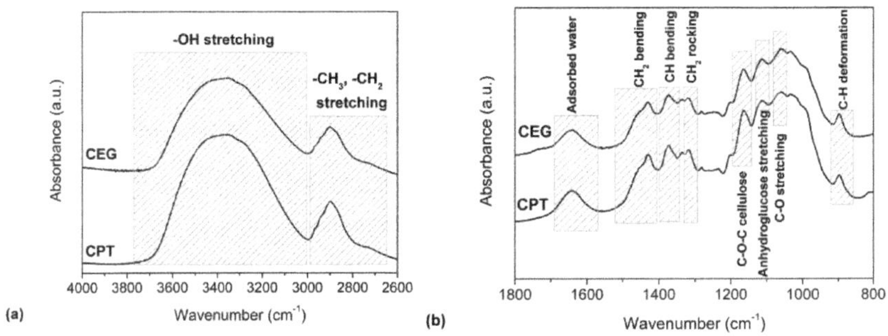

Figure 6. FTIR spectra of celluloses studied in the region between 4000-2800 cm^{-1} (a) and between 1800-800 cm^{-1} (b).

The total crystalline index (TCI), lateral order index (LOI), hydrogen bond energy (E_H), and hydrogen bond intensity (HBI) were calculated from the spectra obtained from FTIR spectroscopy. The obtained results are presented in Table 4.

Samples	IR crystallinity ratio		E_H	HBI
	H1372/H2900 (TCI)	A1429/A897 (LOI)	kJ	A3400/A1320
CEG	0.457 ± 0.020	3.507 ± 0.344	21.133 ± 0.092	1.368 ± 0.014
CPT	0.491 ± 0.010	4.071 ± 0.128	21.630 ± 0.311	1.455 ± 0.002

Table 5. Cellulose infrared crystallinity ratios and hydrogen bond intensity

TCI is proportional to the crystallinity degree of cellulose [14] while LOI is correlated to the overall degree of order in cellulose [17,41]. Based on this fact, CPT showed the higher TCI and LOI value indicating higher degree of crystallinity and more ordered cellulose structure than CEG. On the other hand, for CEG the lower cellulose infrared crystallinity values may indicate that the structure of this cellulose is composed of a larger number of amorphous domains when compared with CPT. The hydrogen bond energy is higher in CPT than in CEG. This is probably associated with higher crystallinity in this sample, as observed in the XRD analysis, which leads to more hydrogen bonds and so higher hydrogen bond energy. The HBI value is higher for CPT than for the CEG sample. This result might indicate that CPT contains much more cellulose chains in a highly organized form which can lead to higher hydrogen bond intensity between neighboring cellulose chains and result in a more packing cellulose structure and higher crystallinity than CEG. The crystallinity of cellulose is closely related to thermal stability [10, 21, 42]. Therefore, it is possible that cellulose samples of higher TCI, LOI and HBI might exhibit higher thermal stability.

5.3. Thermogravimetric analysis

Figure 7 shows the TGA and DTG curves of the two cellulose samples using a heating rate of 10°C min^{-1}. A small weight loss for both samples occurs between 40-70°C which is attributed to the removal of absorbed water in cellulose [20, 43]. As depicted in Figure 7(a), the CEG sample initiates a more pronounced degradation process at around 280°C while for CPT a more pronounced degradation process occurs at 292°C. The main decomposition step occurs in the range of 240°C to 370°C for CEG and 250°C to 375°C for CPT. In this stage the cleavage of the glycosidic linkages of cellulose reduces the polymerization degree leading to the formation of CO_2, H_2O and a variety of hydrocarbon derivatives [44].

According to Figure 7, differences in the decomposition profiles of the two cellulose samples indicate slight thermal stability differences for the samples. The DTG peaks were centered at 353°C and 360°C for CEG and CPT, respectively, as presented in Figure 7(b). The DTG curve for CPT was shifted to higher temperatures with increasing crystallite size. This behavior suggests that higher crystallite size celluloses have higher thermal stability. Kim *et al.* [10] studied different types of cellulose and noted an increase in the crystallite size promoted by higher thermal stability. Therefore, higher TCI, LOI and HBI celluloses have higher thermal stability probably due to a much higher amount of hydrogen bonds between cellulose chains that can lead to more ordered and packed cellulose regions, this in turn possibly increasing the thermal decomposition temperature of cellulose.

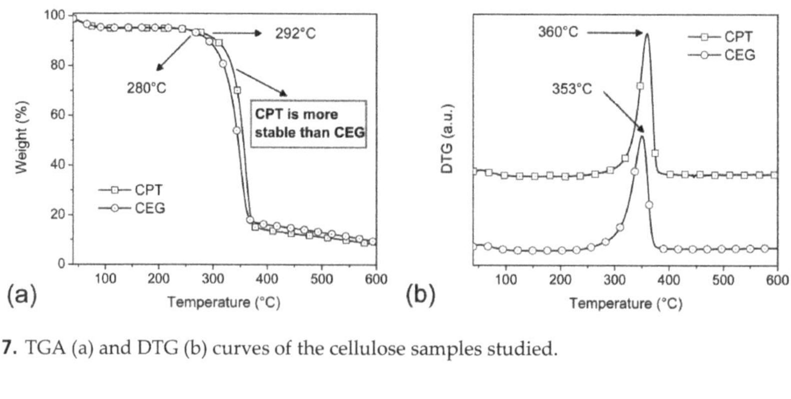

Figure 7. TGA (a) and DTG (b) curves of the cellulose samples studied.

5.4. Activation energy (E_a) in degradation

Figure 8 (A) shows the typical behavior of the thermal analysis conducted at different heating rates for the CPT sample while Figure 8 (B) illustrates the conversion curves determined from Equation (9).

In Figure 8, with the increase in heating rate, the curves show a shift to higher degradation temperatures, *i.e.*, there is a shorter time interval between the amount of heat supplied and absorbed by the sample. The observation of this behavior for the heating rate allows the use of the isoconversional method of Flynn-Wall-Ozawa (FWO) [19-22]. Figure 9 shows the linear fits from the plot of log β versus 1/T in the conversion range of 0.2-0.8 for the determination of the activation energy values obtained using the method proposed by FWO for the CPT sample. The linear fits showed correlation coefficient (r) values close to unity (minimum 0.9878 and maximum 0.9998) with a confidence interval of 95%. Figure 10 shows the activation energy values for the CEG and CPT samples in the conversion range of 0.2-0.8.

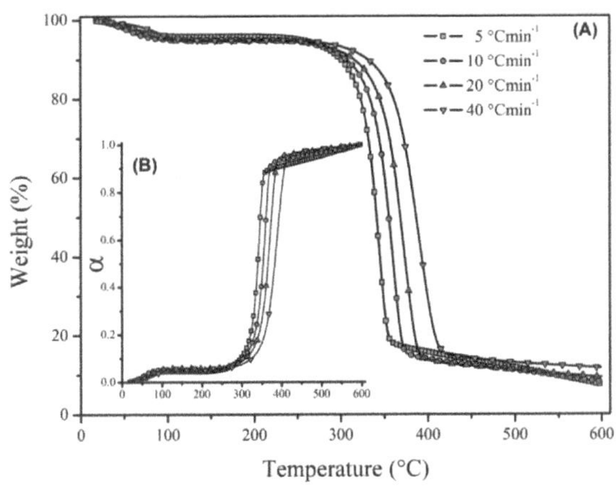

Figure 8. Thermogravimetric curves for several heating rates (A) and conversion curves (B) as a function of temperature for the CPT sample.

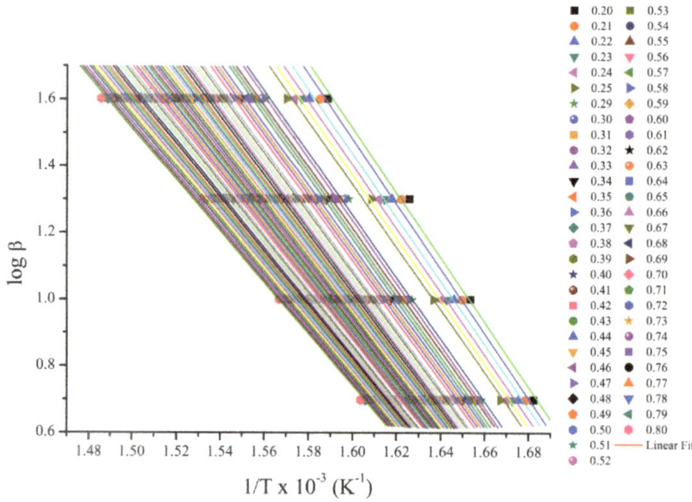

Figure 9. Linear fit determined by the conversion points trough different heating rates for the CPT sample using the FWO method.

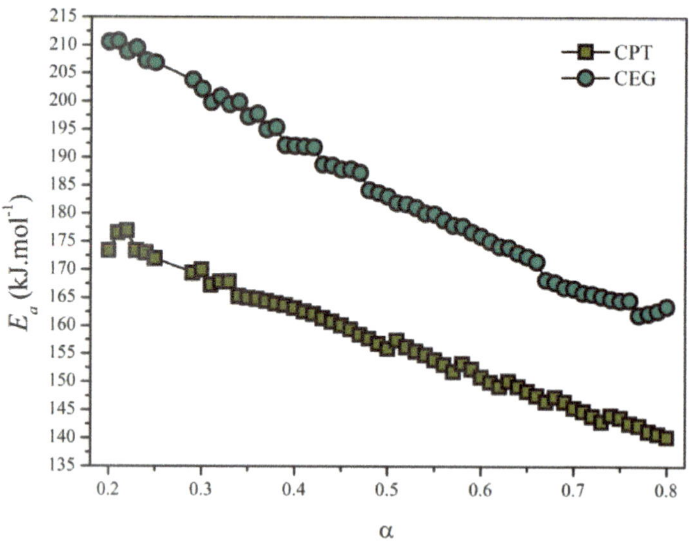

Figure 10. Activation energy values obtained trough the FWO method for both samples

In Figure 10 it is seen that the E_a values decrease progressively as the degradation process occurs for both celluloses studied. The activation energy E_a varies between 210-165 and 178-140 kJ/mol for the CEG and CPT samples, respectively. This range of values is consistent with the literature [35, 45-46]. The reduction in E_a is associated with the breaking of repeating units of ringed glucose molecules and molecular weight reduction of cellulose that may result from autocatalysis in the dehydration process [47]. Then, a mixed mechanism of depolymerization and dehydration can be considered for both pulp samples decomposition, regardless of the kind of pulp treatment. CEG sample exhibited larger values of activation energy than CPT. With respect to thermal stability, the crystal size, as shown in Table 3, may promote an increase in the degradation temperature [10, 21], and by consequence, the CPT cellulose fiber thermal stability may be higher than that of CEG. Moreover, the activation energy E_a value is not affected by the crystallite size [10] and therefore the lower E_a values observed for CPT can be attributed to the thermal decomposition of the sulfite pulp which can be controlled by dehydration, while the higher activation energy E_a values exhibited by CEG may indicate the depolymerization of kraft pulp with the production of levoglucosan [46-47]. Also, when considering the E_a values together with the data in Figure 6(b), the CPT sample showed a band in the 1642 cm^{-1} region, which is to be attributed to the adsorbed water in the cellulose structure in larger amount than in CEG. The higher amount of water adsorbed by the structure of CPT cellulose confirms the fact that the initiation of the CPT degradation can be more directly related to dehydration, which leads to lower activation energy E_a values. Similar behavior was observed by Soares et al. [47] for cellulose powder and kraft paper and by Scheirs et al. [48] for cellulose paper and kraft insulating paper.

Whereas different pulping conditions can affect the crystallinity of cellulose and differences in E_a suggest different relationships between the degradation mechanisms, it is also possible to evaluate the influence of different treatments on the structure of crystalline cellulose with the kinetic mechanism.

The activation energies E_a obtained using the FWO method were used to determine the degradation mechanisms proposed by Criado *et al.* [28]. This method uses reference theoretical curves obtained from Equation (19) that are derivatives of the $f(\alpha)$ and $g(\alpha)$ functions represented in Table 1. These theoretical curves are called master curves and are compared to experimental data obtained from Equation (20) for the determination of a solid-state process mechanism. These mechanisms represent how the solid-state degradation process occurs. The algebraic expressions that represent the theoretical mechanisms are separated into four groups, A_n, R_n, D_n and F_n, as can be seen in Table 1. These mechanisms describe processes of nuclei formation on the propagation of the degradation process; diffusion processes related to the heat transfer capacity along the material structure; reaction mechanisms controlled by the sample surface; and the random degradation of nuclei, respectively. To determine the $Z(\alpha)$ experimental values the heating rate (β) of 10 °Cmin^{-1} was used. The theoretical and experimental curves corresponding to these mechanisms are shown in Figure 11.

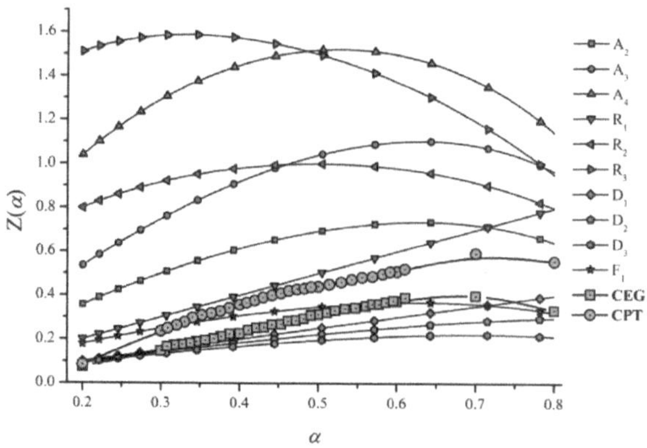

Figure 11. Master curves and experimental data obtained using the Criado method for both CPT and CEG samples.

The experimental data for the CEG sample in the conversion range of α = 0.2 - 0.4 overlapped on the D_n mechanism and according to the literature these degradation mechanisms refer to the diffusion processes in one, two and three dimensions, respectively [23, 28]. Similar results were described by Wu and Dollimore [49].As for the CPT sample, for α values in the range of 0.3 to 0.7 the degradation mechanism corresponded to R_1, *i.e.*, controlled reaction on the surface (motion in one dimension). The R_1 mechanism is also consistent with the results obtained from the XRD analysis, where the larger crystallite size for CPT might be related to the action of temperature on the boundaries due to the larger interfacial perimeter between crystals.

The degradation process is generally initiated in the cellulose amorphous regions, therefore, the smaller the size of the crystalline domains the larger number of amorphous regions which may be present in the structure of cellulose. So, in agreement with the lower crystallinity values found for the CEG sample from FTIR and XRD techniques this sample initiates the degradation process in the cellulose amorphous regions and when the conversion values are around 0.5 the CEG degradation mechanism tends towards F_1, corresponding to random nucleation with one individual particle nucleus. This behavior may be associated with the more pronounced degradation of the cellulose crystallite domains which results in the breakdown of the CEG crystallites and promotes random nucleation of the degradation process. As for CPT, of higher crystallite size, the degradation process is controlled by the degradation on the crystallites surface.

6. Conclusion

The crystallinity and kinetic decomposition of two cellulose samples obtained by two pulping processes were investigated. FTIR results indicated that CPT contains more cellulose chains in a highly organized form which may result in a more packed cellulose structure and higher crystallinity than CEG. Thermogravimetric results confirm that for the

CPT sample the thermal stability was higher than that of CEG probably due to the more ordered cellulose regions. In general, the crystallinity and thermal stability were more affected by the kraft pulping conditions than by those of sulfite pulping.

Through the kinetic parameters it was found that there are differences between the degradation processes of the cellulose fibers studied. For the CEG sample the degradation process occur by a diffusion process and probably starts in the cellulose amorphous domains while for CPT, which exhibits more crystalline regions than CEG, the more densely packed cellulose chains might hinder heat transfer by diffusion through the cellulose chains and then the degradation process may occur by degradation of the cellulose crystallites surface through a phase boundary-controlled reaction.

Science, technology, industry and government continue to move toward renewable, biodegradable, non-petroleum and carbon neutral raw materials. So, more environmentally friendly and sustainable resources and processes are desirable. Therefore, the demand for cellulose and cellulose derivatives is of growing importance in several applications as polymer materials, medical uses, food stuffs and in many other industry fields. However, from the discussion in this work it is obvious that the structure of cellulose is complex and the investigation of the many aspects of cellulose structure should be pursued to better understand this unique material.

Author details

Matheus Poletto* and Ademir J. Zattera
Laboratory of Polymers (LPOL), Center of Exact Sciences and Technology (CCET), University of Caxias do Sul (UCS), Caxias do Sul,RS, Brazil

Vinícios Pistor
Institute of Chemistry, Federal University of Rio Grande do Sul, UFRGS, Porto Alegre, RS, Brazil

Acknowledgement

The authors are grateful to Cambará S.A. and CMPC S.A. for supplying the cellulose samples. The authors also thank to CAPES and CNPq for financial support.

7. References

[1] O'Sullivan A (1997). Cellulose: the structure slowly unravels. Cellulose 4: 173-207.
[2] Klemm D, Heublein B, Fink H-P, Bohn A (2005). Cellulose: Fascinating Biopolymer and Sustainable Raw Material. Angew. Chem. Int. Ed. 44: 3358-3393.
[3] Moon RJ, Martini A, Nairn J, Simonsen J, Youngblood J (2011). Cellulose nanomaterials review: structure, properties and nanocomposites. Chem. Soc. Rev. 40: 3941-3994.

* Corresponding Author

[4] Bledzki AK, Gassan J (1999). Composites reinforced whit cellulose based fibres. Prog. Polym. Sci. 24: 221-274.
[5] Åkerholm M, Hinterstoisser B, Salmén L (2004). Characterization of the crystalline structure of cellulose using static and dynamic FT-IR spectroscopy. Carbohydr. Res. 339: 569-578.
[6] Oh SY, Yoo DI, Shin Y, Kim HC, Kim HY, Chung YS, Park WH, Youk JH (2005). Crystalline structure analysis of cellulose treated with sodium hydroxide and carbon dioxide by means of X-ray diffraction and FTIR spectroscopy. Carbohydr. Res. 340: 2376-2391.
[7] John MJ, Thomas, S (2008). Biofibres and biocomposites. Carbohydr. Polym. 71: 343-364.
[8] Gandini A (2011). The irruption of polymers from renewable resources on the scene of macromolecular science and technology. Green Chem. 13: 1061-1083.
[9] Hult E-I, Iversen T, Sugiyama J (2003). Characterization of the supermolecular structure of cellulose in wood pulp fibres. Cellulose 10: 103-110.
[10] Kim U-J, Eom SH, Wada M (2010). Thermal decomposition of native cellulose: Influence on crystallite size. Polym. Degrad. Stab. 95:778-781.
[11] Koyama M, Helbert W, Imai T, Sugiyama J, Henrissat B (1997). Parallel-up structure evidences the molecular directionality during biosynthesis of bacterial cellulose. Proc. Natl. Acad. Sci. 94: 9091-9095.
[12] Wada M, Chanzy H, Nishiyama Y, Langan P (2004). Cellulose III_I Crystal Structure and Hydrogen Bonding by Synchrotron X-ray and Neutron Fiber Diffraction. Macromolecules 37: 8548-8555.
[13] Wada M, Okano T (2001). Localization of I_α and I_β phases in algal cellulose revealed by acid treatments. Cellulose 8: 183-188.
[14] Popescu M-C, Popescu C-M, Lisa G, Sakata Y (2011). Evaluation of morphological and chemical aspects of different wood species by spectroscopy and thermal methods. J. Mol. Struct. 988: 65-72.
[15] Gümüskaya E, Usta M, Kirei H (2003). The effects of various pulping conditions on crystalline structure of cellulose in cotton linters. Polym. Degrad. Stab. 81: 559–564.
[16] Davidson TC, Newman RH, Ryan MJ (2004). Variations in the fibre repeat between samples of cellulose I from different sources. Carbohydr. Res. 339: 2889-2893.
[17] Carrilo F, Colom X, Suñol JJ, Saurina J (2004). Structural FTIR analysis and thermal characterization of lyocell and viscose-type fibres. Eur. Polym. J. 40: 2229-2234.
[18] Popescu C-M, Singurel G., Popescu M-C, Vasile C., Argyropoulos D.S., Willför S (2009). Vibrational spectroscopy and X-ray diffraction methods to establish the differences between hardwood and softwood. Carbohydr. Polym. 77: 851-857.
[19] Pistor V, Ornaghi FG, Fiorio R, Zattera AJ (2010). Thermal characterization of oil extracted from ethylene–propylene–diene terpolymer residues (EPDM).Thermochim. Acta 510: 93-96.
[20] Poletto M, Dettenborn J, Pistor V, Zeni M, Zattera AJ (2010). Materials produced from plant biomass. Part I: evaluation of thermal stability and pyrolysis of wood. Mat. Res. 13: 375-379.

[21] Poletto M, Pistor V, Zeni M, Zattera AJ (2011). Crystalline properties and decomposition kinetics of cellulose fibers in wood pulp obtained by two pulping process. Polym. Degrad. Stab. 96, 679-685.
[22] Bianchi O, Martins JDeN, Fiorio R, Oliveira RVB, Canto, LB (2011). Changes in activation energy and kinetic mechanism during EVA crosslinking. Polym. Test. 30: 616-624.
[23] Tiptipakorn S, Damrongsakkul S, Ando S, Hemvichian K, Rimdusi S (2007). Thermal degradation behaviors of polybenzoxazine and silicon-containing polyimide blends. Polym Degrad Stab 92: 1265–1278.
[24] Haines PJ (2002). Principles of thermal analysis and calorimetry. United Kingdom: RSC Paperbacks 238 p.
[25] Flynn JH, Wall LA (1966). General treatment of the thermogravimetry of polymers. Journal of Research of the National Bureau of Standards 70A: 487–523.
[26] Ozawa T (1966). A new method of quantitative differential thermal analysis. Bulletin of the Chemical Society of Japan 39: 2071–2085.
[27] Doyle CD (1961). Kinetic analysis of thermogravimetric data. J Appl Polym Sci 5: 285–292.
[28] Criado JM, Malek J, Ortega A (1989). Applicability of the master plots in kinetic analysis of non-isothermal data. Thermochim Acta 147: 377–385.
[29] Pérez-Maqueda LA, Criado, JM (2000). The accuracy of Senum and Yang's Approximations to the arrhenius integral. J. Therm. Anal. Calorim. 60: 909–915.
[30] Wada M, Okano T, Sugiyama J (2001). Allomorphs of native crystalline cellulose I evaluated by two equatorial d-spacings. J. Wood Sci. 47: 124-128.
[31] Howell CL (2008). Understanding wood biodegradation through the characterization of crystalline cellulose nanostructures. Doctoral thesis. University of Maine.
[32] Newman RH (2008). Simulation of X-ray diffractograms relevant to the purported polymorphs cellulose IV_I and IV_{II}. Cellulose 15: 769-778.
[33] Newman RH (1999). Estimation of the lateral dimensions of cellulose crystallites using ^{13}C NMR signal strengths. Solid State Nucl. Magn. Reson. 15: 21-29.
[34] Duchesne I, Hult EL, Molin U, Daniel G, Iversen T, Lennhon H (2001). The influence of hemicellulose on fibril-aggregation of kraft pulp fibres as revealed by FE-SEM and CP/MAS ^{13}C-NMR. Cellulose 8: 103-111.
[35] Nada AMA, Kamel S, El-Sakhawy M (2000). Thermal behavior and infrared spectroscopy of cellulose carbamates. Polym. Degrad. Stab. 70: 347-355.
[36] Quiévy N, Jacquet N, Sclavons M, Deroanne C, Paquot M, Devaux J (2010). Influence of homogenization and drying on the thermal stability of microfibrillated cellulose. Polym. Degrad. Stab. 95: 306-314.
[37] Adel MA, Abb El-Wahab ZH, Ibrahim AA, Al-Shemy MT (2011). Characterization of microcrystalline cellulose prepared from lignocellulosic materials. Part II: physicochemical properties. Carbohydr. Polym. 83: 676-687.
[38] Schwanninger M, Rodrigues JC, Pereira H, Hinterstoisser B (2004). Effects of short vibratory ball milling on the shape of FT-IR spectra of wood and cellulose. Vib. Spectrosc. 36: 23-40.

[39] Chen H, Ferrari C, Angiuli M, Yao J, Raspi C, Bramanti E (2010). Qualitative and quantitative analysis of wood samples by Fourier transform infrared spectroscopy and multivariate analysis. Carbohydr. Polym. 82: 772-778.

[40] Tserki V, Matzinos P, Kokkou S, Panayiotou C (2005). Novel biodegradable composites based on treated lignocellulosic waste flour as filler. Part I. Surface chemical modification and characterization of waste flour. Composites Part A. 36: 965-974.

[41] Corgié SC, Smith HM, Walker LP (2011). Enzymatic transformations of cellulose assessed by quantitative high-throughput Fourier transform infrared spectroscopy (QHT-FTIR). Biotechnol. Bioeng. 108: 1509-1520.

[42] Poletto M, Zattera AJ, Forte MMC, Santana RMC (2012). Thermal decomposition of wood: influence of wood components and cellulose crystallite size. Bioresour. Technol. 109: 148-153.

[43] Yang H, Yan R, Chen H, Zheng C, Lee DH, Liang DT (2006). In-depth investigation of biomass pyrolysis based on three major components: hemicellulose, cellulose and lignin. Energy Fuels. 20: 388-393.

[44] Bourbigot S, Chlebicki S, Mamleev V (2002). Thermal degradation of cotton under linear heating. Polym. Degrad. Stab. 78: 57-62.

[45] Antal MJ, Várhegyi G, Jakab E (1998). Cellulose pyrolysis kinetics: revisited. Ind. Eng. Chem. Res. 34: 1267–1275.

[46] Capart R, Khezami L, Burnham AK (2004). Assessment of various kinetic models for the pyrolysis of a microgranular cellulose. Thermochim. Acta. 417: 79–89.

[47] Soares S, Camino G, Levchik S (1995). Comparative study of the thermal decomposition of pure cellulose and pulp paper. Polym. Degrad. Stab. 49: 275–283.

[48] Scheirs J, Camino G, Tumiatti W (2001). Overview of water evolution during the thermal degradation of cellulose. Eur. Polym. J. 37: 933-942.

[49] Wu Y, Dollimore D (1998). Kinetic studies of thermal degradation of natural cellulosic materials. Thermochim. Acta. 324: 49–57.

Chapter 4

Supra-Molecular Structure and Chemical Reactivity of Cellulose I Studied Using CP/MAS ^{13}C-NMR

Viren Chunilall, Tamara Bush and Per Tomas Larsson

Additional information is available at the end of the chapter

1. Introduction

1.1. Chemical cellulose I

In chemical pulping the components that keep wood cells together, mainly lignin are degraded and dissolved in order to obtain fibres for the dissolving pulp and paper processes. The aqueous solutions of pulping chemicals are transferred from the lumen through the cell walls towards the middle lamella and the lignin rich middle lamella, which actually binds the wood cell wall together is dissolved last [1-3]. Cellulose I is the major component of dissolving pulp and constitutes the cell wall of plants and woods. It is the β-1,4-homopolymer of anhydroglucose [4, 5]. Maximising the commercial use of cellulose I dissolving pulp is dependent on developing a clear understanding of its chemical properties as a function of the structural characteristics [6]. The α-cellulose classification is based on the amount of total hemicellulose and degraded cellulose removed during bleaching. A 96% α-cellulose sample will typically have low amounts of hemicellulose and degraded cellulose.

1.1.1. Methods of isolation – Acid bisulphite pulping

"In a pulping process, wood is converted into fibres. This can be achieved mechanically, thermally, chemically or through a combination of these techniques" [7]. Chemical delignification, an important process during pulping, results in partial or total removal of lignin from wood by the action of suitable chemicals [8]. The lignin macromolecule is depolymerised through the cleavage of the ether linkages to become dissolved in the pulping liquor. The α-hydroxyl and α-ether groups are readily cleaved under simultaneous formation of benzilium ions [9]. The cleavage of the open α-aryl ether linkages represents the fragmentation of lignin during acid sulphite pulping. The benzilium ions are sulphonated by attack of hydrated sulphur dioxide or bi-sulphite ions, resulting in the increased hydrophylic nature of the lignin molecule. The extent of delignification depends on the degree of sulphonation as well as the depolymerisation [9].

Sulphurous acid and bi-sulphite ions are the main ingredients during acid bi-sulphite pulping. The sulphite pulping cycle is divided into three phases; the penetration phase, the pulping phase and the recovery phase. Time must be allowed for the chemicals to penetrate the chips completely. The slowest chemical reaction determines the reaction rate [10]. The temperature in the reaction vessel is raised slowly over a period of about 4 hours to 130°C (Penetration phase). Following the penetration phase, the temperature is raised to the maximum, usually between 135°C and 145°C and pulping commences (Pulping phase). The pressure is allowed to rise until it reaches about 800 kPa and then maintained constant by venting. The pulping phase is varied between 43 minutes to 103 minutes depending on the amount of lignin removal required. When the pulping phase ends, the pressure is reduced below 100 kPa during a period of about 90 minutes in order to recover chemicals (Recovery phase). The pulping finishes during the recovery phase. Thereafter the pulp is washed and screened. The total pulping time for acid bi-sulphite pulping is approximately 8.5 hours [10]. The sulphite pulping process takes longer but has a slightly higher yield than the sulphate/'kraft' pulping process [7]. A key advantage of the the sulphate /'kraft' process over the sulphite process is that chemicals used in the pulping liquor can be economically recovered [7]. The sulphite process is however characterised by its high flexibility compared to the sulphate/'kraft' process. In principle, the entire pH range can be used for sulphite pulping by changing the dosage and composition of the chemicals. Thus, sulphite pulping can be used in the production of many different types and qualities of pulp samples for a broad range of applications. The sulphite process can be categorised according to the pH into four different types of pulping namely Acid bi-sulphite, Bi-sulphite, Neutral sulphite (NSSC) and Alkaline sulphate. Table 1 presents the main pH ranges for different sulphite/sulphate pulping processes.

Process	pH	Base	Active reagent	Pulping temperature °C	Pulp yield%	Applications
Acid bi-sulphite	1-2	Ca^{2+}, Mg^{2+}, Na^{+}	SO_2 H_2O, H^+, HSO_3	125 – 143	40 – 50	Dissolving pulp, tissue, printing paper, special paper
Bi-sulphite	3-5	Mg^{2+}, Na^{+}	HSO_3^-, H^+	150 – 170	50 – 65	Printing paper, tissue
Neutral sulphite (NSSC)	5-7	Na^{+}, NH_4^{+}	HSO_3^-, SO_3^{2-}	160 – 180	75 – 90	Corrugated medium, semi-chemical pulp
Alkaline sulphate	9 – 13.5	Na^{+}	SO_3^{2-}, OH^-	160 – 180	45 – 60	'Kraft' – type pulp

Table 1. pH ranges for different pulping processes [11]

1.1.2. Dissolving pulp

The unbleached pulp that results after acid bi-sulphite pulping is used as raw material for dissolving pulp production. Lignin and hemicelluloses in the unbleached pulp are considered to be contaminants and are removed in order to produce high purity dissolving pulp samples. This can be done using oxygen delignification followed by a 4 step bleaching sequence. The bleaching sequence can either be Chlorine dioxide – Alkali extraction – Chlorine dioxide – Hypochlorite (D_1ED_2H) or Chlorine dioxide – Alkali extraction – Chlorine dioxide – Peroxide (D_1ED_2P) stage depending on the desired end product i.e. 91%α, 92%α or 96%α dissolving pulp samples. "Dissolving pulp is a chemical pulp intended primarily for the preparation of chemical derivatives of cellulose. It is utilized for the chemical conversion into products such as microcrystalline cellulose, cellophane, cellulose acetate, cellulose nitrate" [7].

1.1.3. Characterisation of cellulose I

There are a few traditional methods of analysing the chemical properties of cellulose I. Some of these methods include the Permanganate number determination, which is used to obtain the lignin content of the pulp [12]. The acid insoluble lignin content (Klason lignin) in wood and raw pulp is determined by gravimetric analysis [13]. The viscosity of a pulp sample provides an estimate of the degree of polymerisation (DP) of the cellulose chain. Viscosity determination of pulp is one of the most informative procedures that is carried out to characterise a polymer i.e. this test gives an indication of the degree of degradation (decrease in molecular weight of the polymer, i.e. cellulose) resulting from the pulping [14]. Pulping is known to affect cellulose structure by the generation of oxidised positions and subsequent chain cleavage in pulp samples [15]. The copper number thus gives an indication of the reducing end groups in a pulp sample [16]. Low molecular weight carbohydrates (hemicellulose and degraded cellulose) can be extracted from pulp samples with sodium hydroxide. The solubility of a pulp in an alkaline solution thus provides information on the degradation of cellulose and loss or retention of hemicellulose during the pulping and bleaching processes. S_{10} (%) and S_{18} (%) indicate that proportion of low molecular weight carbohydrates that are soluble in 10% and 18% sodium hydroxide respectively. S_{10} (%) alkali solubility gives an indication of the total extractable material i.e. degraded cellulose/short chain glucan and hemicellulose content in a pulp sample. S_{18} (%) alkali solubility gives an indication of the total hemicellulose content of the pulp sample and is also known as the percentage gamma (γ %) cellulose content of pulp samples [17]. The monosaccharide constituents (glucose, mannose, xylose, arabinose etc.) can be analysed using high performance liquid chromatography coupled with pulsed amperometric detection [18]. The concentrations of the monosaccharide constituents are obtained from the calibration curves of the standards (glucose, mannose, xylose, arabinose etc.). Molecular weight distribution analysis can be performed using Size Exclusion Chromatography coupled with Multi-Angle Laser Light Scattering (SEC-MALLS) on fully bleached 91%α, 92%α and 96%α samples after conversion to cellulose nitrate [19]. Due to its capability of measuring a sample in its native state, CP/MAS ^{13}C-NMR can be applied to investigate both the chemical and physical structure of lignocellulosics [3]. It is characteristic of NMR spectra that chemically equivalent carbons can be distinguished if they reside in magnetically non-equivalent surroundings. So even though the corresponding carbons of different anhydroglucose units of cellulose are chemically similar, they can be distinguished in a CP/MAS ^{13}C-NMR spectrum if they are in different magnetic environments, due to different packing of the cellulose chains or distinct conformations. Separate signals for crystalline and non-crystalline carbons as well as splitting of crystalline signals can thus be detected [20].

1.1.4. The concept of reactivity coupled to the different grades of dissolving pulp i.e. 91%, 92% and 96% α-cellulose for commercial products viz. microcrystalline cellulose, viscose and cellulose acetate respectively.

The 'reactivity' of cellulose can refer to its capacity to participate in diverse chemical reactions. Each anhydroglucose unit in a cellulose polymer has three different hydroxyl groups. The hydroxyl groups at O(2) H, O(3) H and O(6) H are the main reactive groups susceptible to chemical modification [21]. When discussing reactivity of cellulose I, the accessibility of the hydroxyl groups on the surface of fibrils or fibril aggregates to the chemical reagents is a crucial factor [4]. This accessibility is limited by the compact structure of cellulose I, which is determined by the presence of highly ordered regions formed by strong hydrogen bonds [10].

With the advent of infrared spectroscopy, accessibility could be determined by deuteration [22]. The O(3)H can be described as 'unreactive' when the cellulose surface is highly ordered and 'reactive' when the cellulose surface is less ordered [21]. Authors of [21] used the procedure described in [23] to calculate the availability of surface hydroxyl groups. More recently investigators reported a Fourier Transform Infrared (FTIR) spectroscopic method to measure the accessibility and size of cellulose fibrils from the cell wall of *Valonia ventricosa* by investigation of deuteration and rehydrogenation [24]. Investigators proposed that the accessibility of cellulose I depends on the amount of surface that is accessible, as determined by the size of cellulose fibril aggregate, the structure of the cellulose molecules, which will determine which hydroxyl groups are accessible; as well as the size and type of reagent [4]. The method used in this study for assessing cellulose I structure and accessibility was to determine the dissolving pulp reactivity during cellulose derivative formation.

Acid hydrolysis performed on cellulose I rich fibres can serve as one illustration of the relationship between cellulose supramolecular structure and reactivity. Cellulose I isolated from several sources was subjected to acid hydrolysis by hydrochloric acid, 2.5 M HCl (aq) at 100 °C for up to 17 h with reflux [6, 25, 26]. In the case of isolated cellulose I consisting of thinner fibril aggregates, such as cellulose isolated from wood and cotton, a 17 h hydrolysis resulted in the complete conversion to a sol of cellulose nano-particles. This was not observed for cellulose I isolated from *Cladophora sp.*, after a 17 h hydrolysis a fibrous material remained (unpublished data). The lower susceptibility towards acid hydrolysis exhibited by the *Cladophora* cellulose correlates well with the comparatively larger lateral cellulose dimensions. The larger lateral cellulose dimensions result in a low specific surface area which could, at least in part, account for the comparatively low reactivity observed for the *Cladophora* cellulose.

In the case of the 92% α-cellulose pulp the typical final product is synthetic spun textile fibres manufactured after dissolution of the dissolving pulp fibres. Here, the fibre material's 'reactivity' can be considered to be indicative of how easily the fibres can be dissolved in a solvent system. During the initial stage of the dissolution process, solvent needs to be transported into the pore system of the cellulose I rich fibres, and the subsequent dissolution of cellulose liberates polymers or colloidal particles from the fibre material. In such a system both the lateral dimensions of the cellulose (inversely proportional to the specific surface area) and the pore size characteristics of the fibre wall can influence the 'reactivity' or the ease of dissolution of the cellulose fibre. Depending on the lateral dimensions of the cellulose structure to be dissolved and the typical pore size, the dissolution process may become limited by transport processes. If cellulose structural elements are large and the average pore size of the fibre wall is small, the material liberated by the solvent may close up pores needed for transport of the incoming solvent and for transport of dissolved cellulose out of the fibre wall. This can serve as another illustration of how the cellulose supramolecular structure can affect the cellulose 'reactivity'. In this Chapter, cellulose acetates were produced by heterogenous acetylation from bleached acid bisulphite pulp fibres using a mixture of acetic acid and acetic anhydride in the presence of sulphuric acid catalyst. The degree of acetylation was observed as a function of reaction time and characterised by solid state CP/MAS ^{13}C-NMR.

1.2. CP/MAS ^{13}C-NMR for determining cellulose I structure and 'reactivity'

1.2.1. Principles of CP/MAS ^{13}C-NMR spectroscopy

CP/MAS ^{13}C-NMR spectroscopy is a useful technique for studying the structure of semi-crystalline polymorphic solids. CP/MAS ^{13}C-NMR has not always been used for structure determination because of line broadening. Broad lines characteristic of conventional NMR measurements on solid samples is attributed to two causes, static dipolar interactions between ^{13}C and ^{1}H and the chemical shift anisotropy. The strong dipolar interaction between ^{13}C and neighbouring protons can be removed by high-power proton decoupling. The second cause of line broadening, the chemical shift anisotropy is experimentally diminished by Magic Angle Spinning (MAS), which incorporates a rapid spinning (5 - 15 kHz) at an angle of 54.7 degrees with respect to the external magnetic field. Magic angle spinning will also average any residual dipolar broadening [27]. Cross-Polarization (CP) is a pulse technique used to enhance the signal-to-noise ratio of the spectrum, since ^{13}C is a low abundance nuclei with a comparatively small gyromagnetic ratio and its spin-lattice relaxation in solids is long [28]. This enhancement is performed by first exciting the ^{1}H spins and then transferring the magnetism to the ^{13}C-spin system [28]. Cross-polarization in combination with magic angle spinning and high power proton decoupling generates spectra with comparatively high resolution and good sensitivity [28]. Typical CP/MAS ^{13}C-

NMR spectra from cellulose I are made up of six signals from the anhydroglucose unit split into fine structure clusters due to the supra-molecular structure of the cellulose I fibril (Figure 1). The information content in this fine structure is high, but the accessibility of the information is hampered by a severe overlap of the signals [6]. In order to obtain quantitative information on the supra-molecular structure of cellulose, post-acquisition processing (spectral fitting) of the spectra is needed [25].

1.2.2. CP/MAS ^{13}C NMR in combination with spectral fitting

Figure below, shows signification differences are observable in the CP/MAS ^{13}C-NMR spectra of the same polymer, β 1->4, D-glucan isolated from different sources. These differences are attributed to differences in the supramolecular structure or polymer chain packing of the samples. Since there is significant signal overlap within each spectrum the desired information is not directly available. For this reason a post-acquisition processing method was developed based on non-linear least squares fitting of spectra. Using this method at least two things are required; an interpretation of a representative part of the spectrum and a mathematical model describing the functions used to model the recorded signals [25, 26, 29]. A spectral interpretation of the C4 spectral region was made possible by combining a model for the structural elements of cellulose possessing square cross-sections with a mathematical model comprising two distinct kinds of functions; Lorentzian and Gaussian [25]. The basic features of the mathematical model are that all non-crystalline states of polymer order are described as Gaussian functions and polymers located in the crystalline forms of cellulose are assigned Lorentzian functions (Figure 2). The rationale for the need for two kinds of mathematical functions for describing spectra recorded on semi-crystalline cellulose has been described in a recent paper [29].

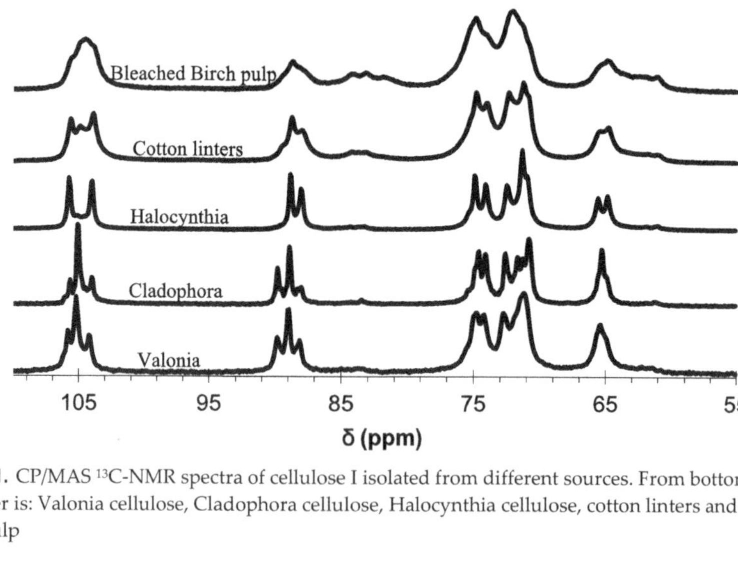

Figure 1. CP/MAS ^{13}C-NMR spectra of cellulose I isolated from different sources. From bottom to top the order is: Valonia cellulose, Cladophora cellulose, Halocynthia cellulose, cotton linters and bleached birch pulp

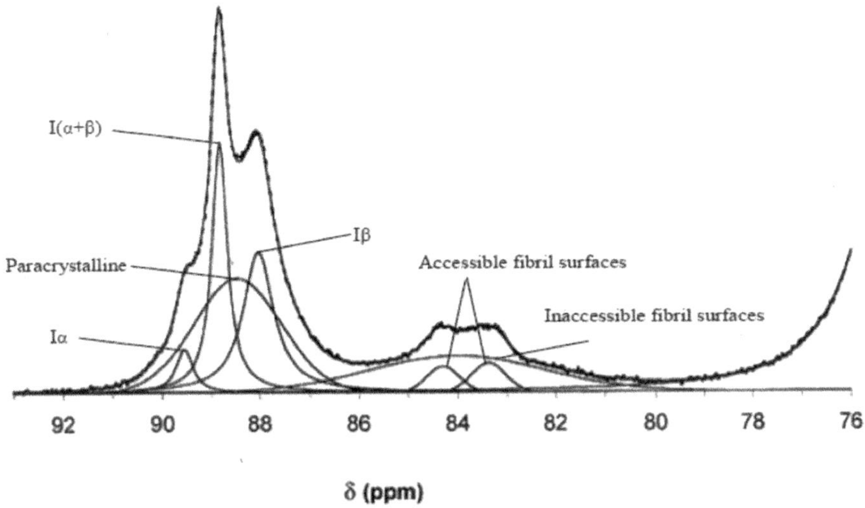

Figure 2. The fitting of the C-4 region of a CP/MAS ^{13}C-NMR spectrum recorded

The interpretation of the CP/MAS ^{13}C-NMR spectra has been further substantiated by ^{13}C enriched Cellulose I samples [30].

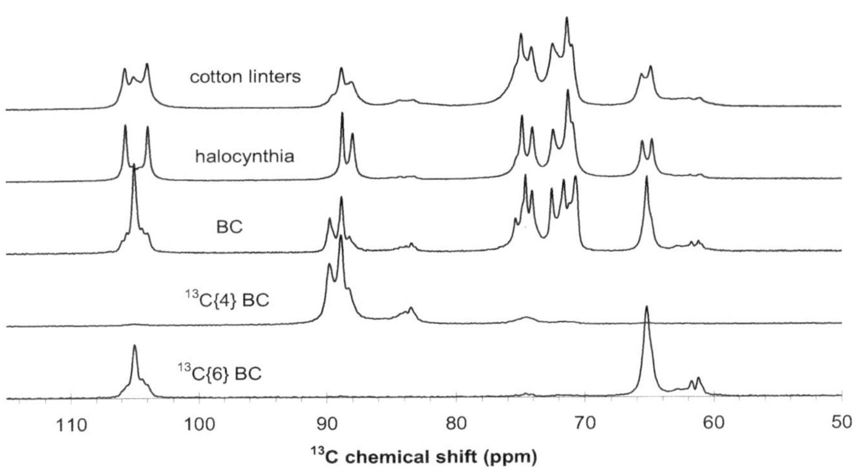

Figure 3. Illustrating the effect of ^{13}C enrichment obtained in bacterial Cellulose I obtained by cultivating bacteria (*Gluconacetobacter xylinus*) in the presence of glucose selectively ^{13}C enriched in the C4 and the C6 position.

Figure 3 shows spectra recorded on Cellulose I isolated from cotton linters, Halocynthia, bacterial cellulose (BC), $^{13}C\{4\}$BC is spectra recorded on bacterial cellulose cultivated from a medium containing glucose selectively enriched in the C4 position, and $^{13}C\{6\}$ BC is spectra recorded on bacterial cellulose cultivated from a medium containing glucose selectively enriched in the C6 position. The occurrence of the enriched signal intensity from the C1 position (about 105 ppm in the bottom spectra is due to scrambling during the biosynthesis [30].

1.2.3. Cellulose I fibril aggregate model

A model of aggregated cellulose I fibrils was constructed based on spectral fitting and spin diffusion experiments (Figure 4) [31]. Two C-4 signals at 84.3 and 83.3 ppm are assigned to cellulose at accessible fibril surfaces in contact with water and the C-4 signal at 83.8 ppm is assigned to cellulose at water-inaccessible fibril surfaces, formed either by interior distortions or aggregation of fibrils [32]. Paracrystalline cellulose can, at least in part, be explained by the presence of phase boundaries, such as the fibril-to-fibril contact surfaces. The finite lateral dimensions of the cellulose I fibril results in the presence of surfaces (phase boundaries) where the polymer conformation deviates from the conformation found for polymers located in the crystalline interior of the fibril. This result in well separated NMR signals from the C4 atoms located in the fibril interior and the C4 atoms located on the fibril surfaces, a 4 to 5 ppm shift difference is typically observed between the two. The responsible conformational differences may disappear gradually towards the fibril centre, resulting in the presence of polymers in a conformation intermediate between that of the fibril surfaces and the fibril interior; denoted para-crystalline cellulose. This is schematically illustrated in Figure 3 where the paracrystalline cellulose is shown to penetrate one layer below the fibril-to-fibril contact surface. This is purely an arbitrary representation since penetration depth is dependent on the size and severity of the distortions induced at the fibril surfaces, and paracrystalline cellulose may also be present beneath the accessible fibril surfaces [31].

It may be pointed out that the terminology used for describing the different states of order of the glucan polymers situated in the cellulose I fibril and fibril aggregate is quite distinct from the terminology used when interpreting results obtained from X-ray diffraction recorded on cellulose samples. However the two sets of terminologies can be reconciled if the degree of crystallinity as determined by X-ray diffraction techniques is compared with the degree of crystallinity computed from the NMR signal intensity originating from the crystalline plus the para-crystalline moieties. This way a reconciliation of the X-ray term "amorphous cellulose" and the corresponding NMR term "non-crystalline cellulose" immediately become obvious by relative signal intensity closure for both methods. The signal intensity detected as amorphous cellulose in the case of X-ray diffraction, is detected and described as non-crystalline cellulose in the case of NMR. This means that the sum of NMR signal intensities from accessible and inaccessible fibril surfaces corresponds to the X-ray estimate of amorphous cellulose. Experimental evidence that supports this reconciliation of NMR and X-ray terminology can be found in [33] where a good agreement was found when the degree of crystallinity and lateral fibril dimensions of cellulose in bleached kraft pulp was determined by both X-ray diffraction and NMR. Such an agreement between results obtained by X-ray and NMR is only possible if the above reconciliation of terminologies and measurements results are correct.

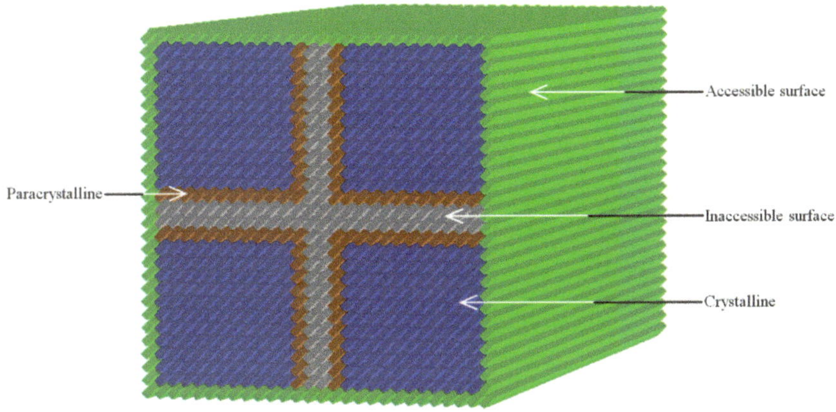

Figure 4. Schematic model of four aggregated cellulose I fibrils

1.2.4. The application of CP/MAS ^{13}C-NMR for determining average lateral fibril dimensions (LFD) and lateral fibril aggregate dimensions (LFAD)

Fibrils have cross-sections of varying shape and width in the range from a few nanometers to a few tenths of nanometers [4, 10, 34]. There is a broad distribution of fibril aggregate structures in pulp probably due to the presence of hemicelluloses and short chain glucan in pulp samples. In order to calculate the average lateral fibril dimensions (LFD) and average lateral fibril aggregate dimensions (LFAD), spectral fitting has to be performed on lignin- and hemicellulose-free pulp samples (glucose content > 95 %) since interfering signals (signal overlap from hemicellulose and spinning side bands from the lignin) influence calculations [32]. From the assignment of the signals in the C-4 region of the CP/MAS ^{13}C-NMR spectra, lateral dimensions can be assigned. Assuming a square cross-section (Figure 4), the fraction of the signal intensity from accessible fibril surfaces (calculation of lateral fibril aggregate dimension) or the fraction of the signal intensity from accessible and inaccessible fibril surfaces (calculation of lateral fibril dimensions) are both denoted q and are given by the equation:

$$q = \frac{(4n-4)}{n^2} \tag{1}$$

where n is the number of cellulose chains on the side of the square fibril or fibril aggregate cross-sections. A conversion factor of 0.57nm width per cellulose polymer has been used in the calculations [35-37]. The final 91%, 92% and 96% α-cellulose dissolving pulp samples were initially run on the CP/MAS ^{13}C-NMR in the wet state (never dried). The samples were then subject to drying via two strategies, viz. oven and condition drying. Oven dried pulp samples were prepared by placing dissolving pulp into the oven at 104°C for 18 hours.

Condition dried pulp samples were prepared by placing pulp in a conditioned room at 23°C with 50% RH for 5 days. The dried pulp samples were re-wetted and analysed by CP/MAS ^{13}C-NMR. Table 2 shows the LFAD measurements of the never dried pulp samples and the pulp samples after condition drying and oven drying.

Pulp sample	S_{10} (%)	S_{18} (%)	S_{10} - S_{18} (%)	$LFAD_{NMR}$ (nm)		
				Never dried	Condition dried	Oven dried
91% α-cellulose	11.4	5.9	5.5	17.5 ± 0.6	21.2 ± 1.6	26.1 ± 1.9
92% α-cellulose	9.1	4.6	4.5	18.3 ± 0.5	22.3 ± 1.6	28.0 ± 2.1
96% α-cellulose	6.7	2.4	3.3	22.5 ± 0.7	28.0 ± 2.6	34.7 ± 2.6

Table 2. Total extractable material S_{10} (%), LFAD (nm), of final dissolving pulp samples subject to different drying strategies

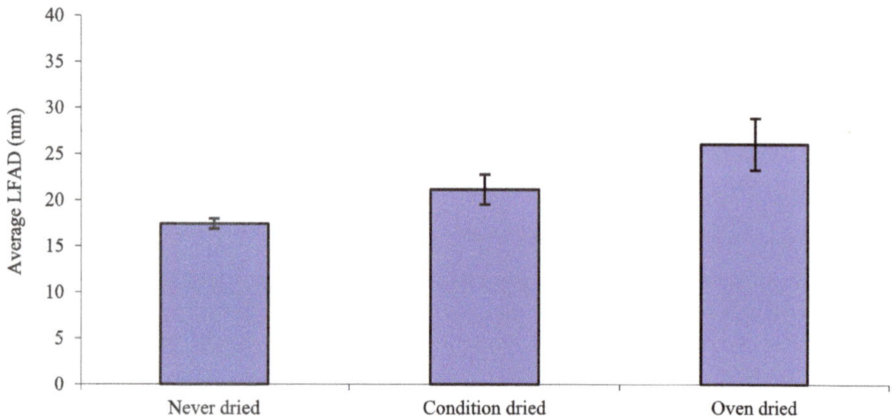

Figure 5. Average LFAD (nm) for the '91% α-cellulose' using the different drying strategies

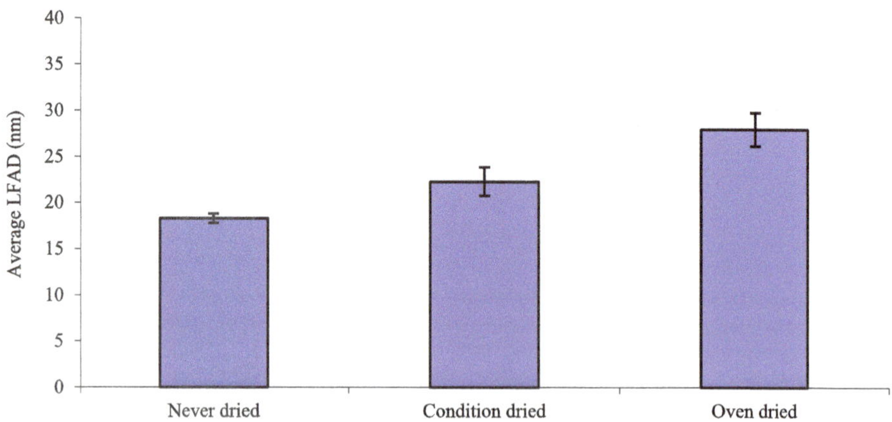

Figure 6. Average LFAD (nm) for the '92% α-cellulose' using the different drying strategies

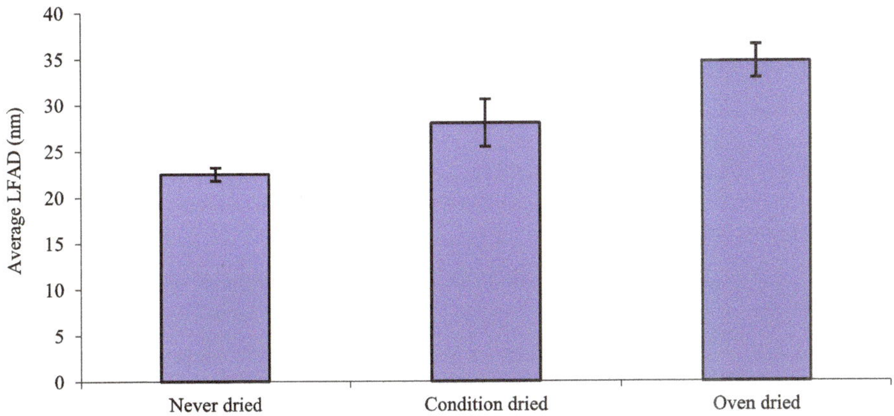

Figure 7. Average LFAD (nm) for the '96% α-cellulose' using the different drying strategies

Inspection of the data in Table 2 shows that there was a change in LFAD upon condition drying e.g. 17.5 ± 0.6 nm to 21.2 ± 1.6 nm for the 91% α; 18.3 ± 0.5 nm to 22.3 ± 1.6 nm for the 92 % α and 22.5 ± 0.7 nm to 28.0 ± 2.6 nm for the 96α. The 91% α and 92% α-cellulose pulp samples (Figures 5 and 6) show an increase in lateral fibril aggregation upon oven drying, 17.5 ± 0.5 nm to 26.1 ± 1.9 nm and 18.3 ± 0.5 nm to 28.0 ± 2.1 nm respectively. The 96 % α pulp sample (Figure 7) shows an increase in aggregate dimensions from 22.5 ± 0.7 nm to 34.7 ± 2.6 nm. Although the three grades of pulp samples show similar trends upon oven drying, the increase in aggregate dimensions is the largest during oven drying. The increase in LFAD was greater for the 96%α-cellulose followed by the 92% and 91% α-cellulose with the aggregate dimensions being 34.7 ± 2.6 nm, 28.0 ± 2.1 nm and 26.1 ± 1.9 nm respectively. First time drying, oven or condition drying, induced a degree of irreversible aggregation of the cellulose fibrils i.e. 'hornification'. Hornification, a term introduced by Jayme [38], is used in wood and pulp literature to describe the 'stiffening of a polymer structure' taking place in lignocellulosic (cellulose containing lignin) material upon drying or water removal [39]. Wetting the samples, prior to running on the CP/MAS ^{13}C-NMR, does not return the LFAD to the original never dried state hence aggregation is irreversible.

The results thus far point to the possibility of controlling LFAD by using different drying strategies. Condition drying the pulp samples is a possible method that can be used to control fibril aggregation. Oven drying presents a relatively rapid form of drying where the high temperature e.g. during oven drying (104°C) increases the rate at which water is removed and the movement of the fibrils. This possibly leads to a random restructure of the fibril aggregates and an increase in lateral fibril aggregate dimension. If LFAD can be controlled then it can be used to provide dissolving pulp samples with pre-defined specific surface area. Since the extractable hemicellulose and degraded cellulose/short chain glucan content influence the LFAD during acid bi-sulphite pulping and bleaching, there is a possibility that it can have an influence on fibril aggregation during drying. This increase in aggregate dimensions upon drying is supported by LFAD results obtained on other

Eucalyptus pulp samples investigated recently [40]. An increase in aggregate dimension due to drying seems to correlate with the total extractable material S_{10} (%) i.e. hemicellulose [(S_{18} (%)] and degraded cellulose/short chain glucan [S_{10} - S_{18} (%)]. During drying, an increased contact between the cellulose fibril surfaces is established [41, 42], and it seems that the dissolving pulp samples with a high total extractable material (i.e. 11.4 - 91% α-cellulose) have a lower tendency for cellulose fibrils to aggregate during drying. A possible explanation for the phenomenon, i.e. an increase in aggregate dimensions upon drying, is shown in Figure 8 and Figure 9 below.

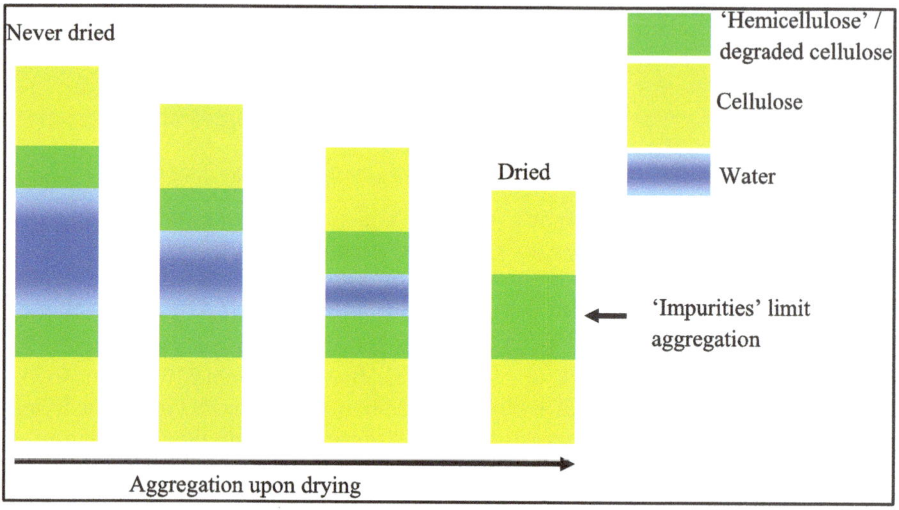

Figure 8. Aggregation of fibrils in the presence of extractable hemicellulose and/or degraded cellulose/short chain glucan.

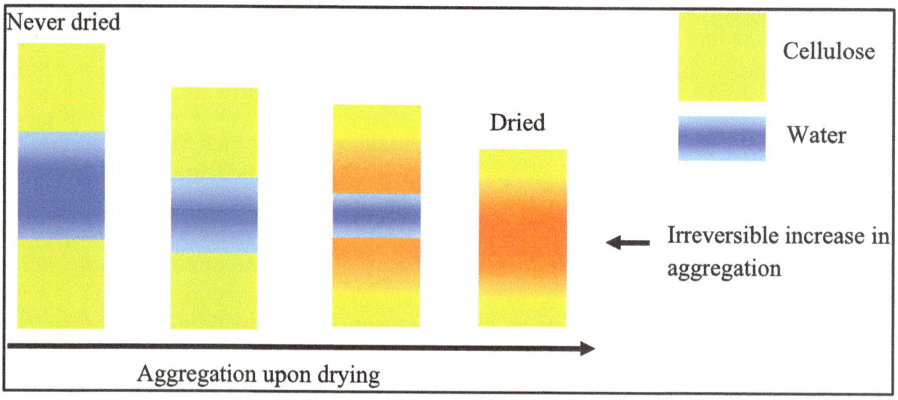

Figure 9. Aggregation of fibrils in the absence of extractable hemicellulose and/or degraded cellulose/short chain glucan

Drying pulp samples results in an irreversible change in lateral fibril aggregate dimensions irrespective of the drying strategies employed. The change can be marginal, as in the case of condition drying, or substantial as in the case of oven drying the pulp samples. Figure 8 shows the aggregation of fibrils upon drying in the presence of the total extractable content S_{10} (%) i.e. hemicellulose and degraded cellulose/short chain glucan ('impurities'). It is however evident, Figure 9, that the measurement of the fibril dimension for the dried material is going to be similar to the never dried material due to the presence of significant amounts of 'impurities'. Investigators recently showed that the presence of 4-O-methylglucuronxylan in the pulp samples diminishes the fibril aggregation and hence hornification during drying [43]. This scenario resembles that prevalent in the 91%α and 92%α pulp samples where the presence of extractable hemicellulose and degraded cellulose/short chain glucan could possibly affect fibril aggregation. The second scenario, Figure 9, shows the aggregation of fibrils upon drying in the absence of 'impurities' or in the presence of small quantities amounts of 'impurities'. Drying removes water from pores between the cellulose molecules facilitating their aggregation, however the absence or presence of small quantities of 'impurities' does not prevent the aggregation of fibrils. This scenario resembles the drying of 96%α pulp where small quantities of extractable hemicellulose and degraded cellulose/short chain glucan do not inhibit the aggregation of fibrils. While this study shows the changes in LFAD upon drying in the radial direction of the fibril, there have been studies that showed that relatively pure forms of cellulose (cellulose nanocrystals) also experience changes in the longitudinal direction upon drying [44].

1.2.5. *Computation of specific surface area from LFAD*

The average density of cellulose molecule is approximately 1500 kg/m³ i.e. $\rho \approx 1500$ kg/m³. The length along one side is taken as a in meters (m) and the longitudinal length of the cellulose molecule is taken as L. The total surface area (A) is thus given as:

$$A = a \times L \times 4 \tag{2}$$

The volume occupied by the cellulose molecule is given as:

$$\text{Volume: } V = a^2 \times L \tag{3}$$

With the mass of the cellulose molecule given as:

$$\begin{aligned}\text{Mass: } M &= V \times \rho \\ &= a^2 \times L \times \rho\end{aligned} \tag{4}$$

Hence the specific surface area (surface area per unit mass) can be calculated as:

Specific surface area

$$\begin{aligned}(\sigma)\ \frac{A}{M} = \sigma &= \frac{a \times L \times 4}{a^2 \times L \times \rho} \\ &= \frac{4}{\rho \times a}\end{aligned} \tag{5}$$

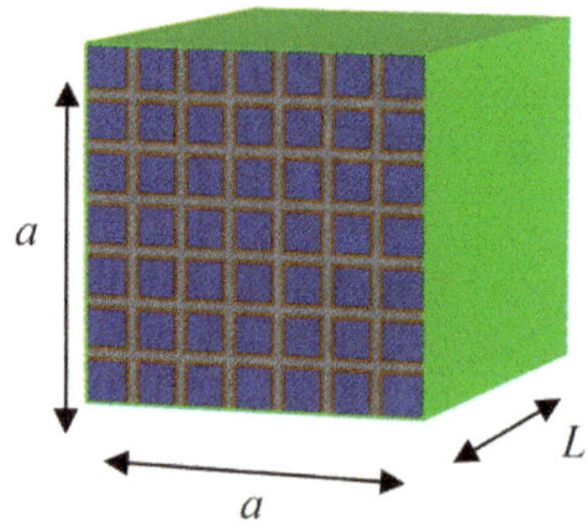

Figure 10. Schematic diagram of the fibril aggregate

Thus fibril aggregates with different lateral dimensions produce different specific surface areas. However this theory is limited since infinitely large objects yield specific surface areas too small to measure. Given that the hydroxyl groups on the fibril aggregate surface are the only functional groups initially available for further reaction, the question that arose from the above computation is whether the specific surface area measured using CP/MAS ^{13}C-NMR could be related to a measure of chemical reactivity. To answer this, a series of reactivity studies were carried out on high purity pulp samples (i.e. 96%α dissolving pulp and cotton linters).

	LFD (nm)	LFAD (nm)	SSA (m^2/g)
Cotton linters	7.1 ± 0.1	47 ± 2	53
Commercial 96%α	4.04 ± 0.04	28 ± 1	89

Table 3. LFD (nm), LFAD (nm) and Specific Surface Area (SSA) in m^2/g for dissolving pulp samples

The LFD for the cotton linters cellulose is 7.1 nm. The native cellulose LFD is larger than the commercial produced dissolving pulp sample (average ca. 4 nm). The LFAD for the cotton linters cellulose was 47 ± 2 nm compared to the commercial dissolving pulp sample i.e. 28 ± 1 nm. For a larger LFAD, evident in Table 3, there is a smaller specific surface area. This implies that there is a smaller specific surface area available for chemical reaction.

1.2.6. The application of CP/MAS ^{13}C-NMR for determining initial reaction rates of dissolving pulp samples

The initial acetylation process depends on the accessibility of cellulose fibres and the susceptibility of individual cellulose I surfaces. Cellulose acetate is obtained from cellulose through the substitution of its cellulosic hydroxyl groups by acetyl groups. The properties of cellulose acetate depend on its degree of substitution, i.e. the average number of acetyl groups per anhydroglucose unit, and on the substituent distribution at three possible sites of anhydroglucose unit and along the length of cellulose chain [45]. The control of the acetylation time can be an important aspect to the variable degree of acetylation and the physical structure of cellulose acetate formed [46]. The fibrous conversion method also provides partially acetylated cellulosic materials. Investigators prepared a paper with improved wet strength and dimensional stability from partially acetylated cellulose fibres [47]. Others studied the homogeneous acetylation process for Valonia and tunicate cellulose samples at the initial un-dissolved stage by CP/MAS ^{13}C-NMR, FT-IR and electron microscopy [48]. As a result, it was found that acetylation proceeds from the surface to the core of each microfibril and the Iα crystals undergo acetylation more rapidly than the Iβ crystals. Since the acetylated surface is expected to provide improved adhesion with hydrophobic matrices, use of partial acetylation of cellulosic fibres has been intensively studied as reinforcing elements for composite materials [49, 50].

To test the hypothesis that the specific surface area determined by the CP/MAS ^{13}C-NMR technique relates to chemical characteristic of the material, an attempt was made to correlate reaction rate ratios to ratios of specific surface areas determined by NMR. Acetylation experiments were performed with commercial 96%α Eucalyptus dissolving pulp and cotton linters. These samples are of high purity with respect to cellulose and differ significantly in their specific surface areas. Both samples were oven dried prior to acetylation. The hypothesis was based on the assumption that during the heterogeneous acetylation of cellulose, only hydroxyl groups situated at the surface of fibril aggregates are directly exposed to the surrounding liquid and hence, initially accessible to the reagent. Other hydroxyl groups reside either in the interior of fibrils or in the interior of fibril aggregates making them less accessible or inaccessible to the reagent.

The reaction scheme is illustrated below

$$ROH + (CH_3CO)_2O \underset{k_{-1}}{\overset{k_1}{=}} ROCOCH_3 + CH_3COOH \tag{6}$$

ROH denotes the accessible hydroxyl groups on the cellulose, $(CH_3CO)_2O$ denotes the acetic anhydride (*AA*), $ROCOCH_3$ denotes the formed cellulose acetate (*CA*) and CH_3COOH denotes the formed acetic acid (*HAc*). k_1 and k_{-1} denotes the forward and reverse reaction rates respectively. The general rate expression for the formation cellulose acetate can be given as (using abbreviations, brackets indicating concentrations)

$$\frac{d[CA]}{dt} = k_1[ROH]^a[AA]^b - k_{-1}[CA]^c[HAc]^d \tag{7}$$

Where $[ROH]^a \propto SSA$

The experimental setup used a constant amount of cellulose for all reaction times, a large excess of acetic anhydride (*AA*) and acetic acid, and measured formed amounts of cellulose acetate (*CA*) only for short reaction times (initial reaction rates). The short reaction time effectively means that only a small fraction of all accessible hydroxyl groups were acetylated and that any reverse reaction rate can be neglected. The large excess of acetic anhydride used makes the concentration of acetic anhydride essentially constant throughout the course of the reaction. The large excess of acetic acid was necessary for two reasons. Firstly, it is necessary for maintaining an approximately constant concentration of acetic acid throughout the reaction time (acetic acid is formed during the course of the reaction), and secondly acetic acid was needed due to its ability to swell cellulose. The swelling action of the acetic acid opens the pore system of the cellulose fibre wall making the specific surface area of the fibre wall cellulose accessible to the reagent acetic anhydride. This allows for the following approximations to be introduced.

$$\frac{d[ROH]}{dt} \approx 0 \tag{8}$$

$$\frac{d[AA]}{dt} \approx 0 \tag{9}$$

$$-k_{-1}[CA]^c[HAc]^d \approx 0 \tag{10}$$

Using these approximations pseudo-zero order reaction kinetics can be realized

$$\frac{d[CA]}{dt} = k_1[ROH]^a[AA]^b \approx k^{'}[ROH]^a \approx k^{''} \tag{11}$$

The connection between the pseudo zero-order reaction rate (k'') and the specific surface area, as determined from NMR LFAD measurements is given below, under the assumption that the exponent a in Equation (11) is equal to one.

$$k^{'}[ROH] \approx k^{'}\frac{\sigma MN}{V} = k^{''} \tag{12}$$

$[ROH]$ denotes the average volumetric concentration of accessible hydroxyl groups, σ denotes the specific surface area, M is the mass of the cellulose sample, N denotes the number of moles of hydroxyl groups per unit surface area and finally V denotes the sample volume. Keeping experimental conditions in agreement with the introduced approximations, keeping all parameters under experimental control equal only changing the cellulose substrate, a meaningful ratio of pseudo-zero order reaction rates can be formed. For substrates labelled A and B

$$\frac{k_A^{''}}{k_B^{''}} = \frac{k^{'}\frac{\sigma_A MN}{V}}{k^{'}\frac{\sigma_B MN}{V}} = \frac{\sigma_A}{\sigma_B} \tag{13}$$

Equation (13) relates the ratio of pseudo-zero order reaction rate determined from the kinetic experiments directly to the specific surface area determined from LFAD measurements. This means that it is possible to test if the LFAD value carries any information that is relevant to the chemical characteristics of the cellulose material, i.e. its behaviour during chemical modification. The system used for the acetylation was formulated in agreement with the theoretical assumptions made to reach pseudo zero-order kinetics (Equations 6-13).

The specific surface area ratio, determined using CP/MAS ^{13}C-NMR, was compared to the pseudo zero order reaction rate ratio which was determined from the degree of acetylation. The question to be addressed was: does the supra-molecular structure of cellulose I, present in the 96%α grade dissolving pulp, influence the reactivity towards acetylation? This study used two pulp samples with different LFAD in the acetylation reaction i.e. 96%α commercial pulp and cotton linters cellulose. They were subject to acetylation at 40°C and 60°C. It was anticipated that the different pulp samples would have different specific surface areas and hence perform differently in the acetylation reactions. The aim of the experiment was to ascertain whether the ratio of specific surface areas for two 'different' pulp samples is similar to the ratio of the pseudo first order initial reaction rate constants. Table 1 presents the lateral fibril dimensions (LFD) and LFAD measurements, specific surface areas, pseudo first order initial reaction rate constant (k) at 40°C and 60°C for the cotton and commercial 96%α dissolving pulp samples. The graph of methyl intensity versus time reveals the pseudo first order initial reaction rate constant for each of the samples.

	SSA (m^2/g)	k at 40°C	k at 60°C
Cotton linters	53	$1.4 \times 10^{-3} \pm 2 \times 10^{-4}$	$6 \times 10^{-3} \pm 6 \times 10^{-4}$
Commercial 96%α	89	$2.4 \times 10^{-3} \pm 3 \times 10^{-4}$	$11 \times 10^{-3} \pm 2 \times 10^{-3}$
Cotton linters / Commercial 96%α pulp (Ratio)	0.59 ± 0.06	0.59 ± 0.07	0.58 ± 0.09

Table 4. Summary of specific surface area ratio compared to initial reaction rate constant ratio computed from LFAD and density of cellulose

The pseudo zero order initial reaction rate constant ratio and specific surface area ratio for the commercial 96%α and cotton linters cellulose are presented in Table 4. The results show that the pseudo-zero order rate constant ratio is 0.59 ± 0.06 with the ratio of specific surface area at 40°C being 0.59 ± 0.07 and at 60°C being 0.58 ± 0.09. This shows that the ratio of initial reaction rate constants, k(cotton linters)/k(commercial 96%α), reproduce the ratio of specific surface are for both 40°C and 60°C reaction temperatures. This implies that:

1. Pseudo zero order initial reaction rate ratio is related to specific surface are ratio for two different substrates and
2. The pseudo zero order initial reaction rate constant ratio is independent of the temperature at which the acetylation reaction is performed.

A further comparison of acetylated sample analysis involved the use of ^{1}H-NMR. It was envisaged that it would provide a rapid result for the initial reaction rate constant compared to CP/MAS ^{13}C-NMR. The dried acetylated pulp samples were placed in deuterated chloroform. In theory, the acetylated surfaces should dissolve in deuterated chloroform with the solid or non-acetylated material filtered. The dissolved acetylated pulp is then analysed by solution state NMR. The graph of acetyl intensity (cellulose triacetate) vs. time gives the initial reaction rate constant [51]. Table 5 shows a comparison of the different processes involved in the analysis of acetylated material by CP/MAS ^{13}C-NMR and ^{1}H-NMR.

CP/MAS ^{13}C-NMR	**^{1}H-NMR**
1. Commercial 96%α pulp, Cotton linters	1. Laboratory 96%α pulp, Cotton linters
2. Acetylated at 40°C for 3, 6, 9 and 12 minutes.	2. Acetylated at 40°C for 10, 15, 20 and 25 minutes, dissolved in deuterated chloroform, undissolved material filtered off [50]
3. Packed in the rotor and run on the CP/MAS 13^{C}-NMR at ca. 6 hours a sample.	3. Run on the Proton NMR at ca. 2 minutes a sample.

Table 5. Comparison of sample preparation and analysis

The preliminary ^{1}H-NMR reactivity study carried out on 96%α pulp samples and cotton linters showed that the short reaction times, 3; 6 and 9 minutes, did not provide any signal intensity. The experiment was thus performed at 10, 15, 20 and 25 minutes. Following first order reaction kinetics at short reaction times, the initial reaction rate constant was determined from a plot of cellulose triacetate signal intensity against time.

	Specific surface area ratio (m^2/g)	**Initial reaction rate constant ratio determined by CP/MAS ^{13}C-NMR**	**Initial reaction rate constant ratio determined by ^{1}H-NMR**
Cotton linters / Commercial 96%α pulp	0.59 ± 0.06	0.55 ± 0.12	0.60 ± 0.05

Table 6. Summary of results for specific surface area ratio vs. initial reaction rate constant ratio determined by CP/MAS ^{13}C-NMR and ^{1}H-NMR at 40°C

Table 6 highlights the results showing the comparison of the two techniques to determine initial reaction rate constant ratios. The specific surface area was determined from LFAD measurements using CP/MAS ^{13}C-NMR. The pseudo initial rate constant ratio determined by CP/MAS ^{13}C-NMR (0.55 ± 0.12) and first order reaction rate constant ratio determined by solution state ^{1}H-NMR (0.60 ± 0.05) is related to the ratio of specific surface area (0.59 ± 0.06) of the acetylated materials. Results indicate that the initial reaction rate constant ratio is proportional to the specific surface area ratio for the cellulose pulp samples. This shows that specific surface area is related to initial reactivity to acetylation. It is thus possible to use solution state ^{1}H-NMR to give an indication of initial reaction rate constants for acetylation.

Author details

Viren Chunilall
Council for Scientific and Industrial Research,
Forestry and Forest Products (CSIR-FFP), Congella, Durban, South Africa

Tamara Bush
Department of Chemistry, University of KwaZulu-Natal, Durban, South Africa

Per Tomas Larsson
Innventia AB, Stockholm, Sweden
Wallenberg Wood Science Centre, KTH Royal Institute of Technology, Stockholm, Sweden

Acknowledgement

The Council for Scientific and Research Council (CSIR) and Sappi Saiccor are acknowledged for financially supporting the project. P.T.L. acknowledges the Wallenberg Wood Science Centre in Sweden for financial support. The host company Innventia AB, Sweden, where the solid state NMR work was conducted is also acknowledged.

2. References

[1] Sjöström E. Wood Chemistry – Fundamentals and Applications, 2nd ed. Academic Press; 1993

[2] Gullichsen J. In: Gullichsen J. (ed) Papermaking Science and Technology, Chemical Pulping 6A. Fapet Oy: Jyväskyla; 2000.

[3] Liitiä T. Application of modern NMR spectroscopic techniques to structural studies of wood and pulp components. PhD thesis. University of Helsinki; 2002.

[4] Krässig HA. Cellulose - Structure, Accessibility and Reactivity. Polymer Monographs M. B. Huglin. Amsterdam, Gordon and Breach science publishers, Vol. 11; 1996.

[5] Sarko A. Cellulose - How much do we know about its structure. In: Kennedy J F. (ed) Wood and Cellulosics: Industrial utilization, biotechnology, structure and properties Chichester, UK: Ellis Horwood; 1987. p55-70.

[6] Larsson PT, Hult E, Wickholm K, Pettersson E, Iversen T. CP/MAS ^{13}C NMR spectroscopy applied to structure and interaction studies on cellulose I. Solid state NMR 1999; 15, 31-40.

[7] Karlsson H. Fibre guide – Fibre analysis and process applications in the pulp and paper industry. AB Lorentzen & Wettre; 2006.

[8] Gierer J. Chemistry of delignification Part 1: General concept and reactions during pulping, Wood Sci. Technol. 1985; 19, 289-312.

[9] Funaoka M, Chang VL, Kolppo K, Stokke DD. Comparison of condensation reactions at C_α positions in kraft and acid sulfite delignification of Western hemlock, Bull. Fac. Bioresources 1991; 5, 37-44.

[10] Fengel D, Wegener G. Wood Chemistry, Ultrastructure, Reactions, Walter de Gruyter; 1984.
[11] Uhlmann T. Ullmann's encyclopedia of industrial chemistry. Paper and Pulp. 1991; 18 (A).
[12] Permanganate number of pulp, Tappi T214 wd-76 (UM-251), Tappi Press, Atlanta.
[13] Acid insoluble lignin in wood and pulp, Tappi TM 222, Tappi Press, Atlanta.
[14] Viscosity of pulp, Tappi T230 om-94, Tappi Press, Atlanta.
[15] Röhrling J, Potthast A, Rosenau T, Lange T, Ebner G, Sixta H, Kosma P. A Novel Method for the Determination of Carbonyl Groups in Cellulosics by Fluorescence Labeling. 1. Method Development, Biomacromolecules 2002; 3(5) 959-968.
[16] Copper number of pulp, paper and paperboard, Tappi T430 OM 94, Tappi Press, Atlanta.
[17] Alkali solubility of pulp, Tappi T235 OM- 60, Tappi Press, Atlanta.
[18] Davis MW. A rapid modified method for compositional carbohydrate analysis of lignocellulosics by high pH anion exchange chromatography with pulsed amperometric detection (HPAEC/PAD). J. of Wood Chem. Technol. 1998; 18(2) 235-252.
[19] Fischer K, Koch R, Fischer M, Schmidt I. Characterization of cellulose and cellulose derivatives by SEC. Das Papier 1999; 53 (12) 722-727.
[20] Mehring M. Principles of high resolution NMR in Solids. Second revised and enlarged edition of NMR - Basic Principles and Progress, Vol 11. Springer-Verlag; 1983.
[21] Verlhac C, Dedier J. Availability of surface hydroxyl groups in Valonia and Bacterial cellulose. Journal of Polymer Science: Part A 1990; 28 1171-1177.
[22] Tsuboi M. Infrared spectrum and crystal structure of cellulose. Journal of Polymer Science 1957; 25 159-171.
[23] Rowland SP, Howley PS. Microstructural order in the developing cotton fibre based on availabilities of hydroxyl groups. J. Polym. Sci. Polym. Chem. Ed. 1985; 23 183-192.
[24] Horikawa Y, Sugiyama S. Accessibility and size of *Valonia* cellulose microfibril studied by combined deuteration/rehydrogenation and FTIR technique. Cellulose 2008; 15 419-424.
[25] Larsson PT, Wickholm K. Iversen T. A CP/MAS ^{13}C-NMR investigation of molecular ordering in celluloses. Carbohydr. Res. 1997; 302 19-25.
[26] Wickholm K, Larsson PT, Iversen T. Assignment of non-crystalline forms in cellulose I by CP/MAS ^{13}C-NMR spectroscopy. Carbohydr. Res. 1998; 312 123-129.
[27] Sanders KM, Hunter BK. Modern NMR Spectroscopy, Oxford/New York/Toronto: Oxford University Press; 1987.
[28] Schaefer J, Stejskal EO. Carbon-13 NMR of polymers spinning at the magic angle. J. Ant. Chem. Soc. 1976; 98 1031-1033.
[29] Larsson PT, Westlund P-O. Line shapes in CP/MAS ^{13}C-NMR spectra of cellulose I, Spectrochemica Acta, Part A 2005; 62 539-546.

[30] Malm E, Bulone V, Wickholm K, Larsson PT, Iversen T. The surface structure of well-ordered native cellulose fibrils in contact with water. Carbohydrate Research 2010; 345 97-100.
[31] Wickholm K. Structural Elements in Native Celluloses. PhD thesis. Royal Institute of Technology (KTH), Sweden; 2001.
[32] Hult E-L, Larsson PT, Iversen T. Cellulose Fibril aggregation-An inherent property of kraft pulps. Polymer 2001; 42 3309-3314.
[33] Larsson PT, Popescu CM, Vasile C. A comparative CP/MAS ^{13}C-NMR and XRD study of the cellulose supra-molecular structure in softwood kraft pulp fibres. Papíripar 2008; 52 235-237.
[34] Atalla RH, VanderHart DL. The role of solid state ^{13}C-NMR spectroscopy in studies of the nature of native celluloses. Solid State Nuclear Magnetic Resonance 1999; 15 1-19.
[35] Sugiyama J, Vuong R, Chanzy H. Electron diffraction studies on two crystalline phases occurring in native cellulose from an algal cell wall. Macromolecules 1991; 24 4168-4175.
[36] Heiner AP, Kuutti L, Teleman O. Comparison of the interface between water and four surfaces of native crystalline cellulose by molecular dynamics simulations. Carbohydrate Research 1998; 306 205-220.
[37] Newman RH. Estimation of the lateral dimensions of cellulose crystallites using ^{13}C-NMR signal strengths. Solid state Nucl. Magn. Res. 1999; 15 21-29.
[38] Jayme G. Mikro-Quellungsmessungen an Zellstoffen. Wochenbl Papierfabr 1944; 6 187-194.
[39] Diniz Fernandes JMB, Gil MH, Castro JAMM. Hornification-its origin and interpretation in wood pulp samples. Wood Sci. Technology 2004; 37 489-494.
[40] Nocanda X, Larsson PT, Spark A, Bush T, Olsson A, Madikane M, Bissessur A, Iversen T. Cross polarisation/ magic angle spinning ^{13}C-NMR spectroscopic studies of cellulose structural changes in hardwood dissolving pulp process. Holzforschung 2007; 61 (6) 675-679.
[41] Oksanen T, Buchert J, Viikari L. The role of hemicelluloses in the hornification of bleached kraft pulps. Holzforschung 1997; 51 (4) 355-360.
[42] Newman RH. Carbon 13 NMR evidence for the co-crystallization of cellulose as a mechanism for hornification of bleached kraft pulp. Cellulose 2004; 11 45-52.
[43] Rebuzzi F, Evtuguin DV. Effect of Glucuronoxylan on the hornification of *Eucalyptus globulus* bleached pulps. Macromol. Symp. 2006; 232 121-128.
[44] Kontturi E, Vuorinen T. Indirect evidence of supramolecular changes within cellulose microfibrils of chemical pulp fibres upon drying. Cellulose 2009; 16 (1) 65-74.
[45] He J, Tang Y, Wang S-Y. Alkaline treatment of diacetate fibers and subsequent cellulase degradation. Journal of Applied Polymer Science 2008; 107, 2466–2474.

[46] Barud HS, de Araujo Junior AM, Santos DB, de Assuncao RMN, Meireles CS, Cerqueira D A. Thermal behavior of cellulose acetate produced from homogeneous acetylation of bacterial cellulose. Thermochimica Acta 2008; 471 61–69.

[47] Herdle LE, Griggs WH. Partially acetylated cellulose-its properties and potential applications. Tappi 1965; 48 103–107.

[48] Sassi J-F, Tekely P, Chanzy H. Relative susceptibility of the Iα and Iβ phase of cellulose towards acetylation. Cellulose 2000; 7 119-132.

[49] Kim D-Y, Nishiyama Y, Kuga S. Surface acetylation of bacterial cellulose. Cellulose 2002; 9 361-367.

[50] Matsumura H, Glasser WG. Cellulosic nanocomposites. II. Studies by Atomic force microscopy. Journal of Applied Polymer Science 2000; 78 2254-2261.

[51] Goodlett VW, Dougherty JT, Patton HW. Characterisation of cellulose acetates by nuclear magnetic resonance. Journal of Polymer science Part A-1. 1971; 9, 155-161.

Chapter 5

Direct Dissolution of Cellulose: Background, Means and Applications

Carina Olsson and Gunnar Westman

Additional information is available at the end of the chapter

1. Introduction

Being the main structural component in plants, cellulose is by far the most abundant organic polymer on earth. The highest quantity of cellulose is found in the secondary walls of higher plants, such as wood, were the polymer is incorporated in a matrix of lignin and shorter heteropolysaccharides such as hemicelluloses and pectins. Cellulose is also found in some fungi and algae, in the marine animal family of tunicates and as an extracellular product of some bacteria. In theory, cellulose is an inexhaustible resource and yields 1.5×10^{12} tons of biomass every year. Despite the fact that cellulose is available in any plant, wood pulp provides the main part of cellulose used today, for example in board, paper and textile. Historically, cellulose has been used for as long as man has existed, first as fuel and shelter and later for clothing and writing material such as papyrus. Cotton was domesticated thousands of years ago and the soft fiber surrounding the cotton seed was used for making textiles in many parts of the world. Cotton is still the main source of cellulose for textile but the demand for wood based raw material is increasing due to the environmental drawbacks associated with cotton cultivation and processing, such as high water and pesticide usage. Both cotton and wood based cellulose can also be used as the basis for cellulose derivatives such as cellulose ethers and cellulose esters. These are widely used, for example in pharmaceuticals, food, construction materials and paint. When it comes to sustainable raw materials for future demands on fuel, chemicals and materials, both bulk and high performance, cellulose is definitely a resource of great importance and still today one that is underutilized.

Corpora non agunt nisi solute (substances do not react unless dissolved) is a statement with roots in ancient alchemy, and still has some validity today. Unlike many petroleum based polymers, cellulose will not melt but decompose at elevated temperatures. To get the cellulose in liquid form, it has to be dissolved or chemically modified. Homogeneous reaction as a route to add functionality to the reactive cellulose hydroxyl groups provides better control of total degree of substitution and better conditions to control not only the substitution pattern along the cellulose chain but potentially even the substitution pattern within each monomer since each part of the polymer is, at least in theory, equally exposed to reagents and solvent. Due to its chemical composition, cellulose is insoluble or only partly soluble in most common solvents. Although quite a few alternatives have proven useful over the years, the search for efficient and non-degrading solvents for cellulose is still ongoing and in the light of current trends within sustainability and exploration of alternatives for oil based products, this work is now more important than ever. Suitable solvents for cellulose are not only essential in homogeneous reactions, but are indispensable in the characterization of the polymer itself, and last but not least in the shaping of cellulose into e.g. textile fibers, films and membranes.

One of the most successful routes to dissolving and shaping of cellulose is via the viscose process. This technique involves conversion of cellulose to cellulose xanthogenate, by a reaction with carbon disulfide in sodium hydroxide first described by Charles Frederick Cross, Edward John Bevan, and Clayton Beadle [1]. The metastable cellulose xanthogenate is soluble in aqueous sodium hydroxide and the resulting viscous liquid is used as a spin dope. This means that the dissolved polymer is not actually cellulose but a cellulose derivative. The polymer is then precipitated in acid, for simultaneous neutralization and regeneration of the cellulose in a wet spinning step. Hence, the resulting regenerated fiber will be pure cellulose. This process is still today widely used to make viscose textile and cellophane. The viscose process yields high performance fibers with excellent properties for various applications, but it has some drawbacks. The sulfuric reagents, their byproducts and the heavy metals used in the regeneration baths are, despite a decrease in usage, still a problem from an environmental point of view. Both air and water streams must be adequately purified to meet today's regulations.

In an attempt to avoid some of the problems associated with the in some sense outdated viscose process, numerous alternatives have been developed. One method where all existing technology can be translated directly from the viscose process to new chemistry is the carbamate process, where urea is used to produce cellulose carbamate, a derivative soluble in aqueous sodium hydroxide that can be fed into the ordinary viscose line [2]. The process utilizes alkalized and pre-ripened pulp in accordance to the viscose process. In this case the spin bath for the wet spinning step can be either acidic or alcoholic to yield the pure cellulose. The urea route to soluble cellulose has a lower environmental impact and lower energy consumption than the traditional viscose process. Commercial plants are not yet in use but Neo Industrial Plc's viscose fiber division, Avilon Ltd, plans to convert a viscose production site to the new urea technology and should open 2013, according to their press release in January 2011 [3]. One major advantage with the carbamate route is that the intermediate cellulose carbamate is stable and may be stored or transported without significant degradation.

Figure 1. Cellulose xhantogenate (left) with R = H or xhanthogenate group and cellulose carbamate (right) with R = H or carbamate group.

Common to the cellulose/solvent systems in both the viscose process and the carbamate process is that they involve an intermediate polymer. As already stated, it is in fact not the cellulose itself but a cellulose derivative that is dissolved. This is true for many other solvent systems for cellulose as well, such as DMF/N_2O_4 or $DMSO/N_2O_4$, CF_3COOH, $HCOOH/H_2SO_4$ and $Cl_2CHCOOH$. This case may or may not be convenient depending on the application. Direct dissolution of cellulose, without derivatization, is not only necessary in some analytical applications but may also simplify the production of regenerated cellulose since many steps are omitted. A direct solvent will also be considerably more straightforward to recycle since no byproducts are generated, at least not in theory. Today, the only really successful direct solvent system for regenerated cellulose on large scale is the NMMO hydrate even if e.g. both alkali and acid systems are proposed for the same use.

When discussing the dissolution of cellulose in general, and the direct dissolution of it in particular, it is necessary to have a proper understanding of the polymer itself. The properties of cellulose are interesting and unique in many ways. Its dissolution process must also be seen as a result of both kinetics and thermodynamics and as a multidisciplinary chemical challenge, where knowledge in both inorganic and organic chemistry as well as physical chemistry and process engineering is required. To understand, and above all control the dissolution process will be of great importance for future applications of the cellulose polymer.

2. Macromolecules

Though the use of both synthetic and modified natural polymers have been of great significance in industry since the 19th century, a reasonably correct understanding of polymer structure was not developed until in the 1920s when Hermann Staudinger proposed a model where polymers, for example cellulose, was composed of covalently linked small repeating units, i.e. the very definition of a polymer today. For this work he received the Nobel Prize in chemistry 1953.

Polymers, with few exceptions, are not homogeneous in length. The degree of polymerization (DP) varies and the molecular weight distribution is one of the major factors influencing the properties of a given polymer.

Some polymers, such as starch, can appear in linear or branched forms. This has great influence on many properties such as crystallinity, solubility, density and as a consequence of this, even thermal and mechanical properties that have direct influence on the industrial applications of the product. The repeating units of any polymers are not restricted to being of the same kind all through the chain length. Nor are the branches restricted to being of the same type as their backbone. Alternating monomers yields so called co-polymers or heteropolymers, that can be randomly organized or well defined. Homopolymers may vary in stereochemical configuration to yield isotactic, atactic or syndiotactic polymers if there are any substituents as seen in Figure 2.

Figure 2. Representation of an isotactic polymer (top) where all monomers have the same stereo/regiochemical configuration, a syndiotactic polymer (middle), where the monomers have alternating positions along the chain and an atactic polymer (bottom) where the configuration of the monomers is random.

Depending on chemical and stereochemical design, polymers experience different interchain forces that, without actual covalent bonding, keep the polymer chains together in the solid state. For electrically neutral polymers, the strongest and most important interchain force is of course hydrogen bond interactions, but dispersion forces may have a significant effect on the polymer bulk too.

3. Solubility of macromolecules

In some aspects macromolecules fall under the same rules as all substances when it comes to solubility, in others not at all. Non-polymeric materials dissolve immediately, and in a good solvent the process is controlled mainly by external mass transfer in a stagnant zone near the solid surface. Polymers, on the other hand, undergo a multistep process going from solid to dissolved state. Mass transfer is bidirectional as solvent penetrates the polymer network and swell its outer layer while polymer disentangles from the outermost surface and transfer into the liquid bulk.

As can be intuitively understood, the dissolution rate of polymers decreases with increasing molecular weight. A more intriguing fact is that polydispersity too has a significant effect on the dissolution rate. This effect was investigated in polymethyl methacrylate solutions by Manjkow and others in the late 80s. Polymers of broad molecular weight distributions ($M_w/M_n \approx 2$) dissolved twice as fast as the monodisperse equivalent with the same M_n [4].

Pure polymer	Infiltration layer	Solid swollen layer	Gel layer	Liquid layer	Pure solvent

Figure 3. Schematic picture of the composition of the surface layer, adapted from Miller Chou [5].

Due to the intrinsic high molecular weight, polymers in general are soluble only in selected solvents. As a result of interchain forces, solubility of polymers is not only dependent on chain length, but also chain-to-chain interaction. Crystallinity and polar groups that may take part in hydrogen bonding thus play a major role in solubility and reactivity of any polymer.

The discussion on thermodynamics of dissolution must begin with Gibbs free energy

$$\Delta G = \Delta H - T\Delta S \tag{1}$$

and the idea of an ideal solution, where the enthalpy of mixing and the volume change upon mixing is equal to zero. In a binary solution this would mean that there is no energy gain or loss for the constituents in the solution to interact with either a molecule of its own kind, or a molecule of the other type. This is of course never really true but has some relevance as a simple model. In the case of polymers in solution, the difference in size leads to a non-ideal mixture. In addition to enthalpy, one must take entropy into consideration. For a macromolecule, the gain in entropy is not as great as for the corresponding small molecule. Furthermore, the monomer-to-monomer covalent bond restricts motion and confines the macromolecule within the solvent. This is especially true for stiff polymers, where the conformational freedom does not increase to the same extent as it does for freely rotating polymers when going from solid to dissolved state.

Dissolution can only take place if the dissolved state represents a lower energy than the solid state, meaning that enthalpy of mixing must be balanced by the entropy term. Again, consider glucose and cellulose. The same type of interactions should take place between a glucose molecule and its solvent as between a polymer of glucose and its solvent. The gain in entropy for glucose monomers is far greater than the entropy gain for the glucose polymer. Hence, glucose dissolves easy in e.g. water, while cellulose does not.

A well needed addition to the Gibbs energy equation is the Flory Huggins model (Equation 2) of the thermodynamics of polymers in solution. The free energy of mixing is described by

$$\Delta G_m = RT[n_1 ln\phi_1 + n_2 ln\phi_2 + n_1\phi_2\chi_{12}] \tag{2}$$

where ΔG_m is the change in Gibbs free energy, R is the ideal gas constant, T is temperature, n_1 is the number of moles of the solvent, Φ_1 is the volume fraction of solvent, n_2 is the number of moles of polymer and Φ_2 is the volume fraction of polymer. The parameter chi (χ) describes the energy of interdispersing polymer and solvent molecules, i.e. both enthalpic and entropic deviations from ideality. The volume fraction Φ is analogous to the molar fraction, but takes into consideration the size of the molecule. For a small solute, with a size comparable to the size of the solvent molecule the volume fraction can be replaced by molar fraction [6].

This model does not take into account a possible change in volume due to mixing, nor is it applicable in the case of very dilute solutions. For these cases, the theory of excluded volume is used. This is based on the simple assumption that one part of a polymer chain cannot occupy the space that is already occupied by another part of the chain. However, the size and shape of dissolved polymers depends on the quality of the solvent. A good solvent is a solvent that coordinates well and strong to the polymer whereas a poor solvent only has weak or almost no coordination to the polymer. The polymer will stretch out in a good solvent and occupy a large volume since interactions with the solvent molecules are favorable. The polymer will contract in a poor solvent, and if the solvent is too poor, eventually precipitate. The so called *theta point* is reached at the exact conditions where the excluded volume effect is balanced by the solvent. In a theta solvent, the polymer chain reverts to its ideal characteristics, and the chain dimensions are no longer dependent on the solvent choice. In other words, the polymer interacts with the solvent as it would interact with itself, and will have the same chain dimensions as it would have in a polymer melt or amorphous solid phase. This means that analytical measurements in a theta solvent will be independent of solvent and give the "true" value of e.g. radius of gyration [6].

Apart from chemical compatibility, i.e. thermodynamic features of the solvent/solute system, there is kinetics to consider. In the case of a macromolecule such as the cellulose polymer, this is a parameter of critical importance since diffusion rate is ideally inversely proportional to the size of the particle, or in this case the molecule. Hence, in the case of entangled polymers with a severe tendency to self assemble the diffusion rate is drastically retarded. It matters not if the solvent is thermodynamically ideal, if the rate of dissolution is too slow and a solution does not form under a desired amount of time the solvent is useless. Kinetics of dissolution can be manipulated by many different means. Diffusion of large molecules is slow, but addition of heat speeds up the process. In general, addition of heat also lowers the viscosity of liquids such as the solvent and this will be advantageous since it will lower the diffusion resistance. Furthermore, mechanical energy added to the system in form of shear stress will decrease the length of the diffusion path from concentrated polymers solution at the particle surface to the bulk solution. This will not affect the rate of diffusion, but the flux of molecules and thereby the overall kinetics of the dissolution process. This can be understood in a simplified way using Fick's first law shown in equation 3 as linear integrated version.

$$J = D\frac{\Delta c}{\Delta x} \tag{3}$$

J is the flux [mol/s], D is the diffusion coefficient [m^2/s], Δc is the concentration gradient from the undissolved particle surface to the bulk solution and Δx is the distance from the particle surface to the bulk, the stagnant layer where mass transport is controlled by diffusion only. Thus, if the distance from the particle surface to the bulk is diminished, the flux of molecules from the solid particle to the bulk solution will increase. It must be stressed that kinetics of polymer dissolution is not by any means as simple as kinetics of dissolution of small molecules, especially when it comes to semi-diluted or concentrated solutions. For example, as the concentration of dissolved polymer increases the viscosity of the solution will increase and this will slow down the process. A basic understanding of the kinetics in a simple system can nevertheless come in handy when it comes to understand the processes in more complex ones.

4. Cellulose morphology

To comprehend the interactions of cellulose and its solvent as well as the dissolution process it is important to have a clear understanding of the polymer itself. Unlike some other biopolymers, like proteins, cellulose in its native form is highly polydisperse and often closely linked to other compounds, e.g. the lignin and hemicelluloses in wood. Apart from this, the polymer structure provides, though at a first glance very symmetric and simple, serious challenges due to its chemical composition and spatial arrangement. From monomer to bulk material, all levels present interesting features that are highly important.

Cellulose was first defined as the remaining part of plant tissue after purification and in the early days of cellulose technology the word cellulose was used also for what is today defined as pulp [7]. Eventually this residue was found to be a carbohydrate very similar to starch, and today it is known that it is a linear syndiotactic polymer (cf. Figure 2) of β-1,4-glycosidic bond linked D-glucopyranose units, where every other monomer is rotated 180 ° with respect to its neighbors. This structure can also be defined, if instead pairs of glucose units, cellobiose, are defined as the repeating unit, as an isotactic polymer. The individual glucose monomers in the cellulose polymer are often referred to as anhydroglucose units, AGU, due to the loss of one water molecule in the polymerization where a hemiacetal reacts with an alcohol to form the acetal bond. From crystallographic [8-9] and NMR spectroscopic data [10] it has been shown that the six membered glucose monomer takes on a chair conformation. Three hydroxyl groups are present in the AGU, secondary on C2 and C3, and a primary on C6. The rotational conformation of the primary alcohol on C6 can be altered, and this has a profound impact on the hydrogen bonding pattern and therefore the crystallinity of the cellulose metastructure [11]. In one end of the polymer the anomeric carbon is involved in a glycosidic bond. This is referred to as the non-reducing end. In the other end of the polymer the anomeric carbon is free to convert to an aldehyde structure, and these two states are in equilibrium. This end is referred to as the reducing end. This chemical polarity of the cellulose chain is important when it comes to its different crystal structures.

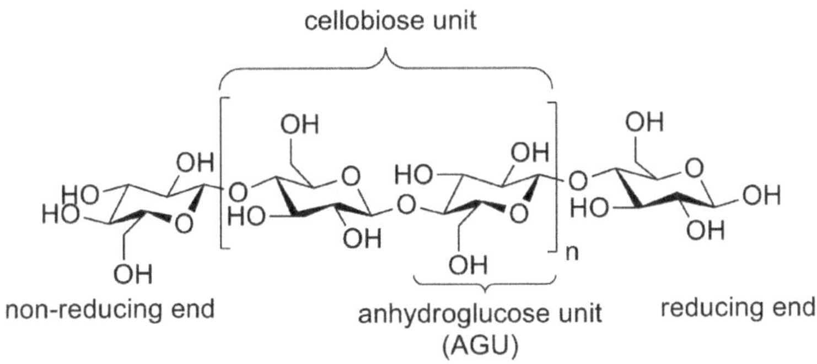

Figure 4. The cellulose polymer chain structure

The number of AGU units defines the degree of polymerization (DP) of the cellulose. Depending on source, this might vary to a great extent. The DP of various sources including native wood was first determined 1962 [12] to between 9000-10000. Extraction methods and treatment of cellulose might degrade the polymer, and the DP of wood pulp varies between 300-1700 whereas regenerated fibers from cellulose have a degree of polymerization no greater than 250-500.

Polymers in native cellulose are densely packed in a semi ordered structure. Intramolecular hydrogen bonds provide chain stiffness, and intermolecular hydrogen bonds allow for the linear polymers to assemble in sheet structures. These sheets are further packed by hydrophobic interactions into crystals. Crystalline regions in silk, bamboo and wood were confirmed by X-ray spectroscopy already 1913 [13] and ever since, the crystalline structure of celluloses has been studied thoroughly. To this day four main different crystal structures of cellulose have been discovered, namely celluloses I, II, III and IV. Based on the X-ray diffraction data for specimens from the sea algae *Valonia ventricosa,* native cellulose I was first said to have a triclinic crystal structure of parallel chains [14] but later work showed by solid state NMR spectroscopy that cellulose I can be subdivided into cellulose Iα and cellulose Iβ [15-16], where the Iα is dominating in bacteria and algae and Iβ is dominating in higher plants. Recent work on X-ray and neutron diffraction data suggests that only cellulose Iα is truly triclinic, while the Iβ form is monoclinic [17]. Cellulose I can easily and irreversible transform into the monoclinic cellulose II by regeneration or alkali treatment, suggesting that the cellulose II is the more stable allomorph [18]. In this transformation, the parallel ordering of polymer chains found in cellulose I is replaced by what seems to be an antiparallel configuration in cellulose II [19]. It is still debated whether this antiparallel ordering is due to chain-folding or interdiffusion of crystallites. Cellulose III and cellulose IV can be reversibly prepared from either cellulose I or cellulose II by various chemical treatments [20].

In its native form, cellulose crystals are never pure. Crystalline regions are interrupted by less ordered structures. Several suggestions on how these crystalline and non-crystalline regions are intermixed have developed over the years, such as single crystals [21] or uniform elementary fibrils [22] but nowadays a so called fringed fibrillar model, developed in the late 50s [23], of the material is widely accepted [24]. In this model the microfibril is not regarded a single crystal but rather a less structured arrangement of non-uniform crystalline segments accompanied by amorphous parts, both longitudinal and lateral.

5. Solubility of cellulose

Being a linear homopolymer, cellulose does not appear as a complex structure but its dissolution is in fact very complicated. Many factors contribute to make the cellulose macromolecule insoluble in most common solvents. Cellulose in its native form is often a very long polymer and high DP inevitably leads to decreased solubility due to decreased entropic gain in the dissolution process as already discussed. Glucose, cellobiose and any oligomer of cellulose structure up to around a DP of less than 10 is soluble in simple solvents such as water and cellulose is not, so this is indeed a relevant factor. Furthermore, the interactions between the cellulose molecules provide a challenge. Three hydroxyl groups per AGU make complex patterns of hydrogen bonds possible. To break these, solvents with high hydrogen bonding capacity are necessary. Quite a few liquids including water are able to swell, but not to dissolve cellulose. Swelling is defined as a process where the liquid molecule penetrates cellulose structure by interacting with the polymer to a certain extent, leaving the volume and physical properties of the cellulose significantly changed but the solid, or semi-solid state remains. Dissolution, on the other hand, completely destroys the supramolecular structure due to superior interaction between solvent and polymer, and results in a clear solution where the polymers are molecularly dispersed. The same liquid may act as a solvent or merely a swelling agent, depending on the conditions used in the experiment. Temperature can be such a determinant parameter.

One way to estimate a new solvent's compatibility is to evaluate the hydrogen bonding and polarizability properties of the solvent. Parameters of the empirical Kamlet–Taft solvatochromic relationship are used to calculate the hydrogen bond donor (α), hydrogen bond acceptor (β), and dipolarity/polarizability (π^*) properties of solvents as contributing to overall solvent polarity. This is done by UV/Vis spectroscopy to monitor the interactions of the substrate with chosen indicators of e.g. the ones shown in Figure 4. Numerous solvatochromic indicator dyes are available [25]. In many reports where new solvents for cellulose are being evaluated, the Kamlet-Taft parameters are held as highly important [26-29]. In Table 1 the solvatochromic parameters α, β and π^* for cellulose, a cellulose non-solvent and a cellulose solvent are compared.

	α	β	π^*
Cellulose (dried)	1.27	0.60	0.41
H_2O	1.17	0.47	1.09
DMAc/LiCl (15%)	0.54	2.0	0.67

Table 1. Kamlet-Taft parameters for cellulose, water and the cellulose solvent system dimethylacetamide (DMAc) with 15 % LiCl [26, 30-31].

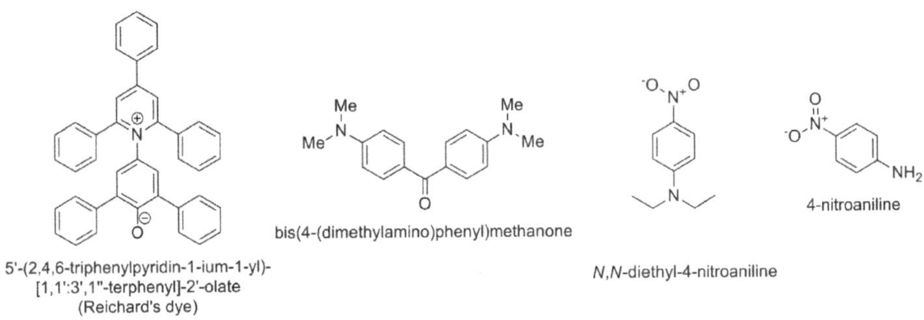

Figure 5. Examples of probe dyes for determination of solvatochromic parameters of cellulose solvents.

Water cannot be used to break the bonds in cellulose, yet both water-water, carbohydrate-water and carbohydrate-carbohydrate hydrogen bond strengths are around 5kcal/mol [32]. Many forget to take this question into account when they reason about the dissolution of cellulose and its relative insolubility. It is obvious that hydrogen bonding is not the only relevant factor in the quest for a suitable solvent, yet the amphiphilic nature of cellulose is still widely overlooked. The last few years, a few eye-opening papers have taken this matter up to discussion [32-37].

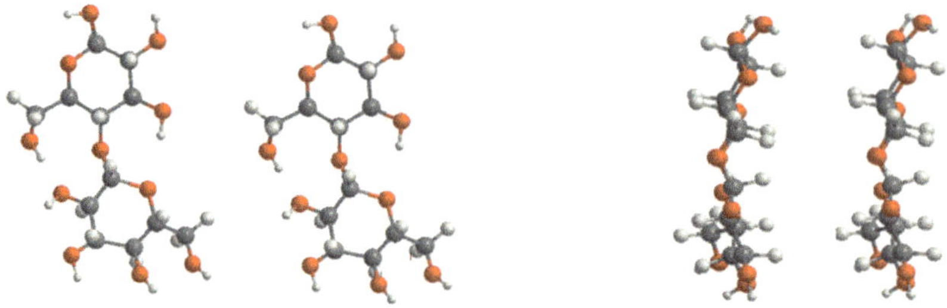

Figure 6. Hydrogen bonds arrange cellulose chains in sheets, and the stacking that forms a three dimensional structure involves hydrophobic interactions.

A relatively new approach when trying to understand molecular interactions is computer simulations and modeling. Some very recently published papers describe this method, showing molecular origins of the recalcitrance of biomass in the terms of decrystallization and dissolution, and showing quantitatively the need for taking cellulose amphiphilicity into account [36-38]. For comparison it may be useful to study cyclodextrins. Typical cyclodextrins are oligosaccharides made up from six to eight 1→4 linked monosaccharides in a torus and are produced from starch but since starch as cellulose is built up from glucose, this may still serve as a model. Cyclodextrins are water soluble, but the arrangements of hydroxyl groups provide a pronounced amphiphilicity.

The inclusion or the interior of the sugar macro ring is if not fully hydrophobic at least significantly less hydrophilic than the outer part of the structure. As a result of this, cyclodextrins may serve as the host in inclusion complexes. Unusual hydrogen bonding patterns take place inside the cyclodextrin, due to the confined space and the hydrophobic nature of the torus, which was shown by Bezel et al. in a neutron scattering study [39]. This relates to and partly explains the concept of "high energy water" i.e. the water trapped in the cyclodextrin ring in a water solution which upon release by addition of a more hydrophobic substance contributes to a favorable enthalpy change. The driving force for inclusion of a hydrophobic substance in a cyclodextrin ring should be considered an "atypical" hydrophobic interaction, since it is not entropy driven [40]. An extensive overview of cyclodextrins as inclusion complexes was published by Clarke et al. and explains many aspects of the amphiphilicity of these substances [41].

The alpha, beta and gamma cyclodextrins with six, seven and eight glucose units in their ring as shown in Figure 6 have all been granted GRAS (generally recognized as safe)-status by the American Food and Drug Administration (FDA). This is useful for many purposes where a hydrophobic compound, or part of a compound, needs to be solubilized in water for example in the drug delivery of lipophilic substances. By considering the uses of cyclodextrins, it is obvious that polymers of glucose should have amphiphilic traits. The difference is that these polymers are linear instead of torus shaped, and make up sheets where the hydrophobic interactions are orthogonal to the hydrophilic.

Coming back to the discussion of polymer solubility, it now seems obvious that any reasonable solvent for cellulose is one that must overcome the low entropy gain by excellent solvent – polymer interactions. For cellulose that has both hydrophilic and hydrophobic features, i.e. the equatorial OH-groups and the axial CH groups, a good solvent should contain both a hydrogen bonding part and a hydrophobic part.

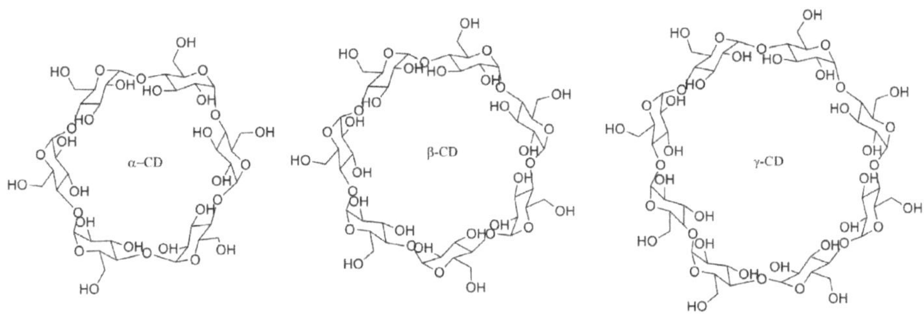

Figure 7. The molecular structure of the six membered alpha cyclodextrin (α-CD), the seven membered beta cyclodextrin (β-CD) and the eight membered gamma cyclodextrin (γ-CD).

Apart from this, the kinetics of cellulose dissolution in a chosen solvent may be controlled by elevated temperatures, at least up to the point where side reactions such as solvent and/or cellulose will start to degrade. In industrial systems, mechanical energy is added as cellulose dissolution takes part in kneaders that speed up dissolution rate and helps in the disintegration of fibers and gel particles that have been formed as a result of initial swelling. Co-solvents may be added to lower the viscosity of the solutions.

6. Solvents for cellulose

Despite the challenges associated with the dissolution of cellulose, numerous options are available. Here the focus lies on direct solvents, i.e. solvents that do not form covalent bonds with the polymer but merely interact by physical means. This distinction between non-derivatizing and derivatizing solvents must be made in order to choose the correct solvent for a specific use. Solvents may further be classified as aqueous or non-aqueous, or as containing halogens or not or whatever else suits the user's purposes.

Solvents can be further classified based on their number of components. Both unicomponent, bicomponent and tricomponent solvent systems are known and including solvent mixtures the range is even broader. Since many non-derivatizing, or direct, solvents for cellulose includes water, as bulk solvent or as bound water in a salt or organic hydrate, it is not fully satisfactory to make a binary distinction between aqueous and non-aqueous solvents. Instead, solvents could be placed on a continuous scale that represents the amount of water in the system. However, herein the binary view is maintained.

6.1. Aqueous media

Being the most green of all chemicals, water is of course an attractive solvent and reaction media. In fact, pure water can actually be used in the field of cellulose technology. Water at around its critical point demonstrates some unique properties and can be an interesting media for various chemistry. Dissolution and hydrolysis of cellulose in subcritical and supercritical water was investigated by Sasaki et al. [42] but the dissolved cellulose in this case was probably shorter fragments and cellooligomers since severe depolymerization must take place before any dissolution can be expected as stated by more recent publications [43-44]. This means that however appealing, it is not yet possible to dissolve high DP cellulose in pure water. Water based solvent systems are plentiful though, and different additives can make water into a very efficient dissolution media. Some of the many alternatives available are listed below.

6.2. Aqueous alkali

Mercerization, the process of soaking cellulose in strong alkali to the extent that the crystal structure changes from cellulose I to cellulose II, is one of the most technically relevant processes in cellulose technology and is used to activate the hydroxyl groups of cellulose for further modification and/or dissolution. Cellulose is partly soluble in an aqueous solution of sodium hydroxide in a concentration of about 10 % w/w. The amount of cellulose that is soluble in $NaOH/H_2O$ depends on degree of polymerization and also mode of crystallinity [45]. Isogai investigated the solubility of cellulose from several sources but never succeeded in preparing cellulose solutions of higher concentrations than 5 % [46]. A method for complete dissolution of cellulose in lye solutions was also patented in 1994 [47]. For cellulose to dissolve in alkaline aqueous media, it needs to be cooled well below room temperature. Soube et al. completed the phase diagram for the ternary system cellulose/$NaOH/H_2O$ [48]. Taking the amphiphilic properties of cellulose into account, it is not surprising that it has been shown that cellulose in $NaOH/H_2O$ is in fact not completely dissolved but forms aggregates [49].

More recently, sodium hydroxide solutions with different additives have turned out to dissolve cellulose more efficiently than the binary $NaOH/H_2O$ system itself. Such additives are for example poly(ethylene glycol) (PEG) [50-51] and urea [52] and/or thiourea [53-55]. The fact that sodium hydroxide, PEG and urea are all environmentally friendly and pose low toxicity towards humans and animals makes these solvent systems interesting for large scale applications. However, mixed systems always pose high demands on recovery systems, and the amount of additives needed in these systems makes recovery and reuse quite necessary.

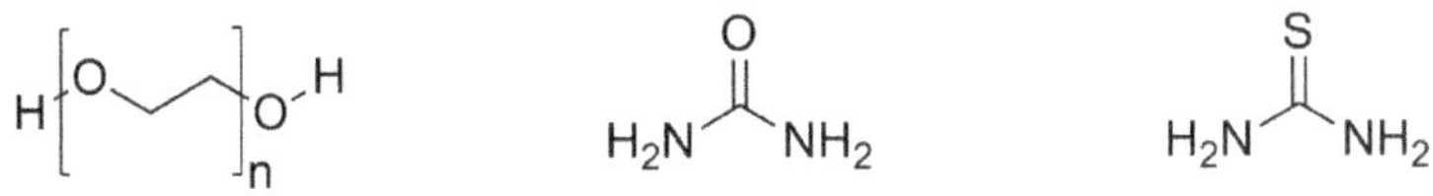

Figure 8. Additives in aqueous alkali solvents for cellulose. From left to right; poly(ethylene glycol), urea, thiourea.

Mechanical or chemical pretreatment is necessary for the dissolution to be efficient enough for industrial needs [56]. Several studies of cellulose in the aqueous NaOH/urea system using e.g. ^{13}C NMR, ^{15}N NMR, ^{1}H NMR, FTIR, small angle neutron scattering and wide angle X-ray scattering suggest that the dissolution mechanism is based on the hydrates of NaOH that in low temperatures are able to form hydrogen bonds with the cellulose chain, while the urea molecules surrounds the cellulose/NaOH/H_2O inclusion complex, screening it from other cellulose molecules and thereby prevent cellulose aggregation [57-59]. Being a hydrotrope, urea is expected to increase the solubility of a poorly water soluble solute in aqueous solutions. This is explained by the ability of hydrotropes to break water structures, i.e. the effect is explained by urea – water interactions, and/or a tendency of hydrotropes to interact with the solute itself by hydrophobic interactions. Interestingly, in the case of cellulose in aqueous NaOH/urea, no direct evidence of interaction between urea and cellulose was found. In the original NaOH/PEG/H_2O publication by Yan and Gao [50] the dissolution mechanism was suggested to be similar to that in NaOH/urea/H_2O. It was proposed that the solution is stabilized by polyethylene glycol chains, here acting as hydrogen bonding acceptors. It has also been suggested that the mechanism for dissolution in these types of solvent mixture relates to a charging up of the cellulose, i.e. turning it into a polyelectrolyte [32].

The solutions of cellulose in aqueous alkali with urea and/or thiourea are being thoroughly investigated for shaping purposes, i.e. regeneration of cellulose into fibers, membranes or similar. Recently, Yang et al. prepared high performance flexible films from different cellulose sources in a solvent of alkali and urea [60]. The films showed good gas barrier properties over the entire relative humidity range, and always one order of magnitude lower gas permeability than cellophane films which are prepared via the viscose process. Compared to conventional films from poly(ethylene) and poly(propylene), all cellulose films show increased water vapor permeability, but this is expected due to the intrinsic hydrophilicity of the polysaccharide structure. Regenerated cellulose fibers similar to viscose or Lyocell fiber have been wet spun from solutions of NaOH/urea [61] and NaOH/thiourea solutions [62-63].

The procedure for dissolving cellulose in aqueous alkali/urea-systems is based on a freeze-thaw method. The solvent is pre-cooled to below the freezing point of water before cellulose is added and kept cold until complete dissolution of the polymer [64]. Regeneration takes place by precipitation in e.g. dilute H_2SO_4, ethanol, butanol or acetone [60]. Lithium hydroxide and potassium hydroxide are expected to have a similar effect as sodium hydroxide. A strong impact of the choice of salt has been observed, and the dissolution power of the different systems is in the order LiOH/urea > NaOH/urea >> KOH/urea. Both LiOH and NaOH in combination with urea are able to rapidly dissolve cellulose, while KOH is not [64].

The possibility to perform homogeneous reaction in aqueous alkali/urea systems has not been neglected and it has been shown to be a stable media for etherification reactions. Carboxymethylation of cotton linters and microcrystalline cellulose was carried out in lithium hydroxide with urea and the degree of substitution could be controlled [65]. The product was water soluble CMC. Another important cellulose derivative, namely hydroxyethyl cellulose, was prepared from cotton linters by homogeneous reaction in sodium hydroxide with urea under mild conditions [66].

6.3. Inorganic metal complexes

One interesting group of solvents is the one of inorganic metal complexes. The complexes are of transition metal ions and nitrous ligands. Some of the most common ones [67] are listed in Table 2.

As in many cases with solutions of transition metal complexes, most of the cellulose solvents in this group are strongly colored. This is true for example with Cuoxam, Nitren and Cuen which all are deep blue, while Pden only displays a weak yellow color. The color of the solvent or in some cases the cellulose – metal complex in itself may cause problems in analytical applications where light scattering or refractive index is used [68].

Cadoxen	$[Cd(H_2N(CH_2)_2NH_2)_3](OH)_2$
Cdtren	$[Cd(NH_2CH_2CH_2)_3N](OH)_2$
Cooxen	$[Co(H_2N(CH_2)_2NH_2)_2](OH)_2$
Cupren	$[Cu(H_2N(CH_2)_3NH_2)_2](OH)_2$
Cuam/Cuoxam/Schweitzer's reagent	$[Cu(NH_3)_4](OH)_2$
Cuen	$[Cu(H_2N(CH_2)_2NH_2)_2](OH)_2$
Nioxam	$[Ni(NH_3)_6](OH)_2$
Nioxen	$[Ni(H_2N(CH_2)_2NH_2)_3](OH)_2$
Nitren	$[Ni(NH_2CH_2CH_2)_3N](OH)_2$
Pden	$[Pd(H_2N(CH_2)_2NH_2](OH)_2$
Zincoxen	$[Zn(H_2N(CH_2)_2NH_2)_2](OH)_2$

Table 2. Transition metal complexes with amines or NH_3, their common names and chemical formula.

Saalwächter et al. compared several metal complexes and found that the coordinative binding metal complexes such as Cuoxam, Nitren and Cdtren were most efficient as cellulose solvents. The coordination takes place at the deprotonated olate anions at C2 and C3 at each AGU. Even though the dissolution appears to be to molecular level and via interaction with the C3 hydroxyl group, the cellulose chain stiffness appears to be considerable. The number of Kuhn segments, i.e. the number of efficient straight segments seconded by kinks, per polymer rarely exceeds 50 [68] which means that several monomers are assembled in rod like structures before an actual bend.

Apart from complex coordination, metal complexes can interact with cellulose via pure Coloumb interactions. This is the case for Cuen [68], but not for Pden which forms square planar complexes with the AGUs of cellulose as with any other ligand [69-70].

Aqueous inorganic salts or metal complexes can also be used for dissolution and regeneration of cellulose. The Cupro process, using cuprammonium hydroxide (Cuam) as solvent, was invented already in the year 1890 and is still used today, although not in a huge scale [71]. The solvents have also proven viable reaction media. For example, completely homogeneous etherification of cellulose may be performed in Nitren to produce carboxymethyl cellulose in a one-phase-reaction with a regioselectivity and substitution pattern similar to the carboxymethyl cellulose obtained from commercial routes in NaOH slurries [72]. Unfortunately, some of the representatives of the inorganic metal complex based solvents initiate severe cellulose depolymerization in the presence of even traces of oxygen [73].

6.4. Molten inorganic salt hydrates/concentrated inorganic salt solutions

Molten inorganic salts have been pursued as cellulose solvents and reaction media for close to 100 years now and some are indeed able to dissolve cellulose without pretreatment and in reasonable concentrations. Both pure salt hydrates and mixtures of different salt hydrates as well as certain concentrated inorganic salt solutions may swell or dissolve cellulose [74]. The most commonly used solvents in this class are $MgCl_2$*$6H_2O$, LiCl*$5H_2O$, $LiClO_4$*$3H_2O$, $ZnCl_2$*$4H_2O$, $ZnCl_2/H_2O$, LiSCN and $Ca(SCN)_2/H_2O$ with or without adition of the sodium or potassium thiocyanate salt [75]. $LiClO_4$*$3H_2O$ is an extremely efficient solvent and cellulose in this particular salt gives clear solutions within a few minutes.

The dissolution process of cellulose in the aqueous $Ca(SCN)_2$ solvent system was examined using IR spectroscopy by Hattori et al. and it seems to be based on complex formation and an addition compound between cellulose and the thiocyanate anion [76] but curiously enough, the crystallinity of regenerated cellulose product appears to depend on what kind of precipitation agent is used. In water, only cellulose II is recovered, but in methanol the product is mainly cellulose I. Upon precipitation in acetone the product is amorphous [77]. The interaction between cellulose and salt hydrates have also been examined using Raman spectroscopy [74], and of course different NMR spectroscopy techniques [78]. The variation in dissolution processes in different molten salt hydrates was again thoroughly investigated by e.g. ^{13}C NMR spectroscopy and discussed by Leipner et al [79].

Homogeneous reactions to obtain cellulose derivatives may be carried out in molten inorganic salt hydrates. Heinze with coworkers reported successful carboxymethylation of cellulose in $LiClO_4$*$3H_2O$ with varying degree of substitution, and a statistical distribution of substituents, showing that no part of the cellulose polymer was inaccessible to the reagents [72, 80]. Esterification in the form of acetylation in the molten inorganic salt $LiClO_4*3H_2O$ and the eutectic mixture of $NaSCN/KSCN/LiSCN*2H_2O$, as well as deacetylation reactions of cellulose triacetate in the concentrated salt solution of $ZnCl_2/H_2O$ has been reported [81]. The acetylation reactions were performed with various acetylating agents such as vinyl acetate and acetic anhydride.

6.5. Acidic solutions of cellulose

It is sometimes claimed that mineral acids are able to dissolve cellulose. However, this must be put in context. Dilute acids swell cellulose, but dissolution can only be achieved using higher concentrations of acids and is expected to be associated with severe, if not complete, chain degradation over time if temperature is not kept very low. The concept of level-off degree of polymerization (LODP) was introduced as the chain length of cellulose after treatment in 2.5 N hydrochloric acid [82-83]. If the hydrolysis is allowed to continue the degree of polymerization will eventually reach the LODP which is thought to reflect the longitudinal size of the native cellulose crystals, since non-crystalline areas are preferably hydrolyzed over the crystallites.

Concentrated mineral acids such as hydrochloric acid, phosphoric acid or mixed acids are utilized as degradation media to prepare cellodextrins, short chained cellulose oligomers [84-87]. Phosphoric acids, with or without additives such as organic acids as a potent solvent for cellulose was patented already back in 1927 [88].

The transition from swelling to dissolution of cellulose in o-phosphoric acid, and its effect on the cellulose accessibility to enzymatic hydrolysis, was studied by Zhang et al. and dissolution took place without severe hydrolysis under the conditions used [89]. As often, water plays a crucial role in the dissolution behavior of cellulose in this solvent. Phosphoric acid is quite unique being triprotic and due to its ability to form dimers and even polymers. The composition of the acid is usually expressed in P_2O_5-concentration, which, at a concentration exceeding 74 % is anhydrous (superphosphoric acid). Thus, mixing different species of phosphoric acids may give a powerful cellulose solvent, claimed to rapidly dissolve up to 38 % w/w cellulose. This was shown recently and from these anisotropic solutions of cellulose with a DP of 700-800 in water free phosphoric acid, spinning dopes for production of textile fibers with only moderate chain degradation could be prepared. Fibers can be spun by air gap spinning in acetone and neutralized by Na_2CO_3. The resulting yarn showed extraordinary tenacity [90-91].

Carboxylic acids are not successfully used as direct dissolution media for cellulose. The acidity is low, and the acids are likely to react with the cellulose, resulting in cellulose derivatives that must be converted to pure cellulose in a second step. Acids in this group that may be used as solvents for cellulose include trifluoroacetic acid, dichloroacetic acid and formic acid with or without addition of sulfuric acid [92]. The dissolution is much faster if sulfuric acid is used as a catalyst. However, the polymers dissolved in these acids are not cellulose but the corresponding cellulose derivative, meaning these are in fact not direct solvents for cellulose but rather derivatizing solvents. Moreover, common for the cellulose derivatives produced via this route is unsurprisingly that they show hydrolytic instability.

6.6. Non aqueous media

In 1934, Charles Graenacher reported the first non-aqueous direct solvent for cellulose in his patent where he applied N-alkylpyridinium salts not only as fairly efficient dissolution media but also for homogeneous reaction [93]. Since then a wide distribution of non aqueous media for cellulose treatment have developed, for uses both within industry and academia.

6.7. DMAc/LiCl

One of the most frequently used solvent systems for cellulose and other polysaccharides when it comes to homogenous modification and analytical work such as size exclusion chromatography is the dimethyl acetamide/lithium chloride mixture [94-96] first patented by McCormic in the early 80s [97]. The DMAc/LiCl solvent system is very efficient and will dissolve even high molecular weight cellulose like cotton linters or bacterial cellulose with negligible chain degradation at moderate temperatures. Besides that, major advantages are that the solutions are colorless and compatible with GPC columns.

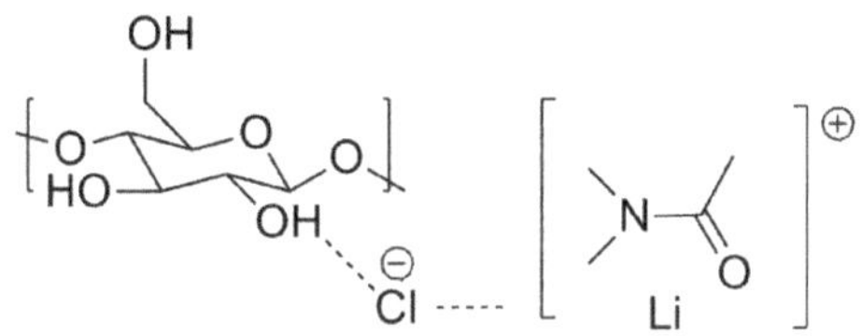

Figure 9. Proposed interaction between solvent and polymer acting as the dissolution mechanism of the system cellulose in DMAc/LiCl, adapted from McCormick, Callais et al. 1985.

The DMAc/LiCl solvent system for cellulose seems to be very specific in its interaction. Neither DMAc with other lithium salts nor DMAc with other chloride salts seem to work in the same way and not even the similar solvent dimethyl formamide with addition of LiCl. The mechanism proposed by McCormick was that hydroxyl groups of cellulose interact with a lithium – DMAc macrocation via hydrogen bonding bridged by the chloride anion. The lithium cation interacts with the carbonyl oxygen via ion – dipole interaction [98]. For this interaction to take place, no cellulose bound water can be present. As a consequence of this, solvent exchange of the cellulose or distillation of the DMAc/LiCl/cellulose system is often necessary. This is a time and solvent demanding step and is one reason why this solvent system may not be preferred outside laboratory scale conditions. Moreover, the solvent itself needs to be dry, since it is easily disturbed by water impurities [99]. Despite early suggestions, the dissolution mechanism is disputed and still not completely understood. However, the DMAc molecule is not able to form hydrogen bonds to any large extent, and that should leave the inorganic ions of the mixtures highly active and free to interact with the polysaccharide instead [94].

This solvent system has been suggested for preparing regenerated cellulose. For example, cellulose beads for support in columns used for liquid chromatography have successfully been produced from DMAc/LiCl solutions and due to the efficiency of the solvent it was possible to optimize the bead performance by adjusting DP and concentration of cellulose [100]. DMAc/LiCl is also suitable for homogeneous derivatization of cellulose. As a "true" solvent where no covalent bonding between solvent and solute takes place, the cellulose is fully susceptible to reaction with other molecules [101-102].

In spite of its many uses, this solvent system experiences some problems. In size exclusion chromatography, unwanted effects have been noticed [103] and under thermal stress the cellulose solutions in DMAc/LiCl may fail to remain inert [104].

6.8. DMSO/TBAF

A new and very powerful solvent for cellulose consists of a mixture of dimethylsulfoxide (DMSO) and tetrabutylammonium fluoride (TBAF). It is an efficient solvent and can dissolve cellulose with a DP of up to 1200 within one hour at 60 °C [105]. So far, the DMSO/TBAF solvent system has been used mostly for analytical purposes e.g. NMR and GPC but also for homogenous reactions for chemical modification of cellulose, both in lab-scale only. Both etherifications and esterifications have been pursued with varying success [106-110].

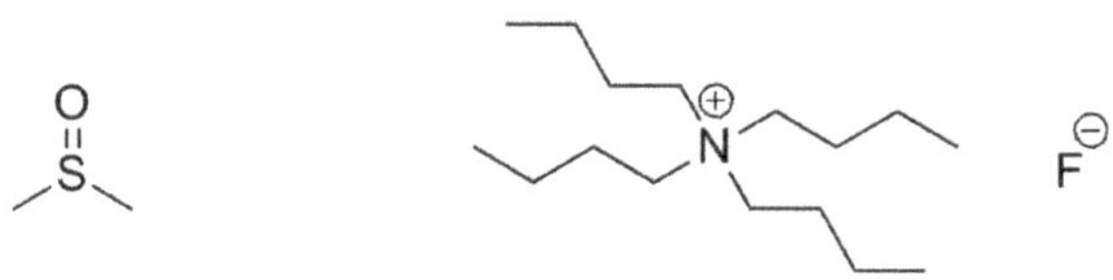

Figure 10. Dimethylsulfoxide (left) and tetrabutylammonium fluoride (right), here represented as anhydrous.

TBAF is a very hygroscopic salt and water content in DMSO/TBAF/cellulose solutions may play a crucial role, depending on circumstances. Anhydrous TBAF is unstable [111-112] but TBAF with water content up to the trihydrate are excellent cellulose solvents, while higher water content is not tolerated. Other ammonium fluorides might be applicable as well [105].

6.9. N-Methylmorpholine-N-oxide: NMMO

Indisputable the most industrially successful of all non-derivatizing cellulose solvents is the N-methylmorpholine–N-oxide, commonly known as NMMO. Due to its ability to dissolve high concentrations of cellulose directly and without alteration of the chemical properties of the cellulose chain, as well as the possibility to recycle more than 99 % of the solvent after usage, this particular solvent has proven to be a viable commercial solvent system. Fibers produced from this solvent also show exceptional mechanical and tactile properties, making NMMO a serious competition to the old but incredibly successful solvent system $NaOH/CS_2$ used in the viscose process. The NMMO solvent system has clear advantages in that it is non-derivatizing, since many process steps may be left out.

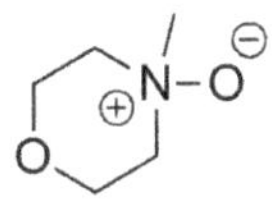

Figure 11. N-methylmorpholine-N-oxide, a cyclic, aliphatic tertiary amine oxide.

NMMO is completely soluble in water, and as a pure substance it is extremely hygroscopic. The high polarity of the N – O bond also results in a pronounced ability to form hydrogen bonds. There are two stable hydrates of NMMO, the monohydrate $NMMO*H_2O$ and the 2.5-hydrate $NMMO*2.5H_2O$. As cellulose solvent the $NMMO*H_2O$ is preferred, and cellulose is soluble all the way up to $NMMO*1.2H_2O$ while for higher amounts of water precipitation of cellulose takes place. The NMMO oxygen is able to form two hydrogen bonds with nearby hydroxyl groups such in water or cellulose [113]. Hence, the monohydrate can dissolve cellulose while the 2.5-hydrate cannot. The competition of water and cellulose for these hydrogen bonds is the central feature of a whole industry based on dissolution and regeneration of cellulose in the NMMO system.

The procedure for dissolving cellulose in NMMO usually includes a first step where a suspension of cellulose in NMMO and a large excess of water is well mixed. The excess water provides low viscosity and thereby superior mixing. Surplus water is then removed by heat between 100 and 120 °C and reduced pressure until the point of complete cellulose dissolution is reached and the spinning dope is in its final form of approximately 14 % cellulose, 10 % water and 76 % NMMO.

The dissolution process of cellulose in NMMO at elevated temperatures is performed under strictly controlled conditions, since NMMO is thermally unstable. The N – O bond is energy rich and upon cleavage releases 222 kJ/mol. Furthermore, NMMO is an oxidant and is sensitive to all forms of catalytic impurities in the dope. NMMO was first considered as an unproblematic solvent without byproduct in the process of dissolving cellulose, but as the knowledge of the possible side reactions in the cellulose/NMMO have increased, the understanding of industrial challenges has improved [114-116]. Due to the reactivity of NMMO and the severe degradation of cellulose that may be the result of this, stabilizers must be added already in the beginning of the dissolution process. Such stabilizers are for example propyl gallate, which is the most known. This compound forms a strongly colored chromophore during usage, and other stabilizers are currently being investigated [117]. The degradation of NMMO and subsequent side reactions may include N-methylmorpholine, morpholine and formaldehyde. Some reactions are induced by transition metal ions such as iron and copper, why these must be strictly excluded from the process [118].

To be able to prepare cellulose solutions in NMMO without stabilizers and without chain degradation, temperature and/or time must be kept low and while this is not doable under commercial conditions, stabilizers are a crucial part of the process. To succeed without stabilizers, high shear must be applied throughout the dissolution and immediately subsequent spinning process [119]. This accelerates the dissolution process and results in a smooth dissolution without undissolved gel particles. Mechanical energy added to the system might have the role of facilitating breakage of cellulose – cellulose intermolecular bonds, enabling solvent molecules to coordinate to the polymer instead. In addition, the apparent viscosity decreases with shear stress due to cellulose alignment, as expected in non-Newtonian polymer solutions. Viscosity is of course also dependent on cellulose concentration and degree of polymerization of chosen substrate.

NMMO is now successfully applied as a non derivatizing cellulose solvent on industrial scale, and a huge amount of research is available on the subtle details on the interactions between polymers and solvent. For example, activation of cellulose seems to play a significant role in the aggregation behavior of cellulose in the NMMO monohydrate solution [120].

6.10. Ionic liquids

This broad class of solvents comprises low melting salts with an organic cation and an organic or inorganic anion. Several comprehensive overviews of the field have been published since 2002 when Swatloski found the neat imidazolium salt 1-butyl-3-methyl imidazolium chloride (BMIMCl) to dissolve cellulose [121]. As previously mentioned, Graenacher found already in 1934 that liquefied quaternary ammonium salts, alone or diluted in suitable solvents, could dissolve cellulose. He worked mostly with pyridinium chlorides, but at the time this did not attract the attention it might have deserved [93]. Ionic liquids (IL) represent chemicals simply defined as organic salts with a melting point below 100 °C and are not limited to being solvents. Room temperature ionic liquids (RTIL) are often considered the second generation of ionic liquids. The possibility to pair anions with cations yields an almost endless library of potential ionic liquids, and the possibilities to adjust chemical and physical properties of the resulting salts are immense. Since the late 1990s, the interest in ionic liquids has grown fast. Ionic liquids have now penetrated many areas of research and industrial applications of ionic liquids can be found in as wide spread areas as pharmaceutics, analytical chemistry, separation and extraction, materials science and as electrolytes in batteries. For cellulose applications the topic is an ever growing area of research. This relatively new solvent class has already shown great versatility in the field of cellulose technology, including dissolution for regeneration purposes [122-123], homogeneous derivatization [124], and biomass processing including wood component separation [125-127].

The ionic liquids that are able to dissolve cellulose include several classes of cations, and a multitude of anions. Some of the most common cations are imidazolium, pyridinium, ammonium and phosphonium derivatives, shown in Figure 12. The most popular cation used today is the imidazolium cation with different alkyl substituents. The effect of alkyl chain length on the cellulose dissolution ability was acknowledged already 2002 by Swatloski and has since then been observed by several groups [128-129]. Studies on the imidazolium cation itself, not taking into account its properties as a solvent, recently explained in detail the effect of cation symmetry and found effects on e.g. glass temperature, viscosity and ion mobility [130]. The asymmetrical cation provides the ionic liquid with lower viscosity, which indeed is an important feature for a solvent.

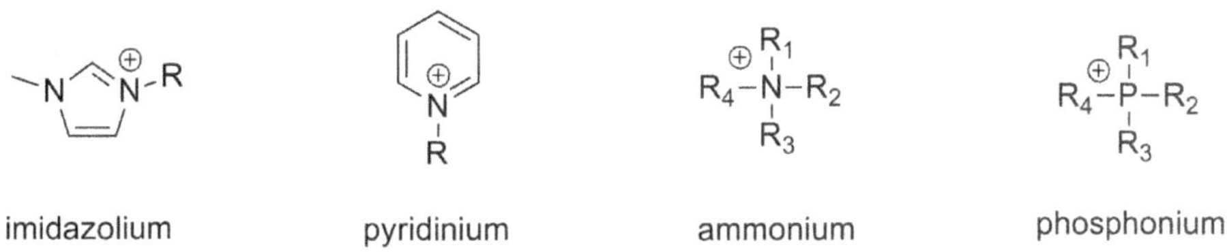

Figure 12. Technically relevant ionic liquid cations for cellulose processing.

The dissolution mechanism of cellulose in ionic liquids has long been argued to be all about hydrogen bond interaction. Several studies have confirmed that the anion of the ionic liquid penetrates the cellulose structure and dissembles the native cellulose structure by competitive hydrogen bonding [131-133]. The anion acts as a hydrogen bond acceptor and the cation as a hydrogen bond donor.

In his original article Swatloski argued that hydrogen bond capability, and therefore the ability to dissolve cellulose, was directly related to the anion concentration in the close proximity of the polymer. This was said to be the reason for the fact that the butyl imidazolium salt but not the corresponding salts with higher alkyl chains could dissolve cellulose. Smaller cations simply allow for more anions to crowd around the cellulose chain [128]. The role of the cation in the dissolution mechanism is still disputed. Some simply leave the cation out of the discussion while others attributes the cation a more prominent role, e.g. as electron acceptor and hydrogen bond donor as according to Feng and Chen, c.f. Figure 12. In that case, both the cation and the anion need to be small enough to reach the hydroxyl groups of the cellulose, forming a electron donor – electron acceptor complex, break the polysaccharide – polysaccharide interactions and finally solubilize the polymer [134]. Proof of this is claimed to be the change in cellulose dissolution capability in imidazolium salts when the acidic proton on C2 is replaced by a methyl group [123].

Viscosity and other rheological properties are always important when working with polymer melts or solutions. One feature to examine is the Mark-Houwink parameters which relate the molecular weight of the polymer with the intrinsic viscosity of the solution according to the simple expression

$$[\eta] = KM^{\alpha} \qquad [4]$$

where $[\eta]$ is the intrinsic viscosity and M is the molecular weight if the polymer. K and α are the so called Mark-Houwink parameters. An indication of the polymer shape and indirectly a measure of the solvent quality can be acquired from these parameters. Cellulose solutions in ionic liquids have been studied by Gericke and co-workers who examined the Mark-Houwink parameters for cellulose/1-ethyl-3-methyl-imidazolium acetate (EMIMAc) solutions, and report α-values of 0.4 – 0.6 in the temperature range of 0 – 100 °C. A value of α around 0.8 indicates a "good" solvent and in general, α-values of 0.65 to ~1 have been reported for other cellulose solutions [135].

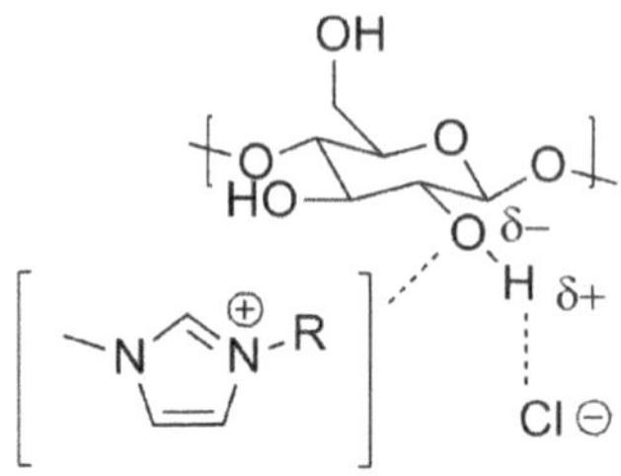

Figure 13. Interaction between cellulose and imidazolium type ionic liquid, adapted from Feng and co workers [134].

It has been shown numerous times that ionic liquids can be used as a reaction media for homogeneous derivatization as well as for dissolution of cellulose. Further, it has been proposed that due to their good dissolution properties, the use of ionic liquids can aid in control of degree of substitution in for example acetylation and tosylation reactions using various ionic liquids, reaction conditions and reactants [136-138]. Unlike in DMAc/LiCl solutions, no catalyst seems to be needed. The degree of substitution be controlled and a very wide range of DS can be achieved [136]. Silylations of cellulose in solutions of BMIMCl and EMIMAc using hexamethyldisilazane as silylating agent with high yields were also recently reported. Degree of substitution was controlled by reaction conditions and DS = 3 were achievable [139].

Ionic liquids as reaction media for homogeneous derivatization of cellulose turn out to be efficient enough to possibly compete with today's heterogeneous reaction routes. With optimization of reaction conditions the possibilities to control DS and possibly even substitution pattern opens up for production of new materials based on cellulose. One of the most important cellulose derivatives is still the cellulose acetate. Acetylation of cellulose can be performed using the ionic liquid not only as a solvent but as the actual reagent [140]. This was first discovered as an unexpected side reaction when in fact the goal was to react cellulose with acid chlorides, trityl chlorides and tosyl chlorides. The resulting polymer was acetylated, meaning that parts of the solvent, in this case the acetate anions of the ionic liquid EMIMAc, are consumed. In large scale applications this might turn out to be a serious problem, since recycling of the solvent requires that it does not degrade or react during the process. Not only will the recycling be more complicated due to side products, it will also be incomplete and in large scale application recycling of the solvent is necessary for both economic and environmental sustainability [141].

Mixed solvent systems provide further possibilities in cellulose processing. Adding a co-solvent might serve as a means to lower the viscosity of the solution and thereby facilitate fast dissolution rate and overall ease of handling. Mixtures of ionic liquids and organic solvents includes BMIMCl in 1,3-dimethyl-2-imidazolidinone (DMI) that, thanks to its low viscosity allows for efficient mixing and no agglomeration. This system is claimed to dissolve 10 % cellulose (Avicel) within a few minutes at 100 °C. This behavior is explained by the fact that only a fraction of IL in the proper molecular solvent may shift the solvatochromic Kamlet Taft parameters α, β and π^* to the point where cellulose is dissolved [142]. Remsing and co-workers studied the molecular interactions of BMIMCl in water and DMSO, respectively, and found that while water completely solvates the ions already at low concentrations high density clusters of ionic liquid were found even at concentrations as low as 10 %. The poor interaction between ions and DMSO thereby showed suggests that DMSO could be used as a rather inert co-solvent, leaving the ionic liquid intact to act as solvent [143].

Recently an interest in so called switchable or distillable ionic liquids has risen. One important reason for this is the need to recycle and also purify the ionic liquid after use. By reintroducing molecular traits of the ionic liquid, i.e. turning it into a molecular solvent (mixture), it may be possible to distill the components and thereby achieve high purities. Thermal instability of the solvent may actually have a profound impact on the choice of recycling techniques. The possibility of irreversible decomposition of the ionic liquid must be taken into account, and there are several routes by which the e.g. the imidazolium cation can decompose [144]. This concept may contradict the often assumed statements that ionic liquids in general are both non-volatile and thermally stable. Imidazolium based ionic liquids are known to decompose under temperatures exceeding 200 °C and reduced pressure. EMIMCl shows fragmentation into 1-methylimidazole, 1-ethylimidazole, chloromethane and chloroethane [145]. The fragments obtained by thermolysis can be further distilled and separated for later use as starting materials in a reaction to regenerate the original ionic liquid [146]. Problems arise when the thermal decomposition occurs unexpected or via unknown routes. Recent investigations have shown that common ionic liquids such as EMIMCl and BMIMCl start to degrade already at temperatures around 120 °C [144]. The free acids resulting from the decomposition may cause problems in carbohydrate processing, since depolymerization can be expected.

In 2009, BASF patented a method for distillation of ionic liquids such as the cellulose solvent EMIMAc. For a successful distillation, the ionic liquid must be converted into volatile compounds, which upon condensation again form the wanted salt. The results were good, but not outstanding, with a yield of > 90 % and a purity of > 95 % at 170 °C [147]. The failure to reach better results is most likely due to side reactions where unwanted decomposition occurs.

Recently, a new class of distillable tetramethylguanidine based ionic liquids was found to dissolve cellulose. In this case, the ionic liquid will dissociate and form the volatile corresponding acid and base pair [29]. Decreased temperature will again form the ionic liquid, according to the equilibrium in Figure 14. In this case, the distillation can be very successful and the reported yield and purity both exceeded 99 %.

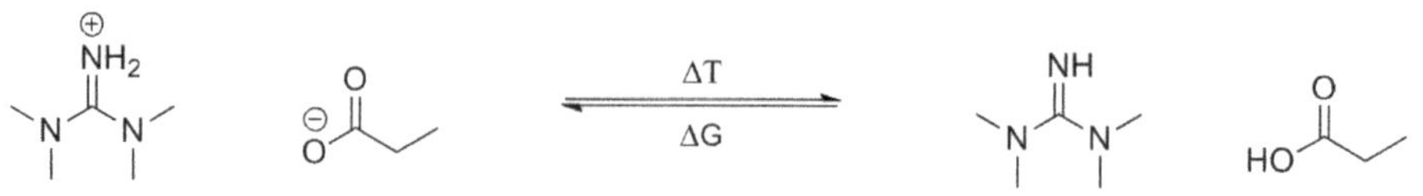

Figure 14. Distillation of a tetramethylguanidine based ionic liquid.

7. Cellulose solutions

Due to the strong tendency to self-aggregation in cellulose and the challenge of finding a proper solvent, most industrial cellulose solutions are actually not true molecular solutions. Knowledge of the character of the solution system is of importance for optimization of industrial systems as well as in academia, e.g. for characterization of the polymer using size exclusion chromatography. Comparisons between the cellulose structures in different solvents may be performed using e.g. small angle X-ray scattering (SAXS), small angle light scattering (SLS), dynamic light scattering (DLS) and depolarized dynamic light scattering (DDLS) [148].

Several different models of partially substituted cellulose in solution have been proposed. By light scattering experiments it is possible to approximate the aggregation number, and also the shape of the cellulose aggregates, if present. Models suggest star-branched macromolecules or fringed micelles [149-150]. However, these models were based on experiments using cellulose derivatives and even low degree of substitution may significantly change the behavior of the polymer in a given solvent.

For high concentration cellulose solutions very different behavior can be observed. Lyotropic mesophases or liquid crystalline phases in solutions of NMMO/cellulose are thought to be possible if the concentration of cellulose exceeds 20 % [151-153]. Comparable solutions can also be formed in an ammonia/ammonium thiocyanate solvent. The nature of the solution can be controlled to selectively produce lyotropic cellulose mesophases having either nematic or cholesteric character. High strength, high modulus fibers can be produced from a solution of cellulose in the nematic phase. Films can be prepared from mesophase solutions of cellulose that are mainly in the cholesteric phase. Material properties vary significant with cellulose concentration [154]. Similar solutions for fiber spinning applications have been prepared and solution anisotropy have been observed using inorganic acids [155] and ionic liquids [121, 133, 156] at cellulose concentrations over 10 %. For DMAc/LiCl solutions of cellulose, the polymer concentrations needs to exceed 15 % [98].

The ordered structure can be induced not only in the liquid state but of course also in the gel like state. Nematic ordered films were prepared from molecularly dispersed DMAc/LiCl solutions by stretching of a water swollen gel phase attained by slow coagulation in water vapor [157]. Still, the mechanisms of liquid crystalline phase transition and gel–sol transition are not yet fully understood and much research is left to do. The question must be raised whether to actually define liquid crystalline phases as true solutions when obvious anisotropy is present.

8. Conclusions

The development and optimization of efficient cellulose solvents have had tremendous impact on cellulose processing for well over a hundred years, and will continue to be important as the interest in cellulose itself as a basis for new materials and other applications is increasing. The most important question when considering the dissolution of cellulose is what the purpose of keeping the cellulose in solution will be. For homogeneous reactions in small scale the requirements on the solvents may be completely different from those in a large scale industrial facility for dissolution and shaping of cellulose where the purpose of regeneration is not only to stabilize the physical structure but also regain the original chemical composition of the polymer. Degree of polymerization, source and purity of the cellulose as well as required concentration in the solution may also have a profound effect on the choice of solvent.

A wide variety of suitable solvents for cellulose are already available, both aqueous and non-aqueous. Despite this, feverish activity goes on in both industry and academia to optimize existing solvent systems and develop new ones. Ideally, such a new cellulose solvent would of course be efficient and inert towards the polymer itself. It would also be 100 % recyclable and non-hazardous to environment and man. To meet future requirements on "green chemistry" and sustainability, it would be produced from non-hazardous starting materials and upon final degradation give non-hazardous, biodegradable products only. Proper life cycle assessment could be conducted to see if any new solvent meets these demands.

New materials produced from cellulose raw material are claimed to fall under the category of sustainable development because the cellulose is a polymer that replenishes itself in plants by air, water and solar energy. However, the chemistry we apply to modify and shape the cellulose to whatever form we desire must also be considered. Solvents are no exception, especially not since they are often used in great excess meaning large quantities. Truly sustainable polymer based materials are only achieved by using a sustainable polymer in combination with a sustainable chemistry to finalize the product. The quest for the cellulose solvent of the future goes on.

Author details

Carina Olsson
Corresponding Author
Department of Chemical and Biological Engineering, Chalmers University of Technology, Gothenburg, Sweden

Gunnar Westman
Department of Chemical and Biological Engineering, Chalmers University of Technology, Gothenburg, Sweden

Acknowledgement

The authors would like to acknowledge Prof. Herbert Sixta and Dr. Merima Hasani for valuable discussions and critical reading of the manuscript. Many thanks also to Södra Cell for financial support.

9. References

[1] Cross, C.F., Bevan, E.T. and Beadle, C. (1893). Thiokohlensäureester der Cellulose. Berichte der deutschen chemischen Gesellschaft, 26(1): 1090-1097.

[2] Loth, F., Schaaf, E., Fink, H.P., Kunze, J. and Gensrich, H.J. Verfahren zur Herstellung von Cellulosecarbamat in Einem Inerten Organischen, mit Wasser Nicht Mischbaren Lösungsmittel. Germany patent (2004).

[3] Rentto, M.E. (2011). Avilon Develops Environmental Friendly Cleantch Innovation. Available: http://www.cisionwire.com/neo-industrial-plc-g/r/avilon-develops-environmental-friendly-cleantech-innovation,e200454. Accessed 2012 April 13.

[4] Manjkow, J., Papanu, J.S., Hess, D.W., Soane, D.S. and Bell, A.T. (1987). Influence of Processing and Molecular Parameters in the Dissolution Rate of Poly-(Methyl Methacrylate Thin Films. Journal of the Electrochemical Society, 134(8): 2003-2007.

[5] Miller-Chou, B.A. and Koenig, J.L. (2003). A Review of Polymer Dissolution. Progress in Polymer Science, 28(8): 1223-1270.

[6] Bovey, F.A. and Winslow, F.H. (1979). Macromolecules. An Introduction to Polymer Science. (New York: Academic press, inc.).

[7] Zugenmaier, P. (2008). History of Cellulose Research. In: Crystalline Cellulose and Cellulose Derivatives. Berlin: Springer Verlag. pp. 7-51.

[8] Chu, S.S.C. and Jeffrey, G. (1968). The Refinement of the Crystal Structures of β-D-glucose and Cellobiose. Acta Crystallographica, 24: 830-838.

[9] Ferrier, W.G. (1963). The Crystal and Molecular Structure of β-D-glucose. Acta Crystallographica, 16: 1023-1031.

[10] Nehls, I., Wagenknecht, W., Philipp, B. and Stscherbina, D. (1994). Characterization of Cellulose and Cellulose Derivatives in Solution by High Resolution ^{13}C-NMR Spectroscopy. Progress in Polymer Science, 19(1): 29-78.

[11] Nishiyama, Y., Langan, P. and Chanzy, H. (2002). Crystal Structure and Hydrogen-Bonding System in Cellulose Iβ from Synchrotron X-ray and Neutron Fiber Diffraction. Journal of the American Chemical Society, 124(31): 9074-9082.

[12] Goring, D.A.I. and Timell, T.E. (1962). Molecular Weight of Native Celluloses. Tappi, 45(6): 454-460.

[13] Nishikawa, S. and Ono, S. (1913). Transmission of X-rays Through Fibrous, Lamellar and Granular Substances. Proceedings of the Tokyo Mathematico-Physical Society, 7: 131-138.

[14] Gardner, K.H. and Blackwell, J. (1974). The Structure of Native Cellulose. Biopolymers, 13(10): 1975-2001.

[15] Atalla, R.H. and VanderHart, D.L. (1984). Native Cellulose: A Composite of Two Distinct Crystalline Forms. Science, 223(4633): 283-285.

[16] VanderHart, D.L. and Atalla, R.H. (1984). Studies of Microstructure in Native Celluloses Using Solid-State Carbon-13 NMR. Macromolecules, 17(8): 1465-1472.

[17] Finkenstadt, V.L. and Millane, R.P. (1998). Crystal Structure of Valonia Cellulose Iβ. Macromolecules, 31(22): 7776-7783.

[18] Langan, P., Nishiyama, Y. and Chanzy, H. (2001). X-ray Structure of Mercerized Cellulose II at 1 Å Resolution. Biomacromolecules, 2(2): 410-416.

[19] Kolpak, F.J. and Blackwell, J. (1978). Mercerization of Cellulose: 2. The Morphology of Mercerized Cotton Cellulose. Polymer, 19(2): 132-135.

[20] Lejeune, A. and Deprez, T. (2010). Cellulose: Structure and Properties, Derivatives and Industrial Uses

[21] Revol, J.F. (1982). On the Cross-Sectional Shape of Cellulose Crystallites in Valonia ventricosa. Carbohydrate Polymers, 2(2): 123-134.

[22] Frey-Wyssling, A. and Muehlenthaler, K. (1963). The Elementary Fibrils of Cellulose. Makromolekulare Chemie, 62: 25-30.

[23] Hearle, J.W.S. (1958). A Fringed Fibril Theory of Structure in Crystalline Polymers. Journal of Polymer Science, 28(117): 432-435.

[24] Fink, H.P., Hofmann, D. and Philipp, B. (1995). Some Aspects of Lateral Chain Order in Cellulosics from X-ray Scattering. Cellulose, 2(1): 51-70.

[25] Reichardt, C. (1994). Solvatochromic Dyes as Solvent Polarity Indicators. Chemical Reviews, 94(8): 2319-2358.

[26] Spange, S., Reuter, A., Vilsmeier, E., Heinze, T., Keutel, D. and Linert, W. (1998). Determination of Empirical Polarity Parameters of the Cellulose Solvent N,N-Dimethylacetamide/LiCl by Means of the Solvatochromic Technique. Journal of Polymer Science Part A: Polymer Chemistry, 36(11): 1945-1955.

[27] Ohno, H. and Fukumoto, K. (2007). Amino Acid Ionic Liquids. Accounts of Chemical Research, 40(11): 1122-1129.

[28] Fukaya, Y., Iizuka, Y., Sekikawa, K. and Ohno, H. (2007). Bio Ionic Liquids: Room Temperature Ionic Liquids Composed Wholly of Biomaterials. Green Chemistry, 9(11): 1155-1157.

[29] King, A.W.T., Asikkala, J., Mutikainen, I., Järvi, P. and Kilpeläinen, I. (2011). Distillable Acid–Base Conjugate Ionic Liquids for Cellulose Dissolution and Processing. Angewandte Chemie International Edition, 50(28): 6301-6305.

[30] Taft, R.W., Abboud, J.-L.M., Kamlet, M.J. and Abraham, M.H. (1985). Linear Solvation Energy Relations. Journal of Solution Chemistry, 14(3): 153-186.

[31] Marcus, Y. (1993). The Properties of Organic Liquids that are Relevant to Their Use as Solvating Solvents. Chemical Society Reviews, 22(6): 409-416.

[32] Lindman, B., Karlström, G. and Stigsson, L. (2010). On the Mechanism of Dissolution of Cellulose. Journal of Molecular Liquids, 156(1): 76-81.
[33] Lindman, B. (2010). Amphiphilic Biopolymers. In: M. Fanun (Ed.), Colloids in Biotechnology. CRC Press. pp. 1-7.
[34] Biermann, O., Hädicke, E., Koltzenburg, S. and Müller-Plathe, F. (2001). Hydrophilicity and Lipophilicity of Cellulose Crystal Surfaces. Angewandte Chemie International Edition, 40(20): 3822-3825.
[35] Bergenstråhle, M., Mazeau, K. and Berglund, L.A. (2008). Molecular Modeling of Interfaces Between Cellulose Crystals and Surrounding Molecules: Effects of Caprolactone Surface Grafting. European Polymer Journal, 44(11): 3662-3669.
[36] Cho, H.M., Gross, A.S. and Chu, J.-W. (2011). Dissecting Force Interactions in Cellulose Deconstruction Reveals the Required Solvent Versatility for Overcoming Biomass Recalcitrance. Journal of the American Chemical Society, 133(35): 14033-14041.
[37] Gross, A.S. and Chu, J.-W. (2010). On the Molecular Origins of Biomass Recalcitrance: The Interaction Network and Solvation Structures of Cellulose Microfibrils. The Journal of Physical Chemistry B, 114(42): 13333-13341.
[38] Beckham, G.T., Matthews, J.F., Peters, B., Bomble, Y.J., Himmel, M.E. and Crowley, M.F. (2011). Molecular-Level Origins of Biomass Recalcitrance: Decrystallization Free Energies for Four Common Cellulose Polymorphs. The Journal of Physical Chemistry B, 115(14): 4118-4127.
[39] Betzel, C., Saenger, W., Hingerty, B.E. and Brown, G.M. (1984). Topography of Cyclodextrin Inclusion Complexes, Part 20. Circular and Flip-Flop Hydrogen Bonding in Beta-Cyclodextrin Undecahydrate: a Neutron Diffraction Study. Journal of the American Chemical Society, 106(24): 7545-7557.
[40] Griffiths, D.W. and Bender, M.L. (1973). Cycloamyloses as Catalysts. Advances in Catalysis, 23: 209-61.
[41] Clarke, R.J., Coates, J.H. and Lincoln, S.F. (1988). Inclusion Complexes of the Cyclomalto-Oligosaccharides (Cyclodextrins). In: R. S. Tipson and H. Derek (Eds.), Advances in Carbohydrate Chemistry and Biochemistry. Academic Press. pp. 205-249.
[42] Sasaki, M., Fang, Z., Fukushima, Y., Adschiri, T. and Arai, K. (2000). Dissolution and Hydrolysis of Cellulose in Subcritical and Supercritical Water. Industrial & Engineering Chemistry Research, 39(8): 2883-2890.
[43] Sasaki, M., Adschiri, T. and Arai, K. (2004). Kinetics of Cellulose Conversion at 25 MPa in Sub- and Supercritical Water. AIChE Journal, 50(1): 192-202.
[44] Yu, Y. and Wu, H. (2009). Characteristics and Precipitation of Glucose Oligomers in the Fresh Liquid Products Obtained from the Hydrolysis of Cellulose in Hot-Compressed Water. Industrial & Engineering Chemistry Research, 48(23): 10682-10690.
[45] Kamide, K., Okajima, K., Matsui, T. and Kowsaka, K. (1984). Study on the Solubility of Cellulose in Aqueous Alkali Solution by Deuteration IR and ^{13}C NMR. Polymer Journal, 16(12): 857-866.

[46] Isogai, A. and Atalla, R.H. (1998). Dissolution of Cellulose in Aqueous NaOH Solutions. Cellulose, 5(4): 309-319.

[47] Isogai, A., Chiba, J., Atalla, R.H. and Verona, W. Alkaline Method for Dissolving Cellulose (1994).

[48] Sobue, H., Kiessig, H. and Hess, K. (1939). The System: Cellulose-Sodium Hydroxide-Water in Relation to the Temperature. Zeitschrift für Physikalishe Chemie, B43: 309.

[49] Roy, C., Budtova, T. and Navard, P. (2003). Rheological Properties and Gelation of Aqueous Cellulose–NaOH Solutions. Biomacromolecules, 4(2): 259-264.

[50] Yan, L. and Gao, Z. (2008). Dissolving of Cellulose in PEG/NaOH Aqueous Solution. Cellulose, 15(6): 789-796.

[51] Han, D. and Yan, L. (2010). Preparation of All-Cellulose Composite by Selective Dissolving of Cellulose Surface in PEG/NaOH Aqueous Solution. Carbohydrate Polymers, 79(3): 614-619.

[52] Zhang, L., Ruan, D. and Zhou, J. (2001). Structure and Properties of Regenerated Cellulose Films Prepared from Cotton Linters in NaOH/Urea Aqueous Solution. Industrial & Engineering Chemistry Research, 40(25): 5923-5928.

[53] Zhang, L., Ruan, D. and Gao, S. (2002). Dissolution and Regeneration of Cellulose in NaOH/Thiourea Aqueous Solution. Journal of Polymer Science Part B: Polymer Physics, 40(14): 1521-1529.

[54] Jin, H., Zha, C. and Gu, L. (2007). Direct Dissolution of Cellulose in NaOH/Thiourea/Urea Aqueous Solution. Carbohydrate Research, 342(6): 851-858.

[55] Zhang, S., Li, F.-X., Yu, J.-y. and Hsieh, Y.-L. (2010). Dissolution Behaviour and Solubility of Cellulose in NaOH Complex Solution. Carbohydrate Polymers, 81(3): 668-674.

[56] Trygg, J. and Fardim, P. (2011). Enhancement of Cellulose Dissolution in Water-Based Solvent via Ethanol–Hydrochloric Acid Pretreatment. Cellulose, 18(4): 987-994.

[57] Egal, M., Budtova, T. and Navard, P. (2008). The Dissolution of Microcrystalline Cellulose in Sodium Hydroxide-Urea Aqueous Solutions. Cellulose, 15(3): 361-370.

[58] Cai, J., Zhang, L., Liu, S., Liu, Y., Xu, X., Chen, X., Chu, B., Guo, X., Xu, J., Cheng, H., Han, C.C. and Kuga, S. (2008). Dynamic Self-Assembly Induced Rapid Dissolution of Cellulose at Low Temperatures. Macromolecules, 41(23): 9345-9351.

[59] Qi, H., Yang, Q., Zhang, L., Liebert, T. and Heinze, T. (2011). The Dissolution of Cellulose in NaOH-based Aqueous System by Two-Step Process. Cellulose, 18(2): 237-245.

[60] Yang, Q., Fukuzumi, H., Saito, T., Isogai, A. and Zhang, L. (2011). Transparent Cellulose Films with High Gas Barrier Properties Fabricated from Aqueous Alkali/Urea Solutions. Biomacromolecules, 12(7): 2766-2771.

[61] Chen, X., Burger, C., Wan, F., Zhang, J., Rong, L., Hsiao, B.S., Chu, B., Cai, J. and Zhang, L. (2007). Structure Study of Cellulose Fibers Wet-Spun from Environmentally Friendly NaOH/Urea Aqueous Solutions. Biomacromolecules, 8(6): 1918-1926.

[62] Ruan, D., Zhang, L., Zhou, J., Jin, H. and Chen, H. (2004). Structure and Properties of Novel Fibers Spun from Cellulose in NaOH/Thiourea Aqueous Solution. Macromolecular Bioscience, 4(12): 1105-1112.
[63] Chen, X., Burger, C., Fang, D., Ruan, D., Zhang, L., Hsiao, B.S. and Chu, B. (2006). X-ray Studies of Regenerated Cellulose Fibers Wet Spun From Cotton Linter Pulp in NaOH/Thiourea Aqueous Solutions. Polymer, 47(8): 2839-2848.
[64] Cai, J. and Zhang, L. (2005). Rapid Dissolution of Cellulose in LiOH/Urea and NaOH/Urea Aqueous Solutions. Macromolecular Bioscience, 5(6): 539-548.
[65] Qi, H., Liebert, T., Meister, F., Zhang, L. and Heinze, T. (2010). Homogenous Carboxymethylation of Cellulose in the New Alkaline Solvent LiOH/Urea Aqueous Solution. Macromolecular Symposia, 294(2): 125-132.
[66] Zhou, J., Qin, Y., Liu, S. and Zhang, L. (2006). Homogenous Synthesis of Hydroxyethylcellulose in NaOH/Urea Aqueous Solution. Macromolecular Bioscience, 6(1): 84-89.
[67] Heinze, T. and Liebert, T. (2001). Unconventional methods in cellulose functionalization. Progress in Polymer Science, 26(9): 1689-1762.
[68] Saalwächter, K., Burchard, W., Klüfers, P., Kettenbach, G., Mayer, P., Klemm, D. and Dugarmaa, S. (2000). Cellulose Solutions in Water Containing Metal Complexes. Macromolecules, 33(11): 4094-4107.
[69] Ahlrichs, R., Ballauff, M., Eichkorn, K., Hanemann, O., Kettenbach, G. and Klüfers, P. (1998). Aqueous Ethylenediamine Dihydroxo Palladium(II): A Coordinating Agent for Low- and High-Molecular Weight Carbohydrates. Chemistry – A European Journal, 4(5): 835-844.
[70] Hanemann, O. and Ballauff, M. (1997). Chain Conformation of Cellulose in a Coordinating Solvent. Macromolecules, 30(24): 7638-7640.
[71] Woodings, C. (2001). Regenerated Cellulose Fibres. Cambridge: Woodhead publishing Ltd.
[72] Heinze, T., Liebert, T., Klüfers, P. and Meister, F. (1999). Carboxymethylation of Cellulose in Unconventional Media. Cellulose, 6(2): 153-165.
[73] Sjöström, E. (1993). Wood Chemistry: Fundamentals and Applications (2nd ed.). San Diego: Academic press.
[74] Fischer, S., Leipner, H., Thümmler, K., Brendler, E. and Peters, J. (2003). Inorganic Molten Salts as Solvents for Cellulose. Cellulose, 10(3): 227-236.
[75] Fischer, S. and Thümmler, K. (2010). Molten Inorganic Salts as Reaction Medium for Cellulose. In: Cellulose Solvents: For Analysis, Shaping and Chemical Modification. American Chemical Society. pp. 91-101.
[76] Hattori, M., Koga, T., Shimaya, Y. and Saito, M. (1998). Aqueous Calcium Thiocyanate Solution as a Cellulose Solvent. Structure and Interactions with Cellulose. Polymer Journal, 30(1): 43-48.
[77] Hattori, M., Shimaya, Y. and Saito, M. (1998). Structural Changes in Wood Pulp Treated by 55 wt% Aqueous Calcium Thiocyanate Solution. Polymer Journal, 30(1): 37-42.

[78] Brendler, E., Fischer, S. and Leipner, H. (2001). ^{7}Li NMR as Probe for Solvent–Cellulose Interactions in Cellulose Dissolution. Cellulose, 8(4): 283-288.
[79] Leipner, H., Fischer, S., Brendler, E. and Voigt, W. (2000). Structural Changes of Cellulose Dissolved in Molten Salt Hydrates. Macromolecular Chemistry and Physics, 201(15): 2041-2049.
[80] Fischer, S., Thümmler, K., Pfeiffer, K., Liebert, T. and Heinze, T. (2002). Evaluation of Molten Inorganic Salt Hydrates as Reaction Medium for the Derivatization of Cellulose. Cellulose, 9(3): 293-300.
[81] Thümmler, K., Fischer, S., Peters, J., Liebert, T. and Heinze, T. (2010). Evaluation of Molten Inorganic Salt Hydrates as Reaction Medium for the Esterification of Cellulose. Cellulose, 17(1): 161-165.
[82] Battista, O.A. (1950). Hydrolysis and Crystallization of Cellulose. Industrial & Engineering Chemistry, 42(3): 502-507.
[83] Battista, O.A., Coppick, S., Howsmon, J.A., Morehead, F.F. and Sisson, W.A. (1956). Level-Off Degree of Polymerization - Relation to Polyphase Structure of Cellulose Fibers. Industrial and engineering chemistry, 48(2): 333-335.
[84] Liebert, T., Seifert, M. and Heinze, T. (2008). Efficient Method for the Preparation of Pure, Water-Soluble Cellodextrines. Macromolecular Symposia, 262(1): 140-149.
[85] Akpinar, O., McGorrin, R.J. and Penner, M.H. (2004). Cellulose-Based Chromatography for Cellooligosaccharide Production. Journal of Agricultural and Food Chemistry, 52(13): 4144-4148.
[86] Miller, G.L., Dean, J. and Blum, R. (1960). A Study of Methods for Preparing Oligosaccharides from Cellulose. Archives of Biochemistry and Biophysics, 91(1): 21-26.
[87] Zhang, Y.-H.P. and Lynd, L.R. (2003). Cellodextrin Preparation by Mixed-Acid Hydrolysis and Chromatographic Separation. Analytical Biochemistry, 322(2): 225-232.
[88] Celanese, B. Improvements in Phosphoric Acid Solutions of Cellulose. GB patent GB 263810 (A) (1927).
[89] Zhang, Y.H.P., Cui, J., Lynd, L.R. and Kuang, L.R. (2006). A Transition from Cellulose Swelling to Cellulose Dissolution by o-Phosphoric Acid: Evidence from Enzymatic Hydrolysis and Supramolecular Structure. Biomacromolecules, 7(2): 644-648.
[90] Boerstoel, H., Maatman, H., Westerink, J.B. and Koenders, B.M. (2001). Liquid Crystalline Solutions of Cellulose in Phosphoric Acid. Polymer, 42(17): 7371-7379.
[91] Northolt, M.G., Boerstoel, H., Maatman, H., Huisman, R., Veurink, J. and Elzerman, H. (2001). The Structure and Properties of Cellulose Fibres Spun From an Anisotropic Phosphoric Acid Solution. Polymer, 42(19): 8249-8264.
[92] Heinze, T. and Koschella, A. (2005). Solvents Applied in the Field of Cellulose Chemistry: a Mini Review. Polímeros, 15: 84-90.
[93] Graenacher, C. Cellulose Solution. U.S. Patent 1,943,176 (1934).
[94] Striegel, A. (1997). Theory and Applications of DMAC/LiCl in the Analysis of Polysaccharides. Carbohydrate Polymers, 34(4): 267-274.

[95] Austin, P.R., Brine, C.J., Castle, J.E. and Zikakis, J.P. (1981). Chitin: New Facets of Research. Science, 212(4496): 749-753.

[96] McCormick, C.L. and Callais, P.A. (1987). Derivatization of Cellulose in Lithium Chloride and N-N-Dimethylacetamide Solutions. Polymer, 28(13): 2317-2323.

[97] McCormick, C.L. Novel Cellulose Solutions. (1981).

[98] McCormick, C.L., Callais, P.A. and Hutchinson, B.H. (1985). Solution Studies of Cellulose in Lithium Chloride and N,N-Dimethylacetamide. Macromolecules, 18(12): 2394-2401.

[99] Potthast, A., Rosenau, T., Buchner, R., Röder, T., Ebner, G., Bruglachner, H., Sixta, H. and Kosma, P. (2002). The Cellulose Solvent System N,N-Dimethylacetamide/Lithium Chloride Revisited: the Effect of Water on Physicochemical Properties and Chemical Stability. Cellulose, 9(1): 41-53.

[100] Oliveira, W.D. and Glasser, W.G. (1996). Hydrogels from Polysaccharides. I. Cellulose Beads for Chromatographic Support. Journal of Applied Polymer Science, 60(1): 63-73.

[101] El Seoud, O.A., Marson, G.A., Ciacco, G.T. and Frollini, E. (2000). An Efficient, One-Pot Acylation of Cellulose under Homogeneous Reaction Conditions. Macromolecular Chemistry and Physics, 201(8): 882-889.

[102] Tosh, B., Saikia, C.N. and Dass, N.N. (2000). Homogeneous Esterification of Cellulose in the Lithium Chloride-N,N-Dimethylacetamide Solvent System: Effect of Temperature and Catalyst. Carbohydrate Research, 327(3): 345-352.

[103] Bikova, T. and Treimanis, A. (2002). Problems of the MMD Analysis of Cellulose by SEC using DMA/LiCl: A Review. Carbohydrate Polymers, 48(1): 23-28.

[104] Potthast, A., Rosenau, T., Sixta, H. and Kosma, P. (2002). Degradation of Cellulosic Materials by Heating in DMAc/LiCl. Tetrahedron Letters, 43(43): 7757-7759.

[105] Heinze, T. and Köhler, S. (2010). Dimethyl Sulfoxide and Ammonium Fluorides Novel Cellulose Solvents. In: Cellulose Solvents: For Analysis, Shaping and Chemical Modification. American Chemical Society. pp. 103-118.

[106] Heinze, T., Dicke, R., Koschella, A., Kull, A.H., Klohr, E.-A. and Koch, W. (2000). Effective Preparation of Cellulose Derivatives in a New Simple Cellulose Solvent. Macromolecular Chemistry and Physics, 201(6): 627-631.

[107] Östlund, Å., Lundberg, D., Nordstierna, L., Holmberg, K. and Nydén, M. (2009). Dissolution and Gelation of Cellulose in TBAF/DMSO Solutions: The Roles of Fluoride Ions and Water. Biomacromolecules, 10(9): 2401-2407.

[108] Chen, Y., Stipanovic, A., Winter, W., Wilson, D. and Kim, Y.-J. (2007). Effect of Digestion by Pure Cellulases on Crystallinity and Average Chain Length for Bacterial and Microcrystalline Celluloses. Cellulose, 14(4): 283-293.

[109] Ass, B.A.P., Frollini, E. and Heinze, T. (2004). Studies on the Homogeneous Acetylation of Cellulose in the Novel Solvent Dimethyl Sulfoxide/Tetrabutylammonium Fluoride Trihydrate. Macromolecular Bioscience, 4(11): 1008-1013.

[110] Hussain, M.A., Liebert, T. and Heinze, T. (2004). Acylation of Cellulose with N,N′-Carbonyldiimidazole-Activated Acids in the Novel Solvent Dimethyl

Sulfoxide/Tetrabutylammonium Fluoride. Macromolecular Rapid Communications, 25(9): 916-920.

[111] Sharma, R.K. and Fry, J.L. (1983). Instability of Anhydrous Tetra-Normal-Alkylammonium Fluorides. Journal of Organic Chemistry, 48(12): 2112-2114.

[112] Köhler, S. and Heinze, T. (2007). New Solvents for Cellulose: Dimethyl Sulfoxide/Ammonium Fluorides. Macromolecular Bioscience, 7(3): 307-314.

[113] Maia, E., Peguy, A. and Perez, S. (1981). Cellulose Organic Solvents. I. The Structures of Anhydrous N-Methylmorpholine N-Oxide and N-Methylmorpholine N-Oxide Monohydrate. Acta Crystallographica Section B, 37(10): 1858-1862.

[114] Rosenau, T., Potthast, A., Sixta, H. and Kosma, P. (2001). The Chemistry of Side Reactions and Byproduct Formation in the System NMMO/Cellulose (Lyocell process). Progress in Polymer Science, 26(9): 1763-1837.

[115] Rosenau, T., Potthast, A., Hofinger, A., Sixta, H. and Kosma, P. (2002). Instabilities in the System NMMO/Water/Cellulose (Lyocell Process) Caused by Polonowski Type Reactions. Holzforschung, 56(2): 199-208.

[116] Adorjan, I., Potthast, A., Rosenau, T., Sixta, H. and Kosma, P. (2005). Discoloration of Cellulose Solutions in N-Methylmorpholine-N-Oxide (Lyocell). Part 1: Studies on Model Compounds and Pulps. Cellulose, 12(1): 51-57.

[117] Rosenau, T., Potthast, A., Adorjan, I., Hofinger, A., Sixta, H., Firgo, H. and Kosma, P. (2002). Cellulose Solutions in N-Methylmorpholine-N-Oxide (NMMO) – Degradation Processes and Stabilizers. Cellulose, 9(3): 283-291.

[118] Klemm, D., Heublein, B., Fink, H.-P. and Bohn, A. (2005). Cellulose: Fascinating Biopolymer and Sustainable Raw Material. Angewandte Chemie International Edition, 44(22): 3358-3393.

[119] Maron, R., Michels, C. and Taeger, E. (1994). Investigations for Preparation of Cellulose Solutions in NMMO and the Following Forming. Lenzinger Berichte, 9: 27-29.

[120] Röder, T. and Morgenstern, B. (1999). The Influence of Activation on the Solution State of Cellulose Dissolved in N-Methylmorpholine-N-Oxide-Monohydrate. Polymer, 40(14): 4143-4147.

[121] Swatloski, R.P., Spear, S.K., Holbrey, J.D. and Rogers, R.D. (2002). Dissolution of Cellose with Ionic Liquids. Journal of the American Chemical Society, 124(18): 4974-4975.

[122] Viswanathan, G., Murugesan, S., Pushpraj, V., Nalamasu, O., Ajayan, P.M. and Linhardt, R.J. (2006). Preparation of Biopolymer Fibers by Electrospinning from Room Temperature Ionic Liquids. Biomacromolecules, 7(2): 415-418.

[123] Kosan, B., Michels, C. and Meister, F. (2008). Dissolution and Forming of Cellulose with Ionic Liquids. Cellulose, 15: 59-66.

[124] Heinze, T., Schwikal, K. and Barthel, S. (2005). Ionic Liquids as Reaction Medium in Cellulose Functionalization. Macromolecular Bioscience, 5: 520-525.

[125] Sun, N., Rahman, M., Qin, Y., Maxim, M.L., Rodriguez, H. and Rogers, R.D. (2009). Complete Dissolution and Partial Delignification of Wood in the Ionic Liquid 1-ethyl-3-Methylimidazolium Acetate. Green Chemistry, 11(5): 646-655.

[126] Ilkka Kilpeläinen, H.X., Alistair King, Mari Granstrom, Sami Heikkinen, Dimitris S. Argyropoulus. (2007). Dissolution of Wood in Ionic Liquids. Agricultural and Food Chemistry, 55(22): 9142-9148.

[127] Fort, D.A., Remsing, R.C., Swatloski, R.P., Moyna, P. and Rogers, G.M.R.D. (2007). Can Ionic Liquids Dissolve Wood? Processing and Analysis of Lignocellulosic Materials with 1-N-Butyl-3-Methylimidazolium Chloride. Green Chemistry, 9: 63-69.

[128] Swatloski, R.P., Spear, S.K., Holbrey, J.D. and Rogers, R.D. (2002). Dissolution of Cellulose with Ionic Liquids. Journal of the American Chemical Society, 124(18): 4974-4975.

[129] Vitz, J., Erdmenger, T., Haensch, C. and Schubert, U.S. (2009). Extended Dissolution Studies of Cellulose in Imidazolium Based ionic Liquids. Green Chemistry, 11(3): 417-424.

[130] Zheng, W., Mohammed, A., Hines, L.G., Xiao, D., Martinez, O.J., Bartsch, R.A., Simon, S.L., Russina, O., Triolo, A. and Quitevis, E.L. (2011). Effect of Cation Symmetry on the Morphology and Physicochemical Properties of Imidazolium Ionic Liquids. The Journal of Physical Chemistry B, 115(20): 6572-6584.

[131] Remsing, R.C., Swatloski, R.P., Rogers, R.D. and Moyna, G. (2006). Mechanism of Cellulose Dissolution in the Ionic Liquid 1-N-Butyl-3-Methylimidazolium Chloride: a ^{13}C and $^{35/37}Cl$ NMR Relaxation Study on Model systems. Chemical Communications (12): 1271-1273.

[132] Zhang, J., Zhang, H., Wu, J., Zhang, J., He, J. and Xiang, J. (2010). NMR Spectroscopic Studies of Cellobiose Solvation in EmimAc Aimed to Understand the Dissolution Mechanism of Cellulose in Ionic Liquids. Physical Chemistry Chemical Physics, 12(8): 1941-1947.

[133] Zhang, H., Wu, J., Zhang, J. and He, J. (2005). 1-Allyl-3-Methylimidazolium Chloride Room Temperature Ionic Liquid: A New and Powerful Nonderivatizing Solvent for Cellulose. Macromolecules, 38(20): 8272-8277.

[134] Feng, L. and Chen, Z.-l. (2008). Research Progress on Dissolution and Functional Modification of Cellulose in Ionic Liquids. Journal of Molecular Liquids, 142(1-3): 1-5.

[135] Gericke, M., Schlufter, K., Liebert, T., Heinze, T. and Budtova, T. (2009). Rheological Properties of Cellulose/Ionic Liquid Solutions: From Dilute to Concentrated States. Biomacromolecules, 10(5): 1188-1194.

[136] Wu, J., Zhang, J., Zhang, H., He, J., Ren, Q. and Guo, M. (2004). Homogeneous Acetylation of Cellulose in a New Ionic Liquid. Biomacromolecules, 5(2): 266-268.

[137] Granström, M., Kavakka, J., King, A., Majoinen, J., Mäkelä, V., Helaja, J., Hietala, S., Virtanen, T., Maunu, S.-L., Argyropoulos, D.S. and Kilpeläinen, I. (2008). Tosylation and Acylation of Cellulose in 1-Allyl-3-Methylimidazolium Chloride. Cellulose, 15: 481-488.

[138] Cao, Y., JinWub, Zhangb, J., Li, H., Zhanga, Y. and Heb, J. (2009). Room Temperature Ionic Liquids (RTILs): A New and Versatile Platform for Cellulose Processing and Derivatization. Chemical Enginering Journal, 147: 13-21.

[139] Mormann, W. and Wezstein, M. (2009). Trimethylsilylation of Cellulose in Ionic Liquids. Macromolecular Bioscience, 9: 369-375.

[140] Köhler, S., Liebert, T., Schöbitz, M., Schaller, J., Meister, F., Günther, W. and Heinze, T. (2007). Interactions of Ionic Liquids with Polysaccharides 1. Unexpected Acetylation of Cellulose with 1-Ethyl-3-methylimidazolium Acetate. Macromolecular Rapid Communications, 28: 2311-2317.

[141] Ebner, G., Schiehser, S., Potthast, A. and Rosenau, T. (2008). Side Reaction of Cellulose with Common 1-alkyl-3-methylimidazolium-based Ionic Liquids. Tetrahedron Letters, 49(51): 7322-7324.

[142] Rinaldi, R. (2011). Instantaneous Dissolution of Cellulose in Organic Electrolyte Solutions. Chemical Communications, 47(1): 511-513.

[143] Remsing, R.C., Liu, Z., Sergeyev, I. and Moyna, G. (2008). Solvation and Aggregation of N,N'-Dialkylimidazolium Ionic Liquids: A Multinuclear NMR Spectroscopy and Molecular Dynamics Simulation Study. The Journal of Physical Chemistry B, 112(25): 7363-7369.

[144] Meine, N., Benedito, F. and Rinaldi, R. (2010). Thermal Stability of Ionic Liquids Assessed by Potentiometric Titration. Green Chemistry, 12(10): 1711-1714.

[145] Jeapes, A.J., Thied, R.C., Seddon, K.R., Pitner, W.R., Rooney, D.W., Hatter, J.E. and Welton, T. Process for Recycling Ionic Liquids. GB patent (2002).

[146] Earle, M.J., Esperanca, J.M.S.S., Gilea, M.A., Canongia Lopes, J.N., Rebelo, L.P.N., Magee, J.W., Seddon, K.R. and Widegren, J.A. (2006). The Distillation and Volatility of Ionic Liquids. Nature, 439(7078): 831-834.

[147] Massone, K., Siemer, M., Mormann, W. and Leng, W. Distillation of Ionic Liquids. Germany patent (2010).

[148] Röder, T., Möslinger, R., Mais, U., Morgenstern, B. and Glatter, O. (2003). Charakterisierung der Lösungsstrukturen in technishen relevanten Celluloselösungen. Lenzinger Berichte, 82: 118-127.

[149] Schmidt, J., Weigel, R., Burchard, W. and Richtering, W. (1997). Methyl-Hydroxypropyl Cellulose - Shear Induced birefringence Measurements in the Semi–Dilute Regime. Macromolecular Symposia, 120(1): 247-257.

[150] Schulz, L., Seger, B. and Burchard, W. (2000). Structures of Cellulose in Solution. Macromolecular Chemistry and Physics, 201(15): 2008-2022.

[151] Gilbert, R.D. (1994). Cellulosic Polymers, Blends, and Composites: Hanser Publishers.

[152] Michels, C. and Kosan, B. (2003). Contribution to Dissolution State of Cellulose in Aqueous Amine Oxide Charachterized by Optical and Rheological Methods. Lenzinger Berichte, 82: 128-135.

[153] Kim, S.O., Shin, W.J., Cho, H., Kim, B.C. and Chung, I.J. (1999). Rheological Investigation on the Anisotropic Phase of Cellulose-MMNO/H_2O Solution System. Polymer, 40(23): 6443-6450.

[154] Cuculo, e.a. Anisotropic Cellulose Solutions, Fibers, and Films Formed Therefrom. Unites States of America patent (1988).

[155] Boerstoel, H.N.K.B., M.; Westerink, J.B. Cellulose Solutions and Products Made Therefrom. (1996).

[156] Song, H., Niu, Y., Wang, Z. and Zhang, J. (2011). Liquid Crystalline Phase and Gel–Sol Transitions for Concentrated Microcrystalline Cellulose (MCC)/1-Ethyl-3-Methylimidazolium Acetate (EMIMAc) Solutions. Biomacromolecules, 12(4): 1087-1096.

[157] Tetsuo Kondo, E.T., and R. Malcolm Br. (2001). "Nematic Ordered Cellulose": A Concept of Glucan Chain Association. Biomcaromolecules, 2(4): 1324-1330.

Chapter 6

Cellulose Microfibril Angle in Wood and Its Dynamic Mechanical Significance

Tamer A. Tabet and Fauziah Abdul Aziz

Additional information is available at the end of the chapter

1. Introduction

The orientation of the cellulose microfibrils in the S_2 layers of the cell walls of softwood has a significant influence on the mechanical properties of wood. The angle between the cellulose fibrils and the longitudinal cell axis, the microfibril angle, MFA was found to be a critical factor in determining the physical and mechanical properties of wood (Cave, 1997). For this reason, considerable effort has been directed towards the measurement of the cellulose MFA. Direct measurement of MFA has been made by highlighting microfibrils in individual cell walls with iodine staining, but the most widely adopted techniques use either wide-angle X-ray diffraction or small-angle X-ray scattering The pioneering work of Cave (1966) and Meylan (1967) led to the use of the '*T*' parameter derived from the curve distribution of the intensity diffracted by the (002) planes of the cellulose fibrils.

MFA has been found to influence shrinkage of wood (Harris and Meylan, 1965). The MFA of the S_2 layer in the tracheid cell wall is known to be one of the main determinants of the mechanical properties of wood (Cave, 1968). Donaldson (1993) reported that the MFA also had a significant impact on paper properties. The MFA has a very strong influence on the stiffness of wood (Walter and Butterfield, 1996). MFA in the S_2 layer of the cell wall of *Picea abies* has been to influence on cambial age and growth conditions (Hakan *et al.*, 1998).

1.1. Problem and background

Much of the future timber supply in Malaysia is expected to come from *Acacia mangium* and *hybrid Acacia* wood grown in managed plantations or from small diameter logs removed during forest management operations. The short age rotation resource will contain higher properties of *Acacia mangium* wood compared to the present resource. Little if any, comprehensive study has been carried out in Malaysia to examine and compare the microstructure of wood and wood quality traits such as microfibril angle (MFA), fibre length, lumen area, surface roughness, mechanical, physical and nanostructural properties in *Acacia mangium* wood. In anticipation of this raw material resource, definitive information is needed on the influence of growth age one the pure crystalline cellulose and the influence of MFA on the mechanical and physical properties of lumber, so that selection and utilization methods can be adjusted sentence structure, accordingly. The angle between the cellulose fibrils and the longitudinal cell axis (MFA) was found to be a crucial factor in determining the mechanical properties of wood (Cave, 1989; Cave and Walker, 1994; Reiterer *et al.* 1999). The traditional methods for determining the mechanical and physical properties of wood are time consuming and tedious therefore, there is little information on microstructure of *Acacia mangium* wood. A much more rapid method of determining mechanical properties, MFA, moisture content and nanostructure of wood is needed.

1.2. Wood microstructure

The primary structural building block of wood is the tracheid or fibre cell. Cells vary from 16 to 42 microns in diameter and from 870 to 4000 microns long. Thus, the cubic centimeter of wood could contain more than 1.5 million wood cells. When packed together they form a strong composite. Each individual wood cell is even more structurally advanced because it is actually a multilayered, closed-end tube (Fig. 1) rather than just a homogeneous-walled, nonreinforced straw. Each individual wood cell is even more structurally advanced because it is actually a multilayered, closed-end tube (Fig. 1) rather than just a homogeneous-walled, nonreinforced straw. Each individual cell has four distinct cell wall layers (Primary, S_1, S_2, and S_3). Each layer is composed of a combination of three chemical polymers: cellulose, hemicellulose, and lignin (Fig. 1). The cellulose and hemicellulose are linear polysaccharides (i.e., hydrophilic multiple-sugars), and the lignin is an amorphous phenolic (i. e., a three-dimensional hydrophobic adhesive). Cellulose forms long unbranched chain and hemicellulose forms short branched chains. Lignin encrusts and stiffens these polymers. Because carbohydrate and phenolic components of wood are assembled in a layered tubular or cellular manner with a large cell cavity, specific gravity of wood can vary immensely. Wood excels as a viable building material because the layered tubular structure provides a large volume of voids (void volume), it has an advantageous strength-to-weight ratio, and it has other inherent advantages, such as corrosion resistance, fatigue resistance, low cost, and ease-of modification at the job site.

1.3. *Acacia mangium*

Acacia mangium is one of the major fast growing species used in plantation forestry programs throughout Asia and the pacific. It was first introduced to Sabah, Malaysia from Australia as an exotic species in 1966. It comes from the family Leguminosae, sub-family Mimosoideae. It has rapid early growth and can attain a height of 30 meters and over 60 cm in diameters (Makdiken and Brewbaker, 1984). There importance as a plantation species can be attributed to a rapid growth, ruther good wood quality, tolerance to a range of soil types and pH values (Latif and Habib, 1994). Little is known about the physical and mechanical properties of *Acacia mangium* wood in general.

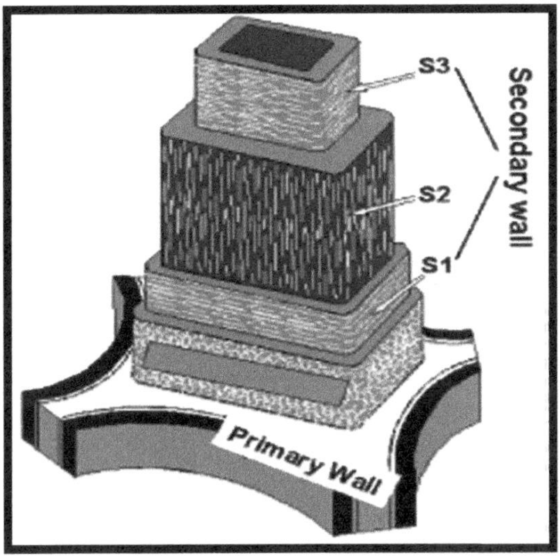

Figure 1. Layers of mature cell ll. Source: Meylan (1972).

1.4. Wood quality

Wood quality results largely from the chemical and physical structure of the cell walls of its component fibres. It can be defined in terms of attributes that make it valuable for a given end use (Jozsa and Middlleton, 1994). In general, density and MFA are indicators of strength and stiffness respectively. They are reputed to be the key determinants of wood quality. For the sawmiller, wood quality is reflected in the value of mill production and this depends on grade out-turn and the value for each grade (Addis Tsehaye *et al.*,1995). Wood quality for the structural engineer means wood with a high stiffness level, an attribute which is most important for beams, joists and purlins. Strong wood is required for studs and trusses (Addis Tsehaye *et al.*, 1995). For the wood technologist, wood density is important, at higher timber strength and greater yield of pulp (Elliot, 1970). The paper and pulp mill requirements for quality wood are long fibre length with low lignin content (Zobel, 1961). A minimum fibre length of 2 mm is necessary to produce acceptable kraft pulp. A reduction in lignin content leads to a considerable savings during the production of bleached kraft pulp (Walker and Butterfield, 1996). The most important parameters of kraft pulping are basic density and fibre length (Cown and Kibblewhite , 1980).

1.5. Chemical composition of wood

Wood material is primarily composed of three organic polymeric components, namely:- cellulose, hemicelluloses and lignin. They primarily determine the chemical and physical properties of wood. Minor amounts of mainly organic inclusions are present collectively called extractives, which are present in wood but are not structural components. The amount of extractives such as gums, fats, resins, sugars, oils, starches, alkaloids and tannins, varies from less than 1% to more than 10% of the oven-dry weight of wood (Tsoumis, 1992).

Cellulose is the main constitutent of wood, occupying 40-45% of the dry wood substance in both softwoods and hardwoods (Watson and Dadswell, 1964). It consist of glucose ($C_6H_{12}O_6$) linked together to form cellulose chain molecule. The structural formula of the cellulose chain molecule is shown in Figure 2 (Cave and Walker 1994) . Each glucose molecule added to the repeating unit of the molecule chain is rotated 180°. The number of the glucose monomers ($C_6H_{10}O_5$) in the cellulose chain is called the degree of the polymerization and in wood it is about 8000-10000, giving the cellulose molecules a length of 4-5 μm (Butterfield and Meylan, 1980). In wood cell walls, cellulose micromolecules are arranged into bundles consisting of a number of cellulose chain molecules called microfibrils, in which cellulose is mainly present in crystalline and in amorphous regions (Tsoumis, 1992). The crystalline regions of cellulose has been investigated by using x-ray scattering methods (Jakob *et al.*, 1995) and atomic force microscopy (Svergun and Stuharmann, 1991).

In addition to cellulose, a number of various polysaccharides called hemicelluloses are present in wood. Hemicelluloses are formed from glucose and other six-carbon and five-carbon sugar molecules and constitute 25-30 % of the dry wood substance in softwoods and 25-35 % in hardwoods (Meylan, 1967). With regard to degree of polymerization, hemicellulosees are quite small macromolecules comparied to cellulose since they seldom have more than 150 -200 monomer units (Wilson and Archer, 1979). However, the clear distinction between cellulose and hemicelluloses is that hemicelluloses are soluble in aqueous alkali but cellulose is not.

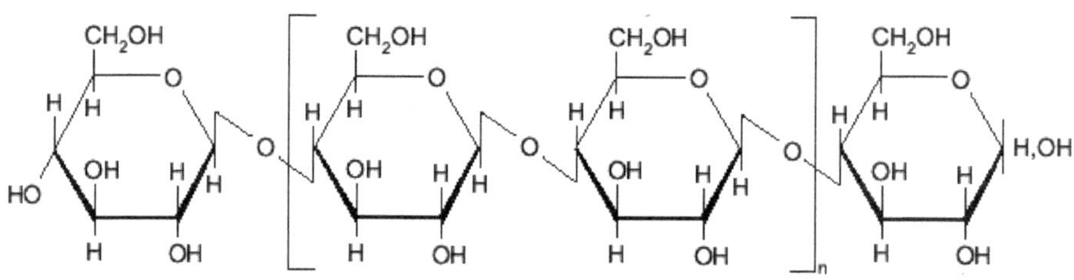

(a) Non-reducing end (b) Repeating cellulose unit (c) Reducing end

Figure 2. The structural formula of cellulose. Source: Cave and Walker (1994).

1.6. The structure of the cell wall

Wood cell walls are structurally complex. It is has a hollow center called lumen and layered cell walls. The cell wall can be divided into separate distinguishable layers, namely, middle lamella (M), primary wall (P) and secondary wall (S_1,S_2 and S_3). These layers differ from one another with respect to chemical composition as well as their structure. The middle lamella is located between the adjacent wood cells and serves the function of binding them together. Sometimes the combine region of the middle lamela and two adjacent primary walls are called the compound middle lamella (CML), which has the most lignin rich region(Batchelor *et al.*, 1997).

The cell wall layers are divided on the basis of how the microfibrils are arranged on the specific layers. The differences in the orientation of the microfibrils help to distiguish the primary wall and the three layers of the secondary wall from each other. The primary wall is very thin, approximately only 0.1 μm, and the microfibrils are randomly orientated on the outer surface and more or less transversally orientated with respect to cell axis on the inner surface near the secondary wall (Cave and Walker, 1994).

In the secondary wall, the microfibrils are closely packed and the differences in the orientation of the microfibrils are quite distinctive. The thin outer S_1 layer adjacent to the CML consist of a few lamella in which the orientation of the microfibrils have an alternating left - handed and right - handed helical arrangement, forming a crossed microfibrillar texture. In each lamela, the microfibrils spiral around the longitudinal cell axis (about 50 - 70°). The thickness of the S_1 layer is about 0.1 – 0.2 μm. The inner S_3 layer is usually even thinner than the S_1 layer and consist of a similar microfibrillar orientation (Cave and Walker, 1994).

The S_2 layer of the secondary cell wall is, by far, the thickest of the cell wall layers and has the most profound effect on the physical properties of wood. The microfibrils in the S_2 layer show a high degree of parallelism in all lamella with only a small dispersion and, in general, are more or less parallel to the longitudinal cell axis. Figure 3 shows the different layers of the cell wall of thin sections through *Populus deltoides* wood (Jean-Paul Joseleau *et al.*, 2004). In hardwood, the microfibril angles are generally lower than in softwood species (Lichtenegger *et al.*, 1999). The S_2 layer can reach a thickness of 5 μm or more and contains, for example in softwoods, up to 80 % or more of the total cell wall material (Tsoumis, 1992 and Donaldson, 1996). Therefore, the magnitude of the MFA in the S_2 layer has been considered as one of the most important elements affecting the different properties of wood. It is defined as the angle between microfibrils of the S_2 layer and the longitudinal axis of the wood cell. In general, when the MFA of wood is discussed in literature, it specifically refers to the angle of microfibrils in the S_2 layer. As the S_2 layer represents the major component of the cell wall, its MFA has a significant effect on the mechanical properties such as modulus of elasticity, strain and extensibility (Reiterer *et al.*, 1999).

1.7. Cellulose microfibrils

Microfibrils are the structural units of plant cell wall. Each microfibril contains a number of cellulose chain molecules bundled together, and is surrounded by low molecular weight hemicelluloses (Tsoumis, 1992). The hemicellulose act as connecting agents that link or of bond the microfibrils together (Hygreen and Bowyer, 1996). The cellulose chain molecules are generally arranged lengthways with regard to the microfibril axis, but run parallel to each other in portions. These portions are called crystalline regions. The molecules in these regions are strongly connected to each other by hydrogen bonding. The crystalline regions are followed by amorphous regions in which the cellulose molecules have no definite arrangement. The transition from crystalline to amorphous region is gradual. Approximately two thirds of the cellulose in the cell wall is crystalline in form while one third is amorphous (Tsoumis, 1992). Microfibrils vary in width from 1 μm in the primary walls to 10 μm in the secondary walls (Zobel, 1961). The angle that the cellulose microfibril make to the axis of the cell wall is known as the microfibril angle (MFA). Microfibrils are present in each of the cell wall layers (Butterfield, 1980). The microfibrils is the smallest component of the cell wall which can be visualized by transmission electron microscopy occurring along fibril 3-4 nm in diameter and it consists of a group of cellulose surrounded by a sheath of hemicellulose (Harris and Meylan, 1965). The cellulose microfibrils are wound helically around the cell wall in the S_2 layer, as shown in Figure 4.

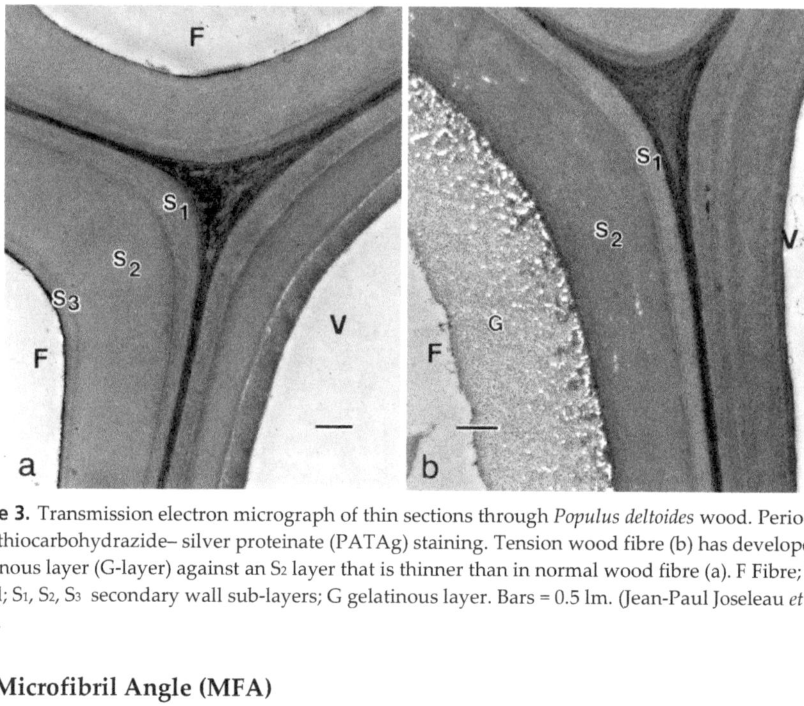

Figure 3. Transmission electron micrograph of thin sections through *Populus deltoides* wood. Periodic-acid–thiocarbohydrazide– silver proteinate (PATAg) staining. Tension wood fibre (b) has developed a gelatinous layer (G-layer) against an S_2 layer that is thinner than in normal wood fibre (a). F Fibre; V vessel; S_1, S_2, S_3 secondary wall sub-layers; G gelatinous layer. Bars = 0.5 lm. (Jean-Paul Joseleau *et al.*, 2004).

1.8. Microfibril Angle (MFA)

One of the primary importance in many investigations of the properties of natural cellulosic fibres is the knowledge of the orientation of the constituent cellulose microfibrils. The orientation of the elementary cellulose fibrils reinforcing the wood cell wall has been a subject of growing interest in recent years. Particular attention has been attracted by the tilt angle of the cellulose fibrils with respect to the longitudinal cell axis called the microfibril angle (MFA) that was found to influence the mechanical properties of wood (Cave, 1997) as well as shrinkage during drying (Meylan, 1967). MFA was also discussed with respect to influences on stiffness and tensile strength of fibres and paper (Cave, 1976). MFA varies from tree to tree, pith to cambium and with height in the stem. It also varies with speed of growth of the tree. Many methods, such polarizing, fluorescence and electron microscopy and iodine staining, used to estimate the MFA are tedious and time consuming because the extreme variability of biological material demands that large numbers of fibre elements be measured to give meaningful average values. In contrast, x-ray diffraction can provide a mean diffraction pattern of several hundred elements in a single exposure, a little cost in preparation and observation time. The main drawback to the x-ray method has been the interpretation of the diffraction patterns in terms of microfibril distribution. Methods to measure MFA include X-ray diffraction (Cave, 1966), Wide angle x-ray scattering (Boyd, 1977), and Small-angle X-ray scattering (Jakob *et al.*, 1994). The recently increased attention on MFA in wood has encouraged researchers to compare different measuring methods (Bertaud and Holmbom, 2004). These approaches are important, as encountered discrepancies foster new research towards a better understanding of MFA in wood.

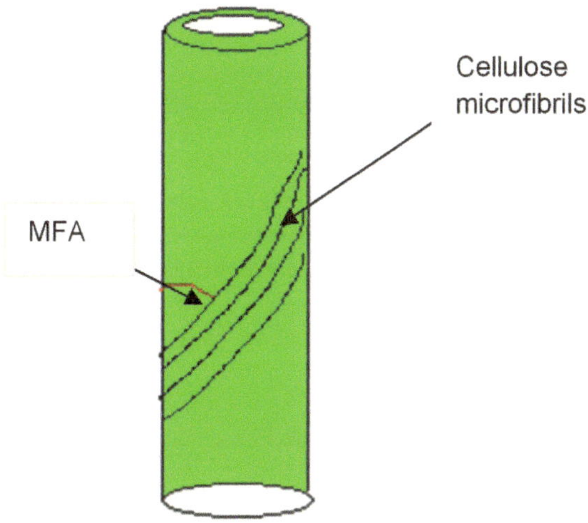

Figure 4. Cellulose microfibrils in wood cell. Source: Fratzl *et al.*, (1997)

In the present study, x-ray diffraction and small-angle x-ray scattering techniques were used to determine MFA as a rapid technique for measuring microfibril angles in *Acacia mangium* wood from pith to bark. The x-ray diffraction technique used the diffraction pattern created by the interaction of x-ray with wood tissue to determine MFA. A group of fibers in the sample is irradiated perpendicular to the fibre length by a narrow, monochromatic x-ray beam. A diffraction pattern is produced by the crystalline cellulose structure and recorded by film or by an electronic detector. The average MFA was calculated using Cave (1966) method. In general, the width of (002) diffraction arc reflects the magnitude of the mean MFA and most methods in use are based on a measurement of the width of the arc. The width of the diffraction arc in the method presented here is determined by the angular separation, *T*, of the intercepts with the zero intensity axis of the tangents at the points of inflexion, of the outer slopes of the intensity curve of the diffraction arc. X-ray diffraction can provide a mean diffraction pattern of several hundred elements in the sample in one single exposure and it has the potential to reduce dramatically the time required to determine MFA. In order to further elucidate to the effect of MFA on the growth age, physical and mechanical properties of *Acacia mangium*, x-ray diffraction (XRD) and small-angle x-ray scattering (SAXS) were used in this study. The combination of XRD measurements with SAXS on the same microtome sections allowed us to establish a direct relationship between microfibril angle and wood behavior.

1.9. Importance of MFA

The term microfibril angle, (MFA) in wood science refers to the angle between the direction of the helical windings of cellulose microfibrils in the secondary cell wall of fibres and the long axis of the cell wall (Cave, 1966). Technologically, it is usually applied to the orientation of the cellulose microfibrils in the S_2 layer that makes up the greatest proportion of the wall thickness, since it is this which most affects the physical properties of wood (Senft Bendetsen, 1985). Figure 5 shows a confocal micrograph of wood fibre under cross-polarisers. It shows the fibre as bundle of helical wound microfibrils composing highly aligned molecules, the MFA in the S_2 layer is the angle of the microfibrils relative to the long axis of the cell wall (Cave, 1966). The MFA of the S_2 layer represent an important ultramicroscopical feature influencing the performance of wood products. It has a major effect on the stability of wood on drying and subsequent manufacturing processes (Zobel, 1961). Orientation of the S_2 MFA has a significant influence on tensile strength, stiffness and shrinkage. Modeling suggests that the relative thickness of the P, S_1 and S_3 layers contributes significantly to the variability of longitudinal shrinkage (Cave, 1976). Both the longitudinal tensile strength and stiffness of wood have been shown to be markedly affected by MFA; as the MFA increases, tensile strength and stiffness quickly decrease (Mary Treacy *et al.*, 2001). Long cells, with low microfibril angles, make for more stable and stronger boards, and the incidence of high MFA is one reason why juvenile wood of loblolly pine (*Pinus taeda*) is weak and somewhat unstable (Cave, 1997). Although other mechanical properties appear to be only slightly affected by MFA (Cave and Walker, 1994). Cave and Walker (1994) concluded that MFA was the only factor that alone could account for the pronounced decrease in stiffness of *radiata pin* wood. The relatively narrow range of variation in microfibril orientation in the S_2 wall layer makes it clear that broad genetic influences are involved (Boyd, 1977). Modifications to the MFA in reaction wood are well known. For example, Addis Tsehaye *et al* (1995) has observed the microfibril angles in the branches of *Radiata pine* trees to be smaller where tensile growth stresses are larger. The within-tree and between-tree variation of MFA in the S_2 layer of the secondary cell wall has attracted the attention of some researchers. Donaldson (1993) found that MFA in *pinus radiata* showed a curvilinear decline from pith to bark, which was more pronounced at the butt end of the stem. The angle also declined rapidly with height within the tree, reaching more or less constant values at heights above 7 m. The tilt angle of the cellulose fibrils versus the longitudinal cell axis, plays an important role in determining the actual stiffness of the material (Meylan and Probine, 1969). Young's modulus in bending of clear small specimens was shown to be closely correlated with both specific gravity and MFA (Cave, 1969). In softwoods the MFA decreased from pith to to bark, there is no consistent explanation for the differences in MFA within a single stem (Preston, 1934). The MFA has already been the subject of extensive investigations (Jakob *et al.*, 1994).

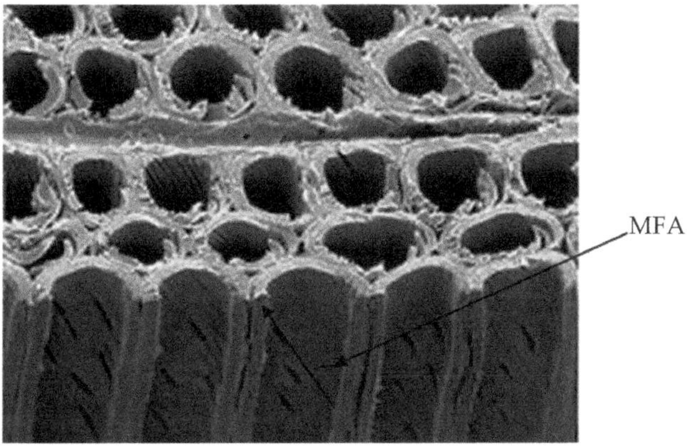

Figure 5. SEM micrograph showing the microfibrils in the S_2 layer of *Norway spruce* wood sample. Seppo Andersson , 2007.

1.10. Environmental impacts on MFA

Growth rate influences MFA in two ways. Firstly, fast growing trees have the largest microfibril angles both in juvenile and mature wood, and secondly, narrow growth rings are formed in some trees when they are suppressed and these rings tend to have tracheids with a large MFA (Brändström, 2002). The results of a trial carried out by Mary *et al.*, (2001) suggest that fast growth may lower the quality of juvenile wood in *Sitka spruce*. Bergander *et al.* (2002) reported that in a comparison of slow-grown versus fast-grown Norway spruce, a significant increase of the S_2 MFA was demonstrated in the fast grown *spruce*. This may change the quality of wood as well as the strength properties of pulp and paper produced from it.

1.11. Methods of measuring MFA

There are essentially four methods for measuring MFA in the cell wall: X-ray diffraction (Cave, 1966; Boyd, 1977), polarized light microscopy (Meylan, 1969; Evans, 1999), direct or indirect observation (Senft and Bendtsen, 1985; Donaldson,1991) and small angle X-ray scattering (SAXS), (Jakob *et al.*, 1995; Reiterer *et al.* 1998; Lichtenegger *et al.* 1999).

X-ray diffraction is the fastest and most modern method of measuring the MFA. This method enables large sample numbers to be measured in a short time. It has been used to determine MFA (Cave ,1966; Harris and Meylan, 1965; Meylan,1967). Evans (1999) used XRD technique for determining the MFA in *Eucalyptus*. Further studies using x-ray diffraction from the 002 as well as the 040 planes of cellulose in wood were carried out by Nomura and Yamada (1972). In this technique, a thin section of wood is irradiated perpendicular to the fibre length by X-ray beam producing a set of diffraction patterns. This pattern consist of a series of arcs that are spaced apart by a number of well-defined concentric circles. The diameters of the concentric circles are indications of the spacing of the crystalline planes with the cellulose crystalline fibrils. Reflections arising from cellulose crystall planes of the type (0K0) give the MFA distribution directly, but generally these reflections are too weak or too poorly resolved to be used and one is forced to use the very strong (002) reflection. Evaluation of the mean MFA then involves an assumption of the form of the microfibril distribution. The angular distance from the equator to the point where the tangent at the point of inflection of the intensity curve cuts the zero intensity axis is *T* (Cave, 1966). The width *T* has been shown to be correlated to the MFA (Meylan, 1967).

1.12. The parameter *T*

In general, the width of the (002) diffraction arc reflects the magnitude of the mean MFA and most methods in use are based on a measure of the width of the arc (Cave, 1966). The width of the diffraction arc in the method presented here is determined by the angular separation, *T*, of the intercepts with the zero intensity axis of the tangents at the points of inflexion, of the outer slopes of the intensity curve of the diffraction arc as shown in Figure 6. The choice of this measure was made originally on practical grounds, but the present study provides further justification by using Scanning Electron Microscope (SEM.

1.13. Angular distribution of microfibrils

The theoretical relation between *T* and the mean MFA is obtained by considering the relationship between the shape of the angular distribution of the microfibrils in the plane of the cell wall and the shape of the intensity distribution of the (002) arc. The shape of the intensity distribution is also dependent to some degree on the cross-sectional shape of the cells.

Tracheids may vary in shape from rectangular through irregular hexagonal to circular. The theory considers two extreme shapes, square and circular, in order to indicate the likely effect of cell cross-sectional shape on the diffraction diagram (Cave, 1966). Wherever possible, a general angular distribution of microfibrils has been assumed in the plane of the cell wall, subject to the following conditions:

i. The microfibril is essentially a single crystall.
ii. All microfibrils are crystallographically identical
iii. The cell wall consist of a single homogeneous layer of microfibrils called S_2 layer embedded in a noncrystalline matrix.

iv. The microfibrils lie strictly in the plane of the cell wall (Cave, 1976). e angular distribution of microfibrils in the plane of the wall. This taken to be approximately that of the normal probability function with a mean MFA, and standard deviation σ (provided σ ≤ 30°) (Meylan, 1972)

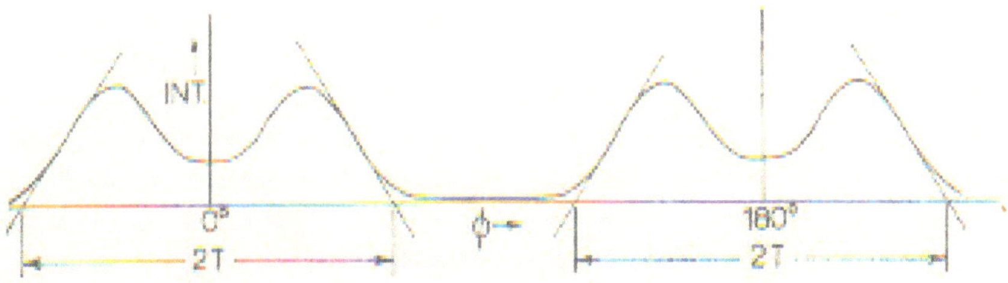

Figure 6. Schematic graph record around the (002) reflection circle of wood showing tangents drawn at the points of the inflexion in the measurement of the parameter *T*. Source: Cave (1966).

2. Materials and methods

The wood samples used in this study were selected from 3, 5, 7, 9, 10, 11, 13 and 15 year-old of *Acacia mangium* tree from Sabah Forestry Development Authority (SAFODA) and Ganui Plantations, Sabah, Malaysia. The standard methodology for the physical characterization followed the International technical standard (ISO standard 4471-1982). The samples were taken out at breast height (1.25m) on each tree. An x-ray diffractometer (Philips X-Pert PRO PW3040/60) was used to determine the average microfibril angle. A point-focused x-ray beam (Cu-Kα x-ray, beam diameter 1 mm) was applied to tangential section, 200 μm thick x 2 mm long, prepared from the pith and bark regions with a sliding microtome as shown in Figure 7. The measurements were made at a speed of 6 degrees per minute, at a Bragg's angle 22.4°, using the 2 mm diverging slit and 1 mm receiving slit. A diffraction pattern is produced by the crystalline structure and recorded by an electronic detector. Parameter *T* defined by Cave, was obtained from the diffraction intensity around (002) arc [3]. Three lines were drawn to derive half the width of the curve. The first was the baseline representing the portion in the curve when the x-ray intensity was more or less minimal. Then, a tangent was drawn to divide the curve in to two equal parts.

Radial slices 50 μm thick were cut by rotary microtome from each trunk and then used for Probe microscope.

Samples for Dynamic Mechanical Analyzer DMA testing were prepared using a table saw. They were further machined down to nominal thickness of 3.0 mm using vertical milling machine. The samples were held in place under controlled humidity and temperature. Care was taken to obtain samples from the same area of the impact region in the wood trunk. Each disc of wood was machined to produce a balance DMA samples desired thickness. The final samples dimensions were 50 mm × 13 mm × 3 mm.

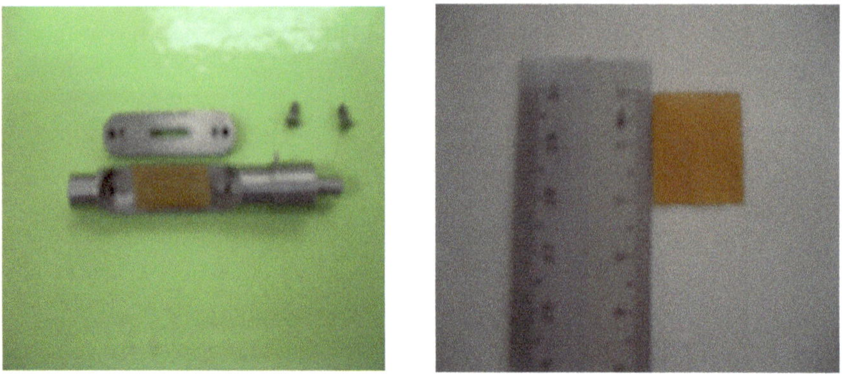

(a) (b)

Figure 7. (a) Sample holder of SAXS machine with wood strip. (b) Wood strip.

3. Determining of MFA

Methods to determine the mean microfibril angle, MFA, crystallinity of wood and the average size of cellulose crystallites were presented. The mean MFA, the crystallinity of wood and the size of the cellulose crystallites were determined as a function of the tree age in *Acacia mangium* wood found in Sabah.

3.1. Calculation of Microfibril Angle (MFA) Using X-Ray Diffraction (XRD)

Microfibril angle, MFA can be defined as the angle between tracheid or fibre axis and microfibril orientation in the S_2 layer. Evaluation of the mean MFA then involves an assumption of the form of the microfibril distribution. It is of interest to examine the differences of the intensity distributions diffracted between the different ages of the real cell wall structure. We can clearly observe the change in the diffraction pattern with increasing the tree age. Figure 14 schematically represents thin sample of wood, which has a rectangular cross-section. The fibre axis or the cell axis is vertical towards the radial direction of the tracheid, which represent the surface of the paper (Figure 8). Two cellulose MFA (Z helix) have been drawn, one in the front cell wall and the other on the back cell wall. When the MFA is determined using x-ray diffraction and a slice of wood, both the front and back cell wall contribute to the intensity curve (Figure 8).

The models, which will be presented in this study, related primarily to studies of wood cell wall structure. An SEM micrograph of a cell wall from *Acacia mangium* wood of age 15-year-old is shown in Figure 8. The model of the cellular geometry shows the microfibrils bundle together in the cell wall, this will subsequently be translated in to a finite element in which each cell wall division as sketched in Figure 9. The orientation of the microfibrils in cell wall layers is illustrated in Figure 10.

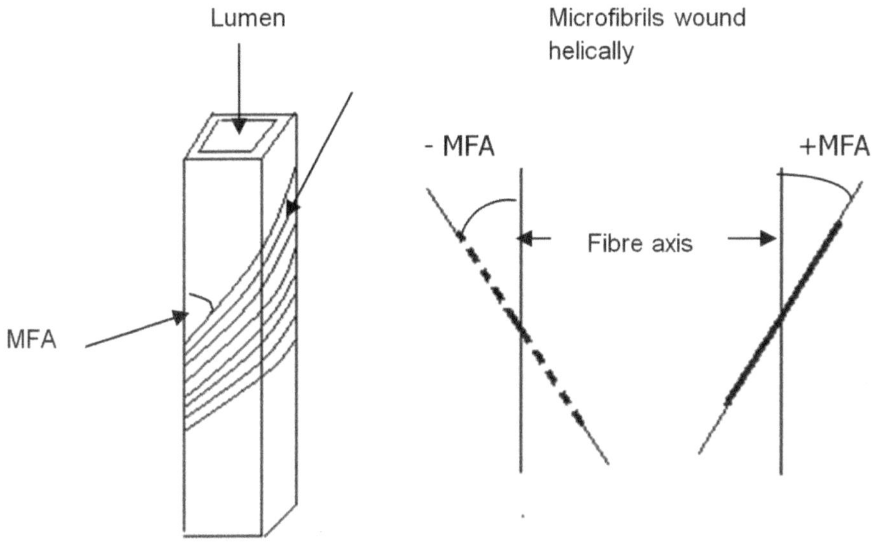

Figure 8. Schematic diagram represents the rectangular cross section of wood sample and the orientation of the cellulose microfibrils in both front, +MFA and back, -MFA around the fibre axis (Drawing not to scale).

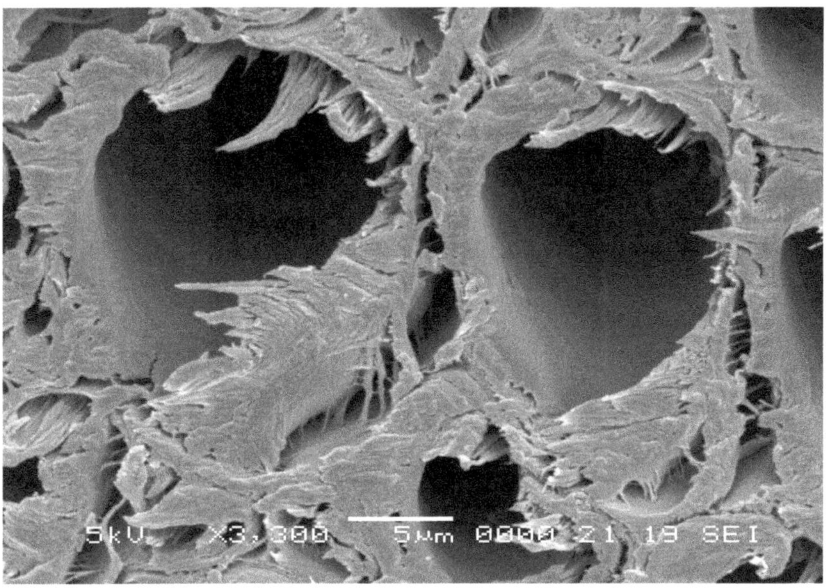

Figure 9. SEM micrograph, X 3300 showing the structure of cell wall *Acacia mangium* wood sample from age 15 year old.

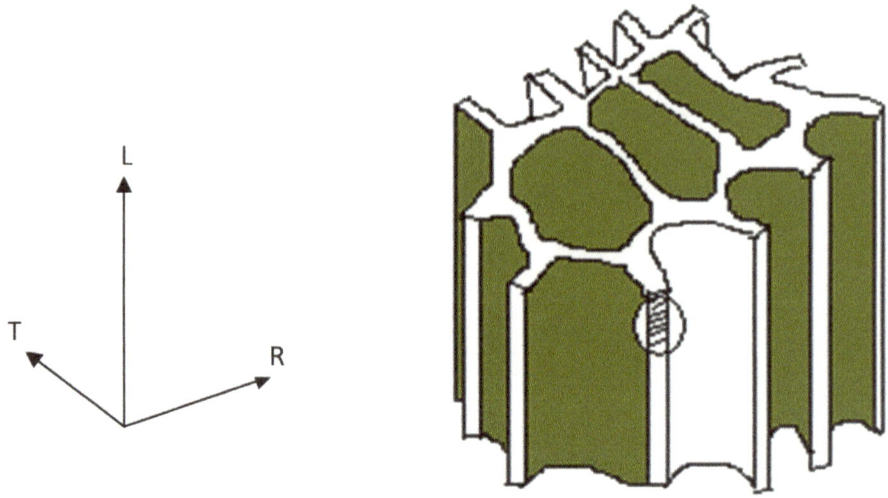

Figure 10. Schematic diagram showing model of the cellular geometry of the cell walls, at radial (R), tangential (T) and longitudinal direction (L).

Figure 11, shows typically diffraction pattern arising from the pith region of *Acacia mangium* wood as a slice samples of thickness 200 μm from growth age 15 year-old. The MFA is determined from the intensity that has the strongest peak (Andersson, 2006). In this work, the intensity peak at $2\theta \approx 22°$ has been used to calculate the parameter T for all growth ages because the peak of the diffraction intensity gives the best Full Width Half Maximum, FWHM value in this case. In case of wood sample from 15-year-old tree. The intensity peak at $2\theta = 22.11°$ gives FWHM about 0.02° was used to calculate the MFA for the wood sample of 15-year-old (Figure 11).

The parameter T was used as an indicator to MFA. T was calculated using Cave method (1966). Mean MFA was calculated using the formula developed by Yamamoto *et al.,* (1993). Three lines were drawn to derive half the width of the curve. The first was the baseline representing the portion in the curve where the X-ray intensity was more or less minimal. Than, a tangent was drawn through the inflection point on one side of the curve. Finally a vertical line was drawn to divide the curve in to two equal parts as shown in Figure 12. The results shows that the MFA ranged from 26.13° at the pith region of *Acacia mangium* wood at age 3 year-old, and decreases to about 0.20° ±0.01° at tree age 15 year-old. It was found that the mean MFA at the bark region of *Acacia mangium* wood behaves the same way. MFA and standard deviation for *Acacia mangium* from 3 year-old at bark region was calculated using the polynomial relationship of Yamamoto *et al.,* (1993).

$$MFA = 1.575 \times 10^{-3} T^3 - 1.431 \times 10^{-1} T^2 + 4.693T - 36.19 \tag{1}$$

$T = 2.3° \pm 0.1°$ for growth age of 3-year-old as shown in Figure 4.9.

$$MFA = -26.13^{o}$$

The (-) signal mean that the microfibrils orient in the back cell wall.

The parameter T given by Cave (1966) by the formula:

$$T = MFA + 2\sigma \quad (2)$$

$$\sigma = 14.21°$$

Where σ = 14.21° represent the standard deviation of the intensity distribution arising from the fibril orientation about the mean value.

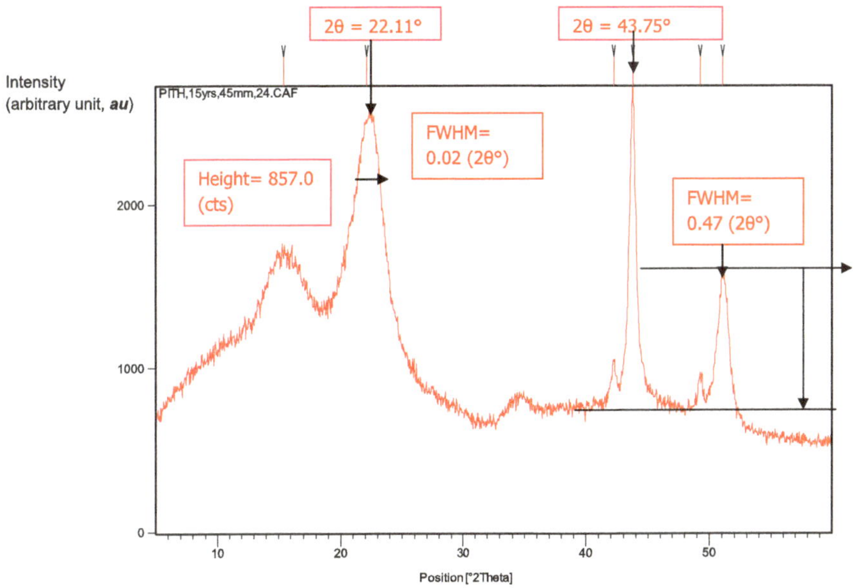

No.	Position 2θ°	Height (cts)	FWHM (2θ°)	d-spacing (Å)	Intensity (%)
1	15.4327	538.24	1.3815	5.74173	28.75
2	22.1110	857.00	0.0200	4.02033	45.78
3	42.2004	245.80	0.2598	2.14149	13.13
4	43.7580	1871.92	0.3247	2.06881	100.00
5	49.2453	209.82	0.3247	1.85037	11.21
6	51.0608	872.48	0.4752	1.78728	46.61

Figure 11. Typical diffracting pattern arising from pith region with peak list of *Acacia mangium* wood of 15-year-old.

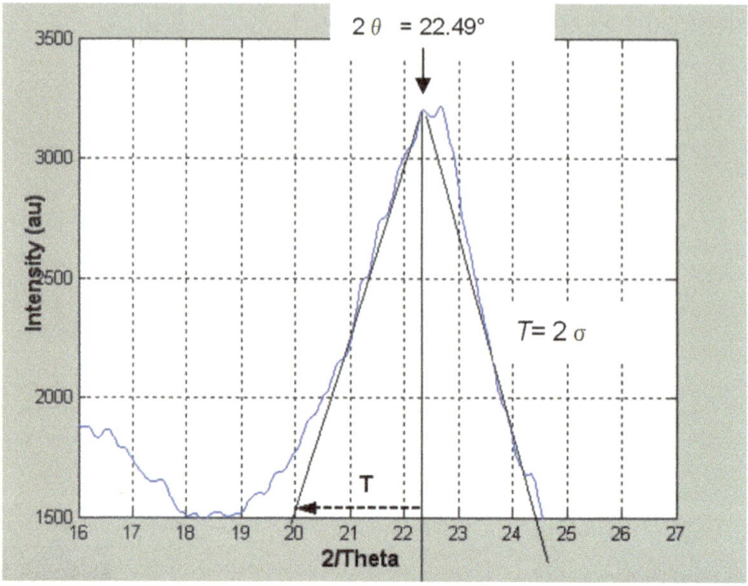

Figure 12. X-ray diffraction intensity used to determine the parameter T in wood model of 3 year-old.

Figure 13 shows SEM micrograph for the wood slice of thickness 50.0 μm taken from the pith region of *Acacia mangium* tree of age 3 year old.

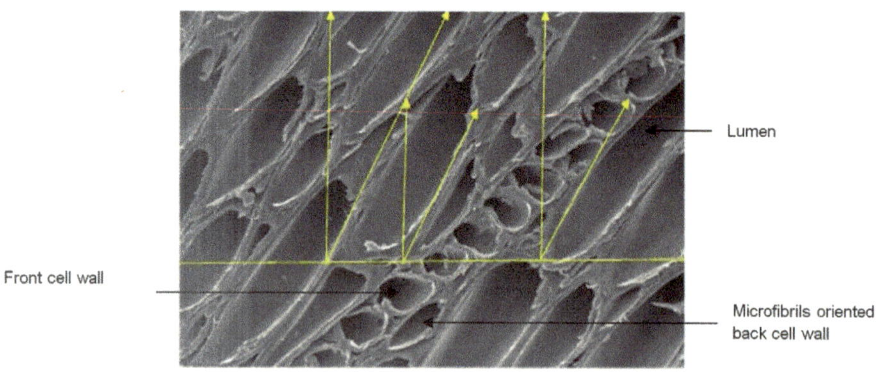

Figure 13. SEM micrograph at magnification x 2000 showing the orientation of microfibrils in the bark region of wood sample at age 3 year old.

4. Results

MFA decreased as the tree age increase. most significant drop occurring from 21.45° at age 5 year-old to 16.14° at age 7 year-old, and from 9.80° at 10 year old to 4.96° at 11 year old at the pith region. The smallest value of MFA was found in the pith center, MFA = 0.20° ± 0.01°. An inverse relationship between MFA and tree age was evident in this study within the pith region (Figure 14). The MFA of *Acacia mangium* in the bark region behaves the same way. The highest rate of decreasing of MFA occurring between ages 5 to 7 year old within the bark region. MFA was found to drop from 27.36° at 5 year old to 17.83°± 0.01° at age 7 year old. An inverse relationship between tree age and MFA was found within the bark region (Figure 15). Tables 1 and 2 are shows the values of the MFA and the standard deviation σ for each age of *Acacia mangium* tree in pith and bark regions respectively.

Sample No.	Tree age (year)	MFA ± 0.01°	σ (°)	Microfibrils orientation
1	3	26.13	14.21	Back cell wall
2	5	21.45	8.97	Back cell wall
3	7	16.14	5.47	Back cell wall
4	9	11.30	2.40	Back cell wall
5	10	9.80	1.40	Back cell wall
6	11	4.96	4.27	Front cell wall
7	13	0.26	0.08	Front cell wall
8	15	0.20	0.07	Front cell wall
9	15 (Pith center)	0.20	0.07	Front cell wall

Table 1. The values of the MFA and the standard deviation σ in the pith region for each age of *Acacia mangium* tree.

Sample No.	Tree age (year)	MFA ± 0.01°	σ (°)	Microfibrils orientation
1	3	31.62	15.13	Back cell wall
2	5	27.36	12.68	Back cell wall
3	7	17.83	6.60	Back cell wall
4	9	14.44	4.47	Front cell wall
5	10	9.87	1.43	Front cell wall
6	11	5.67	1.42	Front cell wall
7	13	3.17	3.16	Back cell wall
8	15	0.47	0.15	Back cell wall

Table 2. The values of the MFA and the standard deviation σ in the bark region for each age of *Acacia mangium* tree.

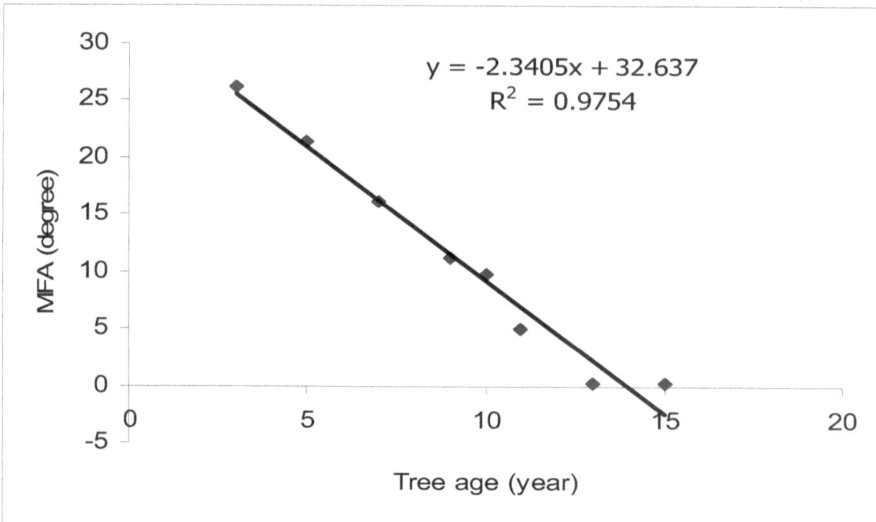

Figure 14. The relationship between MFA and the tree age in the pith region of *Acacia mangium* wood.

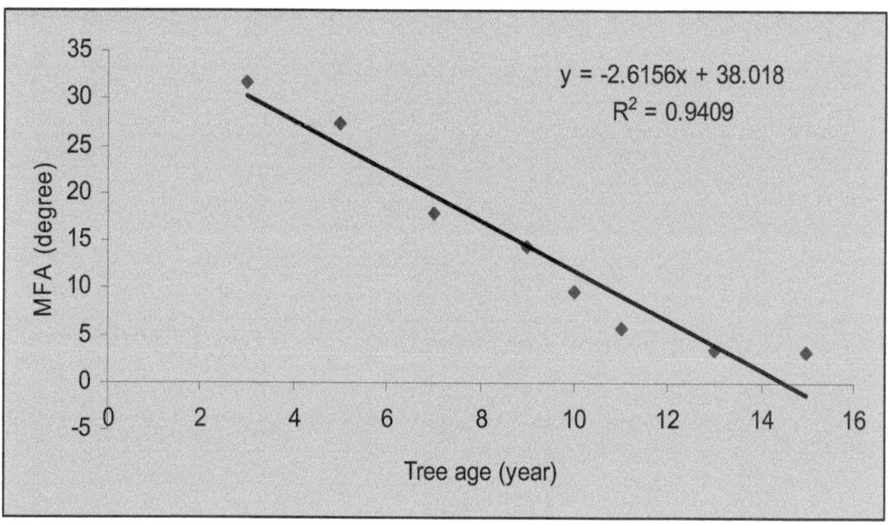

Figure 15. The relationship between MFA versus tree age in the bark region of *Acacia mangium* wood.

5. Thermal and dynamic-mechanical properties of wood

The complicated hierarchical and cellular structure of wood is well known to provide excellent mechanical properties such as stiffness and strength. Wood represents a natural composite with the ability to adapt its structural properties to external mechanical requirements in all hierarchical levels (Cave, 1997a). In the microscopic and nanoscopic scale, the structural optimization of the mechanical behavior of wood is closely related to the cell wall microstructure. The mechanical properties of wood are known to be greatly influenced by its anatomical structure (Boyd, 1977). It is well known that the stiffness of wood is mostly given by the semi-crystalline cellulose microfibrils.

Thermal analysis has been extensively applied to investigate the thermal behavior of various materials as a function of temperature. A number of researches on thermal properties of wood fiber and polymer composites (WFPCs) have been reported (Tamer and Fauziah, 2009). A number of different methods have been used to investigate thermal properties and vescoelastic properties of wood. One such method is dynamic mechanical thermal analysis (DMTA). This has been used to investigate wood from different trees species. The variations of MFA with the tree ages were studied for the real wood cell wall structure of *Acacia mangium* using X-ray diffraction (Tamer and Fauziah, 2009). The results of MFA measurements from the eight *Acacia mangium* trees, 3, 5, 7, 9, 10, 11, 13 and 15-year-old examined in this study showed that differences in MFAs between tree ages were significant. The general trend was for the MFA to be greatest in the young wood of age 3-year-old and decrease gradually with increasing the tree age. This trend has been found for both pith and bark region of the wood samples. It was found that for eight ages of trees examined, the MFA was at its greatest in the sample of 3-year-old, where angle as high as 26.13° were recorded. The lowest for all eight cases were taken from the pith region were found in tree age 15-year-old where the angle varied from 1.99° to 0.20° in the front cell wall direction of the microfibrils. For all cases of the wood samples were taken from the bark region of the eight trees, the results shows that the MFA was its greatest in tree age 3-year-old, were angle as high as 31.62° were recorded. Tables are numbered with Roman numerals. The lowest MFAs for the same samples were found in 15-year-old of the trees were the angle was 0.47° in the back cell wall direction of the microfibrils.

Study by (Tamer and Fauziah, 2009) using small angle x-ray scattering technique shows that MFA in *Acacia mangium* wood increases rapidly as a function of the distance from pith towards the bark in wood models were taken from 10 year-old. The general trend was for the MFA to be lowest near the pith and then to increase gradually towards the cambium. It was found that for the samples taken from 10 year-old trunk, the MFA was at its lowest at a distance 10.0 mm from the pith centre, where angle as 0.46° was recorded. The highest MFA for all samples in the trunk was found at the distance 90.0 mm from the pith where the angle was 14.44°. When a tree is young it needs to be elastic in order to move in the wind. After some decades, however the cells produced by matured trunk cambium have a smaller MFA for stiffness and keeping the trunk upright. By contrast, young woods have to be rather elastic, allowing them to bend. Therefore, MFA need to be larger. The MFA is thus critical to the total mechanical balance of the tree, correct MFAs are essential for its survival. However, the results clearly show that greater variation exists across the distance from pith to bark and between the tree ages.

Results from the same study show a good inversely relationship between MFA and the E' was evidence in *Acacia mangium* wood of 10-year-old. A general declining trend for all curves of E' test is observed when the wood samples go through higher MFA. The only noticeable variation can be detected in the case of MFA about 6.46°. Thus, as the MFA increases the E' decreases. It was found that for a decrease in MFA from 14.44° to 0.46°, the E' increased from 4.76 ×108 GPa to 9.00 ×108 GPa. The results also showed that E'' was most strongly influenced by MFA under the same experimental conditions. The model suggests that a similar reduction in MFA as outlined above, causes E'' to increase about 23,146,225.00 GPa. These results supports those of Cave and Walker (1994) who found that wood stiffness and bending strength are negatively affected by large microfibril angle. Cave and Walker (1994) reported that the tensile strength of the tracheid decreasing with increasing MFA in the cell wall. This behave of dynamic-mechanical properties of wood with MFA may be attributed to the effect of MFA on the density of wood. The density of wood impacts the pulping process, energy consumption and is important for pulp yield (Cave, 1997). High density had a positive correlation with tear strength but the effects diminished when position in the tree was taken in to account indicating that it is not the density in itself that is important but ruther the properties of the fibres which are mirrored by the density (Cave, 1976).

Differences in MFA have a profound effect on the properties of wood, in particular its stiffness. The large MFA in juvenile wood confers low stiffness and gives the sapling the flexibility it needs to survive high winds without breaking. It also means, however, that timber containing a high proportion of juvenile wood is unsuitable for use as high-grade structural timber. This fact has taken on increasing importance in view of the trend in forestry towards short rotation cropping of fast grown species. These trees at harvest may contain 50% or more of timber with low stiffness and therefore, low economic value. Although they are presently grown mainly for pulp, pressure for increased timber production means that ways will be sought to improve the quality of their timber by reducing juvenile wood MFA (Barnett *et al.*, 2004). Glas transition (T_g) determined by the onset of the E' change and the maximum of E'' at MFA = 18.0°.

5.1. Determination of glass transition of *Acacia mangium* Wood by Dynamic Mechanical Thermal analyzer (DMTA)

In this study, the DMTA technique is used in glass transition, T_g calculation. The procedures described in several standards and recommendations can result in significantly different values of the same data, as discussed in detail by Yamamoto *et al.*, (1993). This is due, in part, to the fact that T_g is a single temperature that represents a range over which the glass transition takes place (Work Item Wk278, 2003). In an effort to simplify the determination of T_g , it is commonly defined as the maximum of the damping ratio E' / E'', ($tan\ \delta$) or the maximum of E'' . The T_g also can be derived from the onset of the change in the slope of E' curve (Akay, 1993). A comparison of these calculation techniques is shown in Figures 16, 17 and 18 If T_g is used for engineering design, that is the determination of maximal end-use temperatures, there is a strong believe that the conservative estimate of T_g is warranted (Wong *et al.*, 1993). Therefore, in this work, one method of T_g calculation was based on the temperature at which mechanical properties began to be compromised, that is the maximum damping ratio. This method is used to present the glass transition of *Acacia mangium* wood as a function to the MFA, because this is the method commonly found in the literature. The thermal and mechanical properties of *Acacia mangium* wood under analysis are presented in Table 3. The values of T_g were calculated based on the maximum damping ratio. The results from pith region of the wood trunk are summarized in Table 4.

MFA (°)	E′ (GPa)	E″ (GPa)	Glass transition (C°)	*tanδ*
18.0	9.00×10^8	55129999	102.583	0.175
19.8	7.79×10^8	46028906	103.213	0.145
19.2	7.16×10^8	48354672	92.694	0.166
22.2	7.13×10^8	45717485	88.899	0.165
24.0	7.14×10^8	44826412	99.669	0.151
25.8	6.47×10^8	45953213	84.633	0.165
29.4	6.24×10^8	42198028	84.113	0.153
28.8	6.71×10^8	37456490	83.688	0.172
30.6	4.76×10^8	31983774	83.895	0.147

Table 3. Thermal and mechanical properties of *Acacia mangium* wood at bark region with respect to the MFA. The values of T_g were calculated based on the *tanδ*.

MFA (°)	E′ (GPa)	E″ (GPa)	Glass transition (°C)	*tanδ*
0.46	9.00×10^8	55129999	146.825	0.175
3.17	7.79×10^8	46028906	103.213	0.145
4.90	7.16×10^8	48354672	92.694	0.166
6.46	7.13×10^8	45717485	88.899	0.165
7.68	7.14×10^8	44826412	99.669	0.151
8.66	6.47×10^8	45953213	84.633	0.165
11.91	6.24×10^8	42198028	84.113	0.153
12.84	6.71×10^8	37456490	83.688	0.172
14.44	4.76×10^8	31983774	83.895	0.147

Table 4. Thermal and mechanical properties of *Acacia mangium* wood at the pith region with respect to the MFA. The values of T_g were calculated based on the *tanδ*.

The general declining trend for all curves of E'' test is observed when the wood samples go through higher temperatures. The only noticeable transition can be detected at around 41.173 C° (indicated by an arrow). As it seen in Figure 16, wood sample at MFA 18.0° have higher T_g based on the maximum value of E'' about 77.64 C°. No significant difference between the temperatures based on the change of storage modulus is observed. Results also show that, in the case of the temperature based on the *tanδ*, a significant different due to the increase of MFA can easily be seen. It was found that T_g ranging between 102.583 C° at MFA about 18.0° and 83.895 C° at MFA about 30.6°. Figure 17 and Figure 18 shows the determination of T_g in wood sample at MFA about 19.8° and 0.46° based on E'' and E' respectively. Note that the highest value of T_g was recorded at MFA about 0.46° which is 146.825 C°.

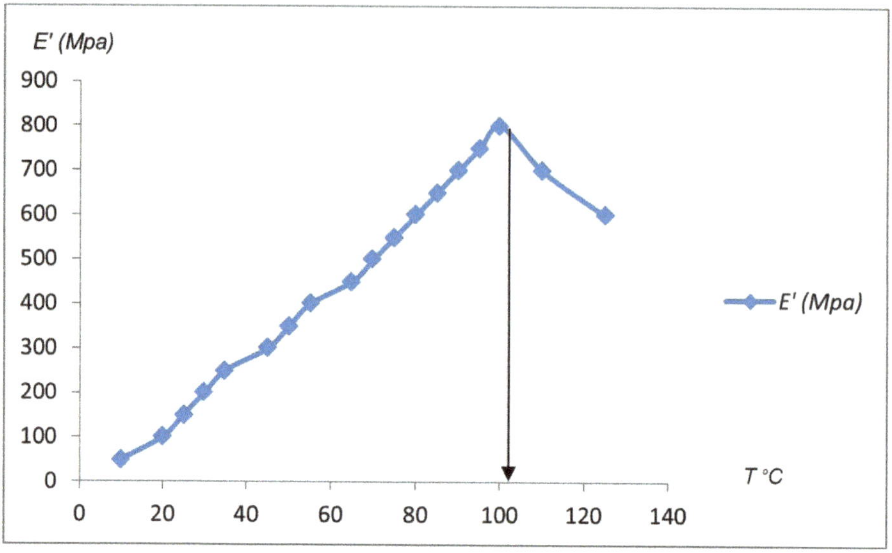

Figure 16. T_g determined by the maximum value of E' at MFA = 18.0°. The T_g value based on the change of E' = 102.583 C° while T_g based on the maximum of E'' = 77.647 C°. The Frequency = 1 Hz.

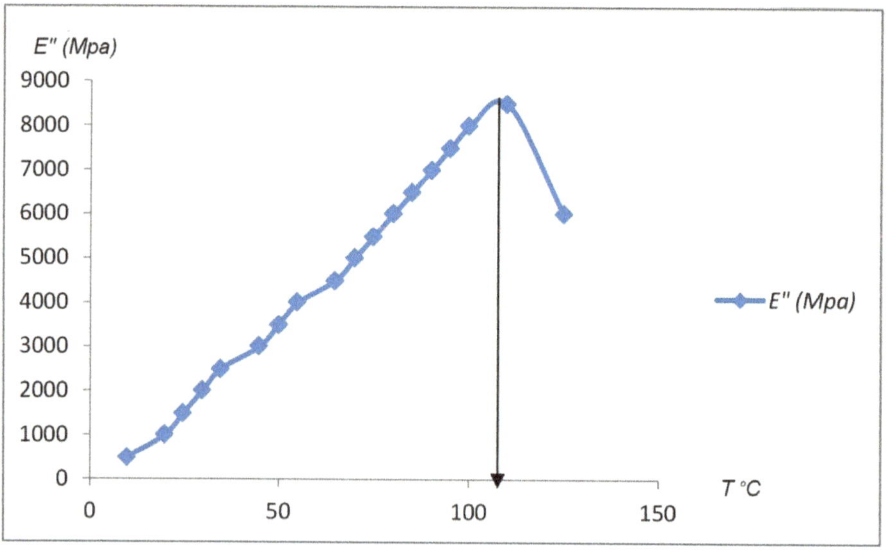

Figure 17. The T_g value based on the change of E'' = 103.213 C° at MFA = 19.8°. Frequency = 1 Hz.

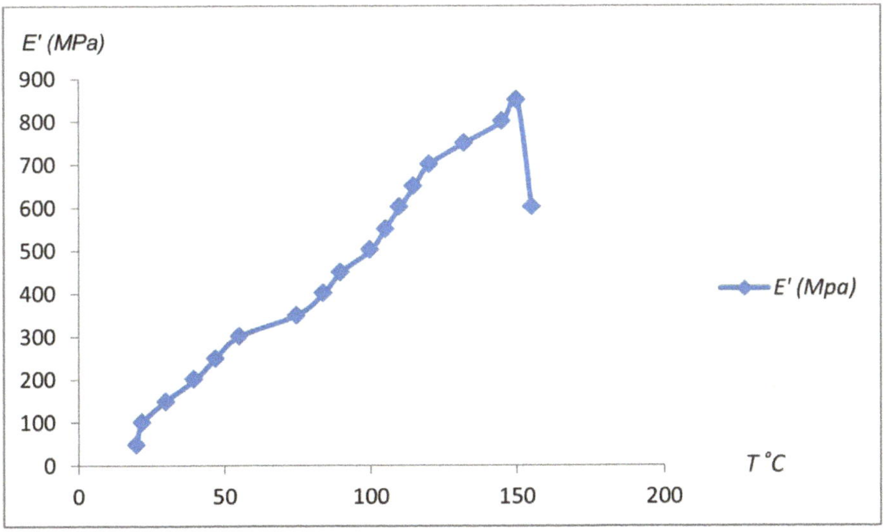

Figure 18. The T_g value based on the maximum of E' = 146.825 C° at MFA = 0.46°. The Frequency = 1 Hz.

5.2. The relationship between microfibril angle and storage modulus, loss modulus and glass transition

The relationship between MFA and the thermal and mechanical properties in *Acacia mangium* wood was studied in this research. Results show, the storage modulus, E' decreased with the increase of MFA. At low MFA about 18.0°, the wood sample shows the highest value of E' about 9.0×10^8 Gpa. When the MFA increased over 30.0° the E' decrease to 4.76×10^8 Gpa. A statistical analysis was conducted on these data to determine the relationship between MFA in the cell wall and the thermal and mechanical properties of *Acacia mangium* wood. Table 5 present the descriptive statistics of the data under analysis. The regression analysis of MFA and E' shows that 74% of variation in storage modulus was explained by MFA. This showed that one regression equation with intercept and slope was required to describe the relationship between MFA and E' (Figure 19).

	N Statistic	Range	Minimum	Maximum	Mean	Std. Error	Std.	Variance
MFA	9	12.60	18.00	30.60	24.20	1.57	4.7244	22.320
Storage	9	4.2×10^8	4.8×10^8	9.0×10^8	6.9×10^8	3.8×10^7	1.2×10^8	1.3×10^{16}
Loss	9	2.3×10^7	3.2×10^7	5.5×10^7	4.4×10^7	2187580	6562740	4.3×10^{16}
Damping	9	0.3	0.14	0.18	0.16		0.1226	0.000
Trans	9	19.52	83.69	103.21	91.48			69.442
Valid N	9							

Table 5. The descriptive statistics of the thermal and mechanical data under Analysis.

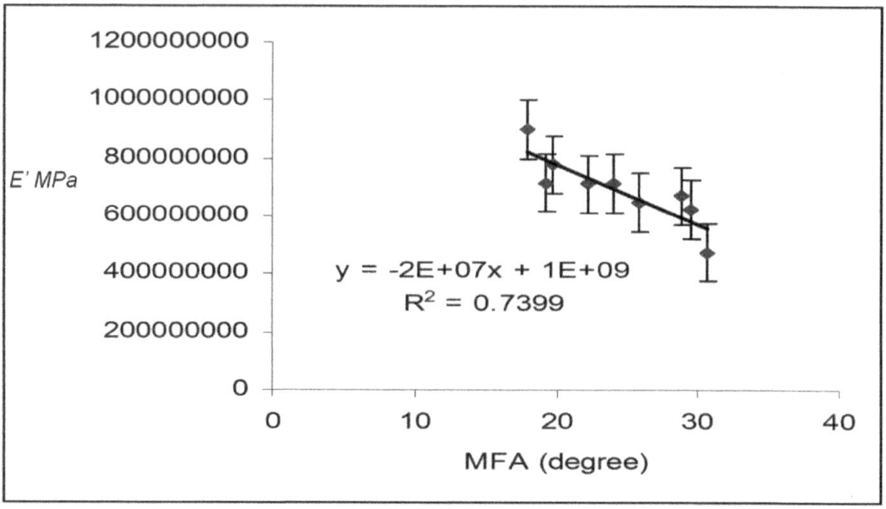

Figure 19. The relationship between MFA and *E'* for wood of 10-year-old.

The relationship between MFA and the loss modulus was discussed. The statistical analysis showed that a straight line fit the data very well. Thus, as the MFA increases the *E″* decreases (Figure 20). The regression analysis of MFA and *E″* shows that 76.2% of variation in *E″* was explained by MFA and the analysis of variance shows the impact of MFA on *E″*.

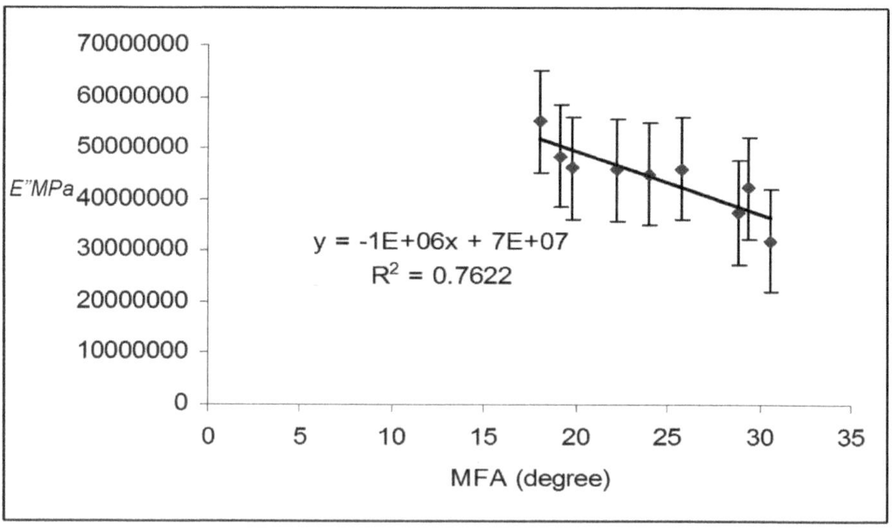

Figure 20. The relationship between MFA and *E″* in *Acacia mangium* wood of 10- year-old.

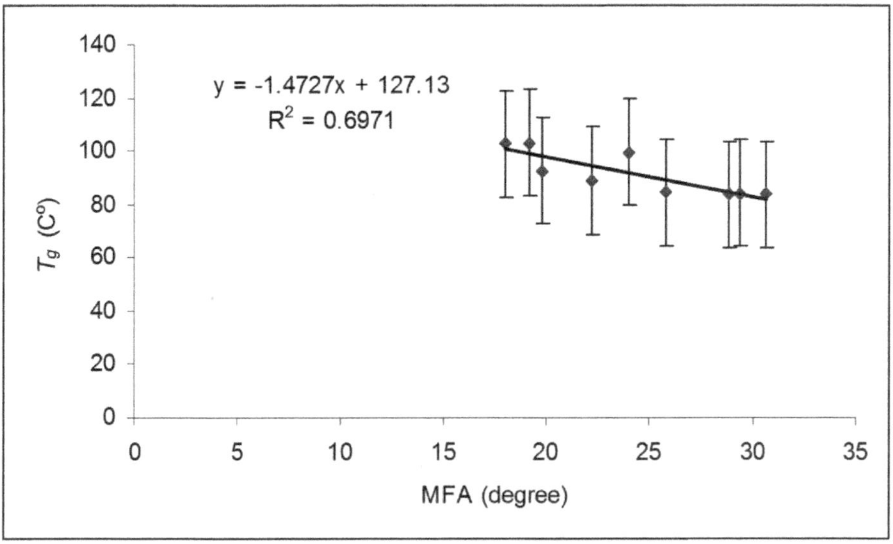

Figure 21. The relationship between MFA and T_g in *Acacia mangium* wood of 10 - year-old.

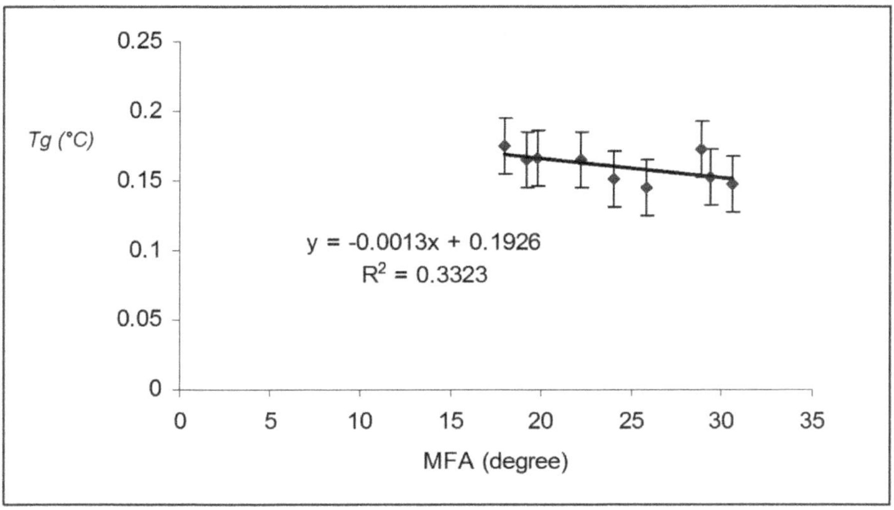

Figure 22. The variation of MFA with *tan δ* in *Acacia mangium* wood of 10-year-old.

The effect of MFA on the glass transition is studied in this research. Figure 21 present the relationship between MFA and T_g. A similar trend is seen for both E' and E'' with a noticeable reduction in the value of R^2 of the graph. As a comparison with the regression analysis of MFA and E'', where 76.2% of variation in E'' was explained by MFA, the MFA accounts for 69.7% of variation in T_g (Figure 21). In contrast to the regression analysis of MFA and E'', no clear relationship between MFA and the damping ratio $tan\delta$ is observed (Figure 22). Inversely correlation was evident between T_g and MFA in *Acacia mangium* wood.

As a result, the mechanical properties of *Acacia mangium* wood can be expected to weaker with high MFA. Glass transition in highly crystalline polymers is difficult to identify (Barnett *et al.*, 2004). This is true because in such cases T_g is a minor event, masked by crystallinity, and because crystalline polymers frequently have multiple transitions arising from relaxations associated with amorphous phase, or both. The effect of MFA on the glass transition was discussed in this study. T_g was calculated based on the temperature at which mechanical properties began to be compromised, that is the maximum damping ratio (Tamer and Fauziah, 2009).

6. Conclusion

In this work, the variations of MFA with the tree ages were studied for the real wood cell wall structure of *Acacia mangium* using X-ray diffraction. The results of MFA measurements from the eight *Acacia mangium* trees, 3, 5, 7, 9, 10, 11, 13 and 15-year-old examined in this study showed that differences in MFAs between tree ages were significant. The general trend was for the MFA to be greatest in the young wood of age 3-year-old and decrease gradually with increasing the tree age. This trend has been found for both pith and bark region of the wood samples. It was found that for eight ages of trees examined, the MFA was at its greatest in the sample of 3-year-old, where angle as high as 26.13° were recorded. The lowest for all eight cases were taken from the pith region were found in tree age 15-year-old where the angle varied from 1.99° to 0.20° in the front cell wall direction of the microfibrils. For all cases of the wood samples were taken from the bark region of the eight trees, the results shows that the MFA was its greatest in tree age 3-year-old, were angle as high as 31.62° were recorded. Tables are numbered with Roman numerals. The lowest MFAs for the same samples were found in 15-year-old of the trees were the angle was 0.47° in the back cell wall direction of the microfibrils. Results shows that MFA in *Acacia mangium* wood increases rapidly as a function of the distance from pith towards the bark in wood models were taken from 10 year-old. The general trend was for the MFA to be lowest near the pith and then to increase gradually towards the cambium.It was found that for the samples taken from 10 year-old trunk, the MFA was at its lowest at a distance 10.0 mm from the pith center, where angle as 0.46° was recorded. The highest MFA for all samples in the trunk was found at the distance 90.0 mm from the pith where the angle was 14.44°. Obviously, the MFA has a strong correlation with the wood quality as well as with the physiomechanical function of tree (Cave, 1997)]. When a tree is young it needs to be elastic in order to move in the wind. After some decades, however the cells produced by matured trunk cambium have a smaller MFA for stiffness and keeping the trunk upright. By contrast, young woods have to be rather elastic, allowing them to bend. Therefore, MFA need to be larger. The MFA is thus critical to the total mechanical balance of the tree, correct MFAs are essential forits survival. However, the results clearly show that greater variation exists across the distance from pith to bark and between the tree ages. Effect of MFA on the storage modulus, loss modulus and glass transition of *Acacia mangium* wood is discussed in this study. Results show a good inversely relationship between and the E' was evidence in *Acacia mangium* wood of 10-year-old. A general declining trend for all curves of E' test is observed when the wood samples go through higher MFA.

The only noticeable variation can be detected in the case of MFA about 6.46° (Table 4). Thus, as the MFA increases the E' decreases. It was found that for a decrease in MFA from 14.44° to 0.46°, the E' increased from 4.76 ×108 GPa to 9.00 ×108 GPa. The results also showed that E'' was most strongly influenced by MFA under the same experimental conditions. The model suggests that a similar reduction in MFA as outlined above, causes E'' to increase about 23,146,225.00 GPa. These results support those of Cave and Walker (1994) who found that wood stiffness and bending strength are negatively affected by large microfibril angle. Cave (1976) and Mary Treacy (2001) reported that the tensile strength of the tracheid decreasing with increasing MFA in the cell wall. This behaves of dynamic-mechanical properties of wood with MFA may be attributed to the effect of MFA on the density of wood. The density of wood impacts the pulping process, energy consumption and is important for pulp yield (Donaldson, 1993). High density had a positive correlation with tear strength but the effects diminished when position in the tree was taken in to account indicating that it is not the density in itself that is important but the properties of the fibres which are mirrored by the density (Cave, 1976). Farther more has been reported that these two properties have been successfully combined to predict the dynamic-mechanical properties of small specimens (Cave, 1997).

Author details

Tamer A. Tabet*

School of International Tropical Forestry, University Malaysia Sabah, Kota Kinabalu, Sabah, Malaysia

Fauziah Abdul Aziz

Physics Department, Centre for Defence Foundation Studies, National Defence Universiti of Malaysia, Kem. Sg. Besi, Kuala Lumpur, Malaysia

Acknowledgement

We greatly appreciate the professional co-operation and assistance of Universiti Malaysia Sabah. We also wish to thank Universiti Putra Malaysia in accessing and using the X-Ray

* Corresponding Author

diffraction equipment. Finally, we would like to thank fully Universiti Kebangsaan Malaysia for the professional cooperation.

7. References

[1] Addis, A., Buchanan, B.H. & Walker, J.C.F. 1995. Stiffness and Tensile Strength Variation Within and Between *Radiata pin* Trees. *Journal of Institute of Wood Science* 13(5):513-518

[2] Andersson, A.D. 2006. A Study of the Nanostructure of the Cell Wall of the Tracheids of Conifer Xylem by X-Ray Scattering. *University of Helsinki, Report Series in Physics. HU-P-D* 135: 18-25.

[3] Arntzen and Charles J. 1994. Encyclopedia of Agricultural Science. Orlando, FL; Academic Press. Vol 4. Pp 549-561.

[4] Barnett J.R., Victoria A. and Bonham. 2004. Cellulose microfibril Angle in the Cell Wall of Wood Fibres. *Biological Reviews Journal*. Vol. 79 Issue 2, pages461-471.

[5] Bertaud, F. & Holmbom, B. 2004. Chemecal Composition of Earlywood and Latewood in Norway spruce Heartwood, Sapwood and Transition Zone Wood. *Wood Science and Technology*, 38 (4):245-256.

[6] Batchelor, W.J., Conn, A.B. & Barker, I.H. 1997. Measuring the Fibril Angle of Fibres Using Confocal Microscopy. *Appita J.* 50:377-380.

[7] Bergander, A., Brändström, J., Daniel, G. & Salmén, L. 2002. Fibre Angle Variability in Earlywood of Norway spruce Using Soft Root Cavities and Polarization Confocal Microscopy. *J. Wood Sci*, 21(3): 209-211.

[8] Boyd, J.D. 1977. Interpretation of X-Ray Diffractograms of Wood for assessments of Microfibril Angles in Fibre Cell. *Wood Science and technology* 11(2): 93-114

[9] Brändström, J. 2002. Morphology of *Norway spruce* Tracheids with Emphasis on Cell Wall Organization. Department of Wood SLU. *Sueciae. Silvestria Vol.* 237: 403-415.

[10] Butterfield, B.J. 1980. *Handbook of Three-Dimensional Structure of Wood.* 2nd ed. London, United kingdom: Chapman and Hall Publishers, pp. 99 -115.

[11] Cave, I.D. 1966. X-Ray Measurement of Microfibril Angle. *For. Prod. J.* 44: 37 – 4.

[12] Cave, I.D. 1997a.Theory of X-Ray Measurement of Microfibril Angle in Wood. Part 1. The Condition for Reflection. X-Ray Diffraction by Materials with Fibre Type Symmetry. *Wood Science and Technology*, 31 (3): 143-152.

[13] Cave, I.D. 1976. Modeling the Structure of the Softwood Cell Wall for Computation of Mechanical Properties. *Wood Science and Technology*. 10:19-28.

[14] Cave, I.D. & Walker J.F.C. 1994. Stiffness of Wood in Farown Plantation Softwood: The Influence of Microfibril angle. *Forest Product Journal*, 44 (5): 43-48.

[15] Cown, D.J. & Clement, B.C. 1983. A Wood Densitometer Using Direct Scanning with X-Ray. *Wood Science and technology* 17: 91-99.

[16] Dadswell, H.E., Nicholls, J. W. P. and Echols R. M. 1959. Assessment of Wood Qualities for Tree Breeding in *Pinus elliottii* from Queensland. Queensland, Australia: *Div. of For. Prod,* 12 (4): 98-112.

[17] Donaldson, L.A. 1991. The Use of Pit Apertures as Windows to Measure Microfibril Angle in Chemical Pulp Fibres. *Wood and Fibre Science.* 23: 290-295.

[18] Elliot G.K. 1970. *Handbook of Wood Density in Conifers.* Technical Communication No.8, Commonwealth Forestry Bureau: Oxford, England. pp. 88-101.

[19] Hakan, H., Jemes W., Evans, R. & Steve, P. 1998. Influence of Cambial Age and Growth Conditions on Microfibril Angle in Young *Norway spruce* and *Picea abies. Holzforschung.* 52: 573-581.

[20] Haygreen, J.G. & Bowyer J.L. 1996. *Forst Product and Wood Science* (An introduction). 3rd edition, PP. 490-513.

[21] Harris, J.M. & Meylan, B.A. 1965. The Influence of Microfibril Angle on Longitudinal and Tangential Shrinkage in *Pinus radiate. Holzforschung* 19:144-153.

[22] Latif, M.A., & Habib, M.A. 1994. Biomass Tables for *Acacia mangium* Grown in the Plantation in Bangladesh. *J. Trop. For. Sci.* 7 (2): 296-302.

[23] Meylan, B.A. 1967. Measurement of Microfibril Angle by X-Ray Diffraction, *Forest Prod. J.* 17: 15-58.

[24] Meylan, B.A. 1972. The Influence of Microfibril Angle on the Longitudinal Shrinkage-Moisture Content Relationship. *Wood Science and Technology* 6: 293-301.

[25] Reiterer, A. Lichtenegger, H., Tschegg S. & Fratzl 1999. Experimental Evidence for Mechanical Function of the Cellulose Microfibril Angle in Wood Cell Walls. *Philosophical Magazine A* 79 (9): 2173-2184.

[26] Svergun, D.I. & Stuharmann H.B. 1991. New Developments in Direct Shape Determination from Small-Angle X-Ray Scattering, 2. Theory and Model Calculations. *Acta. Cryst.* A 47: 736-744.

[27] Tamer A. and Fauziah Aziz. 2009. Influence of Microfibril Angle on Thermal and Dynamic-Mechanical Properties of *Acacia mangium* Wood Using X-Ray Diffraction and Dynamics-Mechanical Test. *Journal of Nuclear and Related Technologies. Vol. 6,* No. 1. P.P 71-86 (Special edition).

[28] Tsoumis, G. 1992. *Science and Technology of Wood Structure, Properties and Utilization.* New York: Van Nostrand Reinhold: London; Chapman and Hall, PP 494-501.

[29] Jakob, H. F., Tschegg, S. E. and Fratzl, P. 1994. Size and Arrangement of Elementary Cellulose Fibril in wood Cells: A small-angle X-ray scattering study of *Picea abies, J. Struct. Biol.* 113: 13-22.

[30] Jean-Paul Joseleau, Takanori Imai and Katsushi Kuroda Katia Ruel. 2004. Detection in situ and characterization of lignin in the G-layer of tension wood fibres of Populus deltoids. Planta (2004) 219: 338–345.

[31] Jozsa, L.A. & Middleton G.R. 1994. *A Discussion of Wood Quality Attributes and their Practical Implications.* Forintek, Canada special publication No. SP-34. pp. 113-123.

[32] Wilson, B.F. & Archer R.A. 1979. Three Designs: Some Biological Solutions to Mechanical Problems. *Forestry J.*, 9: 293-298.

[33] Yamamoto H., Okuyama T. & Yoshida M. 1993. Method of Determining the Mean Microfibril Angle of Wood Over a Wide Range by the Improved Cave's Method. *Mokuzai aaagkkaishi* 39: 11-1.

[34] Zobel, B.J. 1961. Inheritance of Wood Properties in *Conifers. Silvae Genet*. 10: 65-70.

Chapter 7

Rapid Dissolution of Cellulose in Ionic Liquid with Different Methods

Wu Lan, Chuan-Fu Liu, Feng-Xia Yue and Run-Cang Sun

Additional information is available at the end of the chapter

1. Introduction

Cellulose, the most widespread natural polymers, can be used as a green source for fabricating biodegradable and biocompatible materials with attractive properties by chemical modification or mixing with other components (Potthast et al., 2006). Effective utilization of cellulose not only mitigates the fossil resources crisis but also protects the environment of the Earth. However, cellulose is extraordinarily difficult to be dissolved due to its considerable hydrogen bonds and partially crystalline structure.

Ionic liquids (ILs), also known as room temperature ionic liquids (RTILs), are organic salts comprised completely of ions. They are regarded as novel green solvents and have attracted tremendous attentions in analysis chemistry, electric chemistry, catalysis chemistry and organic synthesis (Forsyth et al., 2002; Turner et al., 2003) because of their unique physico-chemical properties, such as chemical and thermal stability, non-flammability and immeasurably low vapor pressure (Feng & Chen, 2008; Cao et al., 2009). Furthermore, it is feasible to modify the physicochemical properties such as viscosity, melting point, polarity and solubility of the ILs through changing the structure of cations and anions (Hermanutz et al., 2008). Recently, ILs were applied in the field of biomass research, such as dissolution and the following homogeneous modification of cellulose (Forsyth et al., 2002; Cao et al., 2010). Swatloski and his group (Swatloski et al., 2002) reported the pioneer study on the dissolving capability of a series of ILs containing 1-butyl-3-methylimidazolium ($[C_4mim]^+$) as cation for cellulose. Among this kind of ILs, 1-butyl-3-methylimidazolium chloride ($[C_4mim]Cl$) exhibited the best solubility, which could dissolve up to 10 wt % of pulp cellulose with a degree of polymerization of (DP) of 1000 at 100 °C. Soon after, 1-allyl-3-methylimidazolium chloride ([Amim]Cl) was synthesized and found to be a powerful solvent of cellulose (Zhang et al., 2005). Cotton cellulose could be dissolved in [Amim]Cl within 120 min at 110 °C. Heinze and his co-workers (Heinze et al., 2005) tested the solubility of $[C_4mim]Cl$ for cellulose and the results showed that $[C_4mim]Cl$ could dissolved cotton linter (DP=1198) with the concentration of 10 wt%, though a relatively long time (12 h) was required.

In view of long treatment duration, some pretreatment or auxilary methods should be introduced to reduce the dissolution time of cellulose in ILs. Ultrasonic and microwave are the most popular activation methods or "unconventional" physical agents in the concept of "Green Chemistry" (Kardos & Luche, 2001). Ultrasonic treatment has been well established in the separation of plant materials, particularly for extraction of low molecular weight substances (Tang et al., 2005; Gadhe et al., 2006). The mechanical and chemical effects of ultrasound are believed to accelerate the extraction of organic compounds from plant materials due to disruption of cell walls and to enhance mass transfer in cell walls. Microwave heating, as an alternative to conventional heating technique, has been successfully applied in organic synthesis and carbohydrate degradation (Satge et al., 2002; Ma et al., 2009; Ramier et al., 2012) due to its thermal effects and non-thermal effects, raised from the heating rate, acceleration of ions and molecules collision, and rapid rotation of dipoles. In addition, microwave energy can easily penetrate to particle inside hence all particles can be heated simultaneously, which resolve the heat transfer resistance problem and make it more uniform and efficient than conventional heating methods (Hou et al., 2008). Consequently, it is reasonable to assume that cellulose could be rapidly dissolved in ILs using ultrasonic and microwave as pretreatment or auxilary methods to active cellulose structure. Actually Mikkola et al (Mikkola et al., 2007) and Swatloski et al (Swatloski et al., 2002) have already mentioned that ultrasonic and microwave irradiation can significantly accelerate dissolution of cellulose in ILs. Besides ultrasonic and microwave, ethanol treatment and mechanical milling were also believed to enhance cellulose dissolution rate in ILs. Mechanical milling was one of the most common mechanical pretreatment methods which can cut down the particle size and reduce molecular weight. Previous studies (Evtuguin et al., 2001; Fujimoto et al., 2005) have already applied mechanical milling to break lignocellulose structure in order to dissolve lignin from biomass matrix with organic solvents. Ethanol treatment was a new kind of pretreatment method to increase cellulose dissolution rate in our lab because it was considered to be capable of preventing cornification of cellulose fiber.

In this article, different pretreatment methods were applied to accelerate dissolution of cellulose in ILs, including mechanical milling, ethanol treatment, ultrasonic irradiation and microwave irradiation. Moreover, the effects of different pretreatment conditions on dissolution rate were also investigated, for example, the ethanol treatment time, mechanical milling time, ultrasonic and microwave irradiation power and duration. In addition, ultrasonic and microwave irradiation were also applied during the dissolution process of cellulose samples. We also investigated the influence of various reaction conditions on acceleration of cellulose dissolution and physicochemical properties of regenerated cellulose, such as degree of polymerization (DP) and crystallinity.

2. Experimental

2.1. Materials

Sugarcane bagasse, obtained from a local sugar factory (Guangzhou, China), was first dried in sunlight and then cut into small pieces. The cut SCB was ground and screened to prepare

40-60 mesh size particles. The ground SCB was dried again in a cabinet oven with air circulation for 16 h at 50 °C.

1-Butyl-3-methylimidazolium chloride ([C_4mim]Cl) was provided by the Chemer Chemical Co., Ltd., Hangzhou, China, and used as received. All of other chemicals used were of analytical grade and obtained from Guangzhou Chemical Reagent Factory, China.

2.2. Isolation of cellulose from bagasse

The dried SCB powder was first dewaxed with toluene-ethanol (2:1 v/v) in a Soxhlet apparatus for 6 h. The dewaxed SCB was treated with distilled water (800 mL) for 2 h at 80 °C. Further the insoluble residue was delignified with sodium chlorite at pH 3.8-4.0, adjusted by 10% acetic acid, at 75 °C for 2 h. After filtration, the residue was sequentially washed with distilled water and 95% ethanol, and then dried in a cabinet oven at 50 °C overnight. Finally the holocellulose obtained was extracted with 10% KOH (600 mL) at 20 °C for 10 h to release the hemicelluloses. The residue was collected by filtration and washed thoroughly with distilled water until the filtrate was neutral, then dried in an oven at 50 °C for 16 h.

2.3. Pretreatment of cellulose

2.3.1. Mechanical ball milling

Dry cellulose samples were subjected to ball milling for 6, 12, 18, and 24 h, respectively, on PM 0.4L planet ball mill machine (PM 0.4L Nanjing Chishun Science & Technology Co., Ltd, China). Ball milled samples were dried in an oven at 50 °C for 6 h.

2.3.2. Ethanol treatment

Residue collected from alkali treatment of holocellulose was washed with distilled water and 95% ethanol and then dispersed in 95% ethanol and subjected to mechanical agitation for 4, 6, 8, and 10 h, respectively, to make cellulose adsorb sufficient ethanol. Then the sample was filtrated by Buchner funnel and dried in cabinet oven with air circulation for 8 h at 55 °C

2.3.3. Ultrasonic irradiation

10 g of cellulose sample was suspended in 95% ethanol solvent with mechanical agitation. It was irradiated with ultrasound at the power of 30W, 50W, and 70W for 10, 20, 30 and 40 min, respectively, at room temperature (25 °C). Ultrasonic treated samples were filtrated and dried in oven for 8 h at 55 °C.

2.3.4. Microwave irradiation

10 g of cellulose sample was suspended in 95% ethanol solvent with mechanical agitation and then subjected to microwave oven (XH-100B, Beijing Xianghu Technology Development

Co., Ltd., China) for 200, 400, 600 W microwave irradiation with 20, 30, 40 min at room temperature and 50 °C, respectively. Microwave treated samples were filtrated and dried in oven for 8 h at 55 °C.

2.4. Dissolution and regeneration of cellulose

2.4.1. Dissolution of cellulose without auxilary method

Cellulose or pretreated cellulose sample was added into a flask containing [C_4mim]Cl with a weight ratio of 2% (cellulose to [C_4mim]Cl). The mixture of cellulose/[C_4mim]Cl was stirred continuously in an oil bath at 110 °C with N_2 atmosphere to form clear solution.

2.4.2. Dissolution of cellulose with ultrasonic assisted

2% dried cellulose sample with ethanol pretreatment was added into a flask with [C_4mim]Cl at 110 °C. N_2 was purged continuously and the mixture was stirred. Then the mixture was irradiated at 110 °C for 5, 10, 15, and 20 min, respectively, with ultrasound provided with a horn at sonic power of 20, 30, 40, 50, 60 and 75 W, respectively. After ultrasound irradiation, the mixture was stirred at 110 °C until cellulose was completely dissolved.

2.4.3. Dissolution of cellulose with microwave irradiation

Molten [C_4mim]Cl was mixed with cellulose sample with ethanol pretreatment (2 wt%) in a three-neck flask. The flask was positioned in the microwave reactor (XH-100B) and heated by the microwave irradiation of 300W, 400W, 500W, 600W, and 700W, respectively, to a specific temperature of 85 °C, 90 °C, 95 °C, 100 °C, 105 °C, 110 °C, and 115 °C, respectively, and stirred until a clear cellulose/IL solution was formed. The temperature was real-time monitored by a sensor directly contacted with the reaction medium. When a required temperature was reached it was maintained by impulse microwave irradiation at low heating power. The accuracy of temperature was ±0.2 °C for measurement and ±1°C for controlment.

2.4.4. Regeneration of cellulose

After complete dissolution, the cellulose/IL mixture was slowly poured into 250 mL ethanol with vigorous agitation. The precipitate was collected by filtration and washed thoroughly with ethanol to eliminate [C_4mim]Cl, and then freezen-dried for 24 h. The yield was determined from the regenerated cellulose on the basis of initial freezen-dried measurements. To reduce errors and confirm the results, all experiments were performed at least in duplicate, and the yield represents the average value.

2.5. Characterization of the native, pretreated and regenerated cellulose

2.5.1. Degree of polymerization

The cupriethylene-diamine method (British standard) was used for determining the limiting viscosity number [η] and DP of native, pretreated, and regenerated cellulose. The viscosity

DP was calculated from the following equation, $DP^{0.90} = 1.65\ [\eta]$. Molecular weight (Mw) of cellulose was then calculated from DP by multiplied by 162, the Mw of an AGU.

2.5.2. Fourier transform infrared

The FT-IR spectra of the native and regenerated cellulose were obtained on an FT-IR spectrophotometer (Nicolet 510) using a KBr disc containing 1% finely ground samples. Thirty-two scans were taken for each sample in the range 4000~400 cm^{-1} with a resolution of 2 cm^{-1} in the transmittance mode.

2.5.3. Wide angle X-ray diffraction

The XRD spectra were recorded at room temperature with a Rigaku D/MAX-IIIA X-ray diffractometer (Japan) in the scattering angle range from 5° to 45° with 8°/min scanning speed and a 2θ step interval of 0.02°. Cellulose samples were cut into small pieces and laid on the glass sample holder, analyzed under plateau conditions. Ni-filtered Cu $K\alpha_1$ radiation (λ=0.154 nm) generated at a operating voltage of 40 kV and current of 30 mA was utilized.

2.5.4. Solid-state CP/MAS ^{13}C Nuclear Magnetic Resonance

The solid-state CP/MAS ^{13}C NMR spectra were measured on a Bruker DRX-400 spectrometer with 5mm MAS BBO probe at the frequency of 100 MHz employing both Cross Polarization and Magic Angle Spinning and each experiment was recorded at ambient temperature. Acquisition time was 0.034 s, the delay time 2 s, and the proton 90° pulse time 4.85 μs. Each spectrum was obtained with an accumulation of 5000 scans.

2.5.5. Thermogravimetric analysis

Thermal analysis of the native and regenerated cellulose was performed on thermogravimetric analysis (TGA) on a thermal analyzer (SDT Q500, TA Instrument). The sample weighed between 7 and 11 mg and the scans were run from room temperature to 600 °C at a rate of 10 °C/min under nitrogen flow.

3. Results and discussion

3.1. Dissolution and regeneration of cellulose with different pretreatment methods

It took quite long to completely dissolve cellulose in ILs because of the high crystallinity and tremendous hydrogen bonding in cellulose. When cellulose dissolved in ILs its crystalline structure was totally destroyed thus no bright eyeshot, which represented the crystalline structure of cellulose, could be found under polarizing microscope. Consequently, in this study we employed polarizing microscope to monitor the dissolution process and defined that dissolution of cellulose was completed when polarizing microscope was in black eyeshot. Ethanol treatment, mechanical milling, ultrasonic irradiation and microwave irradiation were applied to pretreat cellulose for enhancement of cellulose dissolution in ILs, and the results are showed in Table 1.

No.	Pretreatment Condition				Dissolution Time	Yield
	Pretreatment	Power	Duration	Temperature		
1	-	-	-	-	11.1 h	1.18
2	Mechanical milling	-	6 h	25 °C	8.9 h	0.95
3	Mechanical milling	-	12 h	25 °C	7.6 h	0.94
4	Mechanical milling	-	18 h	25 °C	6.8 h	0.95
5	Mechanical milling	-	24 h	25 °C	6.4 h	0.94
6	Ethanol treated	-	4 h	25 °C	4.5 h	1.04
7	Ethanol treated	-	6 h	25 °C	3.8 h	1.06
8	Ethanol treated	-	8 h	25 °C	3.1 h	1.05
9	Ethanol treated	-	10 h	25 °C	2.6 h	1.06
10	Ultrasonic irradiation	30 W	20 min	25 °C	2.2 h	1.05
11	Ultrasonic irradiation	30 W	30 min	25 °C	1.5 h	1.06
12	Ultrasonic irradiation	30 W	40 min	25 °C	0.9 h	1.04
13	Ultrasonic irradiation	75 W	20 min	25 °C	1.4 h	1.02
14	Ultrasonic irradiation	75 W	40 min	25 °C	0.8 h	1.01
15	Microwave irradiation	200 W	30 min	25 °C	2.6 h	1.06
16	Microwave irradiation	400 W	30 min	25 °C	2.4 h	1.02
17	Microwave irradiation	600 W	30 min	25 °C	1.6 h	1.04
18	Microwave irradiation	200 W	20 min	50 °C	1.3 h	1.03
19	Microwave irradiation	600 W	20 min	50 °C	0.9 h	0.98

Table 1. Dissolution of cellulose in [C_4mim]Cl pretreated with different methods.

Yield of regenerated cellulose was approximately between 94%~118%. On one hand, cellulose would probably absorb inorganic salt, such as impurity of IL, during dissolution process. In this case yield would increase or even beyond 100%. On the other hand, cellulose was also likely to be degraded and produced oligosaccharide or even monosaccharide which could hardly be recovered. In addition, washing step and filtration process also contributed to the losses. According to the yield of regenerated cellulose, it was reasonable to deduce that cellulose did not degraded significantly during dissolution process in ILs at 110 °C. However, yields of mechanical milling samples were relatively lower, because mechanical milling was able to degrade cellulose molecular and break the macromolecules into small fragments. In the following dissolution and regeneration process cellulose molecules were further destroyed and fragments with relatively lower molecular weights were produced which cannot be precipitated and recovered. The same results were also discovered for samples pretreated with relatively high power of microwave.

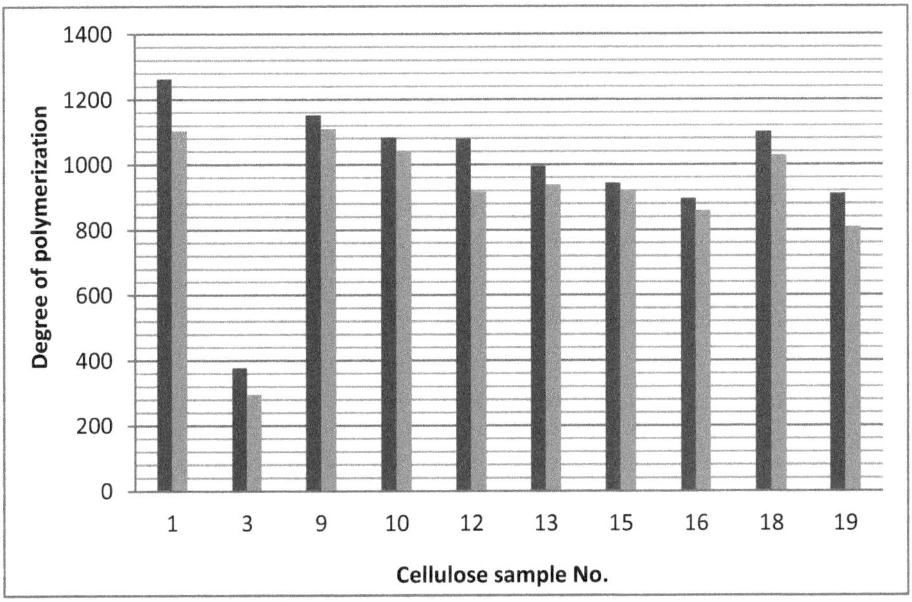

Figure 1. Degree of polymerization of pretreated and regenerated cellulose. Sample No. is in accordance with Table 1. Left bar is DP of pretreated sample, right bar is DP of regenerated sample.

[C_4mim]Cl was one of the most widely studied ILs for cellulose dissolution, but it took 11.1 h to totally dissolve unpretreated cellulose sample at 110 °C, which significantly limited the large-scale utilization of ILs in biomass. Commercial application would be promising if the dissolution of cellulose was improved. As shown in Table 1, dissolution time decreased from 11.1 h to about 6.4~8.9 h when cellulose were subjected to mechanical milling for 6, 12, 18, and 24 h (samples 2~5), indicating the great effect of particle size on cellulose dissolution. However, the fine cellulose sample after mechanical milling would be easily agglomerated during mixing cellulose in [C_4mim]Cl because of its high viscosity. Outside of agglomerates were swollen cellulose macromolecules and inside was dry sample. This kind of particles would seriously prolong the dissolution time and make it difficult to operate during dissolution process. Ethanol pretreated cellulose (samples 6~9) could be dissolved in [C_4mim]Cl at 110 °C within 2.6~4.5 h. During ethanol pretreatment, free water and bound water in cellulose could be replaced by ethanol, which was able to evaporate more quickly during drying, therefore inhibiting high moisture sample become cornification. Cellulose samples after ethanol pretreatment were loosen and porous, which enabled it to open the structure and enlarge superficial area, facilitating penetration of ILs into cellulose. So ethanol pretreatment was one of the effective and feasible methods to improve cellulose dissolution. In addition, dissolution time would be further reduced when ethanol pretreated samples were subjected to ultrasonic and microwave irradiation. It needed only 2.2 h at 110 °C in [C_4mim]Cl to entirely dissolve 2% cellulose pretreated with 30 W ultrasound for 20 min. When the power increased to 75 W, the sample dissolution time was reduced to 0.9 h. Further prolonging irradiation time to 40 min also decreased dissolution time to 0.8 h. The cellulose samples pretreated with microwave irradiation exhibited the similar results. It should be noted that cellulose irradiated with high microwave power at high temperature was more likely to be degraded.

Fig. 1 shows the DP of pretreated and regenerated cellulose. DP of cellulose samples after ethanol treatment, mechanical milling, ultrasonic irradiation and microwave irradiation were obviously lower than the DP of native cellulose. For ultrasonic and microwave pretreatment, enhancement of power and irradiation time further decreased the DP of cellulose, even though the decrement was marginal. For example, DP was reduced from 1083 to 994 when ultrasonic power rose from 30 W to 75 W. In addition, mechanical milling for 12 h significantly brings down DP from 1263 to 378, demonstrating the serious degradation of cellulose during mechanical milling. Moreover, DP of all the samples declined slightly after dissolution in [C_4mim]Cl at 110 °C and regeneration, indicating that cellulose molecules were broken into relatively low molecular fragments.

3.2. Dissolution and regeneration of cellulose assisted with ultrasonic

To further study the influence of ultrasonic irradiation on cellulose dissolution in ILs, we tried to apply ultrasonic irradiation directly during the dissolution process as not a pretreatment but an auxilary method. Results showed that it was more effective to utilize ultrasonic in dissolution process than in pretreatment, as listed in Table 2. Cellulose could be dissolved in [C_4mim]Cl within 1 h when irradiated by 30W ultrasonic for 20 min. With the same irradiation power and duration, it required 2.2 h to entirely dissolve sample when the ultrasonic was employed as pretreatment method. Enhancement of cellulose dissolution with ultrasonic was probably because ultrasound irradiation provided a greater penetration of ionic liquid into cellulose and improved mass transfer. When the dissolving system was irradiated with ultrasound, the extreme condition was probably obtained. It was said that transient temperature of at least 5000 K and pressure up to 1200 bar could be achieved, accompanied by vigorously physical agitation and shock (Kardos & Luche, 2001; Adewuyi, 2001). All of the abovementioned effects provided excellent conditions for cellulose dissolution in IL. This beneficial effect of ultrasound techniques has also been reported in isolation of cell wall polymers (Aliyu & Hepher, 2000). Cellulose dissolution time decreased from 147 (sample 24) to 120 (sample 20), 110 (sample 25) and 100 min (sample 26) with the improvement of ultrasonic power from 20 to 30, 40 and 50 W. However, further improvement of ultrasonic power (samples 26–28) did not result in the decrement in cellulose dissolution time. It seems that increase of power within small range was not as efficient as extension of duration. Compared with sample 3 obtained with one-step ultrasound irradiation, sample 11 with two-step irradiation with 5 min interval needed the same time to totally dissolve sample as well, indicating that the same ultrasound energy obtained from same power and total irradiation duration did not result in different effect on cellulose dissolution.

No.	Irradiation condition		Dissolution Time/min	Yield /%	DP
	Power /W	Duration /min[a]			
20	30	5	150	102.70	799
21	30	10	120	100.32	800
22	30	15	100	101.34	744
23	30	20	60	100.77	682
24	20	10	147	97.86	946
25	40	10	110	99.11	962
26	50	10	100	97.29	1022
27	60	10	100	101.29	-
28	75	10	99	105.19	1019
29	30	10	120	103.63	-

[a] Samples 20-28 were obtained with one-step ultrasound irradiation, while sample 29 was irradiated for 5 min, stirred for 5 min in interval, and then irradiated for another 5 min.

Table 2. Dissolution and regeneration of cellulose in [C_4mim]Cl assisted with ultrasound irradiation

No.	Microwave irradiation condition		Dissolution time/min	Yield/%
	Maximum power/W	Temperature/°C		
30	300	90	50	93.30
31	400	90	20	86.05
32	500	90	17	88.11
33	600	90	13	87.42
34	400	85	37	87.38
35	400	95	14	-
36	400	100	12.5	-
37	400	105	10.5	-
38	400	115	8.5	-

Table 3. Dissolution and regeneration of cellulose in [C_4mim]Cl assisted with microwave irradiation

DP of native and regenerated cellulose is also listed in Table 2. After dissolution and regeneration, DP decreased from 1263 for native cellulose to 1103 for regenerated cellulose sample obtained without ultrasound irradiation. Ultrasound irradiation for 5, 10, 15, and 20 min led to a slight reduction of DP to 799, 800, 744 and 682, respectively. Therefore, direct utilization of ultrasonic in cellulose dissolution degraded cellulose molecular more significantly than using it as pretreatment method. However, the increase of ultrasound power from 20 to 40, 50, and 75 W resulted in an improvement of DP from 946 to 962, 1022, and 1019. The reason for this improvement is not clear and perhaps due to the reassociation and polymerisation of the radicals formed at higher ultrasound power.

3.3. Dissolution and regeneration of cellulose assisted with microwave

Microwave was also applied as a heating method instead of pretreatment to improve cellulose dissolution. Table 3 shows the results of cellulose dissolution and regeneration under microwave heating with different power and temperature. Obviously, cellulose dissolved in [C_4mim]Cl far more quickly with microwave heating than with traditional heating, like in oil bath. Cellulose could be dissolved in [C_4mim]Cl within 50 min at 90 °C under 300 W microwave heating. Dissolution rate would further increase with the increment of microwave power and reaction temperature. Heating with 400W microwave at 115 °C, ethanol treated sample was completely dissolved in only 8.5 min. It should be noted that, however, the yield of regenerated cellulose were about 86~94%, lower than that heating in oil bath. The low yield suggested that degradation of cellulose was more severe under microwave heating than in oil bath.

3.4. FT-IR spectra

FI-IR characterization was employed to measure the changes of the structure of regenerated cellulose samples obtained after pretreatment and dissolution in [C_4mim]Cl assisted with or without ultrasonic and microwave irradiation. Fig. 2 illustrates FT-IR spectra of native cellulose (spectrum 1) and regenerated cellulose sample 17 (spectrum 2, pretreated with 600 W microwave irradiation for 20 min at 25 °C). In these two spectra, the absorbances at 3416, 2916, 1638, 1417, 1323, 1161, 1046, and 895 cm^{-1} are associated with native cellulose (Oh et al., 2005; Liu et al., 2006). Bands in the two spectra were rather similar and no new band was displayed in spectrum of regenerated cellulose, indicating that the structure of both samples are similar and [C_4mim]Cl is direct solvent for cellulose. However, there were still slightly differences in the bands in the spectra of samples dissolved with ultrasonic irradiation and microwave heating (Figures not showed). The absorption from O-H stretching moved to higher wavenumbers from 3407 cm^{-1} to 3423 cm^{-1} after regeneration, indicating the broken hydrogen bonds in cellulose. In addition, the peak for CH_2 bending shifted to a lower frequency from 1431 cm^{-1} to 1419 cm^{-1}, which was a hint for the splitting of hydrogen bonds in C_6-OH (Zhang et al., 2005).

The value of N_O'KI index, proposed by Nelson and O'connor, represents the relative changes of cellulose crystallinity. It can be calculated using the formula N_O'KI $=a_{1372}/a_{2900}$ (Liu & Sun, 2010), in which a_{1372} and a_{2900} are the absorbance intensities of the bands at 1372 and 2900 cm^{-1} for O–H bending and C–H stretching, respectively, in FT-IR spectra of cellulose. In present study, N_O'KI of regenerated cellulose was calculated from FT-IR by the ratio of two absorbances at 2900 cm^{-1} and 1372 cm^{-1}, and the results are listed in Table 2. As seen in Table 2, N_O'KI of regenerated cellulose from ultrasonic dissolution sample increased from 0.930 to 0.963 and 1.084 when the irradiation time increase from 0 to 5 and 10 min. However, it was reduced to 0.975, 0.966 as prolonging the irradiation duration to 15 and 20 min. The N_O'KI of regenerated cellulose changed irregularly with the enhancement of ultrasonic power, and the reason need to be further studied. On the other hand, N_O'KI of regenerated cellulose from microwave heating samples rose with the enhancement of heating power and increase of temperature. It was probably because that cellulose samples with lower molecular fragments were more likely to recrystallize during regeneration. In addition, N_O'KI of the regenerated cellulose from mechanical milling samples dissolved in different temperature (110, 120, 130, and 140 °C) were also measured. Results demonstrated that the crystallinity increase with the enhancement of temperature, similar to that of microwave heating samples.

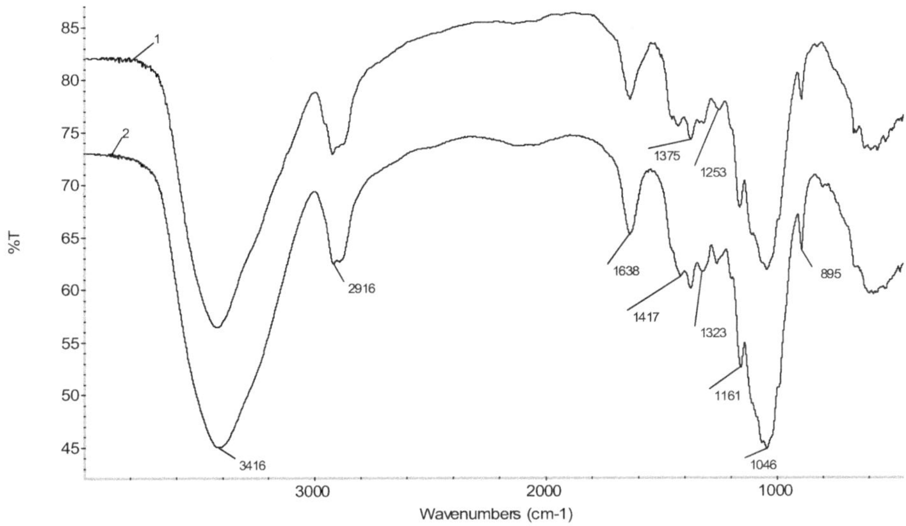

Figure 2. FT-IR spectra of native cellulose (spectrum 1) and regenerated cellulose sample 17 (spectrum 2).

No.	Treatment condition	a_{2900} (C-H)	a_{1372} (O-H)	N_O'KI
9	ultrasonic 0 W, 0 min	0.271	0.252	0.930
20	ultrasonic 30 W, 5 min	0.271	0.252	0.963
21	ultrasonic 30 W, 10 min	0.190	0.183	1.084
22	ultrasonic 30 W, 15 min	0.154	0.167	0.974
23	ultrasonic 30 W, 20 min	0.151	0.147	0.966
30	microwave 300W, 90 °C	0.188	0.191	1.016
31	microwave 400W, 90 °C	0.229	0.240	1.048
33	microwave 600W, 90 °C	0.196	0.217	1.107
34	microwave 400W, 85 °C	0.202	0.200	0.991
35	microwave 400W, 95 °C	0.370	0.383	1.035
36	microwave 400W, 100 °C	0.248	0.257	1.036
38	microwave 400W, 115 °C	0.487	0.528	1.084

Table 4. Relative changes of crystallinity of regenerated cellulose dissolved with ultrasonic and microwave assistance, calculated from FT-IR spectra.

Temperture/°C	a_{2900} (C-H)	a_{1372} (O-H)	N_O'KI
110	0.263	0.243	1.082
120	0.362	0.327	1.107
130	0.363	0.313	1.160
140	0.468	0.396	1.182

Table 5. Relative changes of crystallinity of regenerated samples from dissolution of mechanical milling cellulose at different temperature.

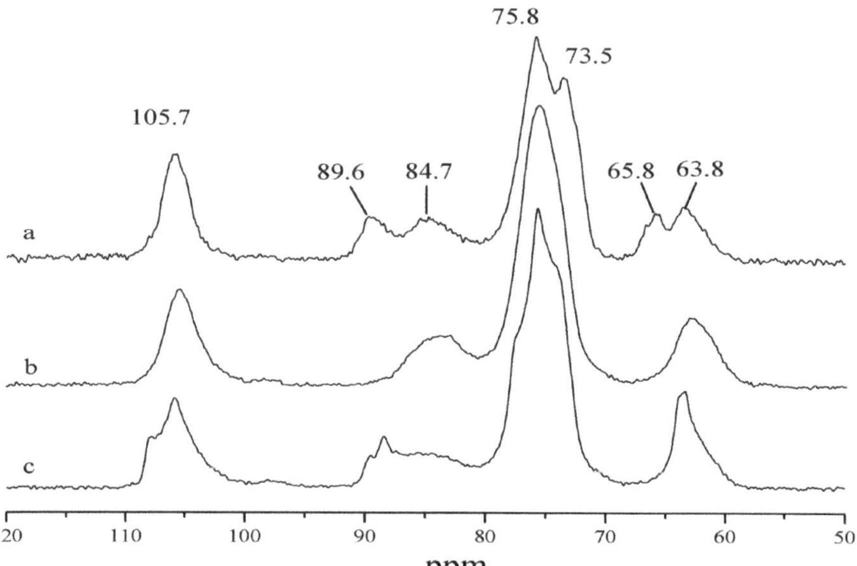

Figure 3. Solid state CP/MAS ^{13}C-NMR spectra of native cellulose (spectrum a) and regenerated cellulose sample 23 (spectrum b) and sample 38 (spectrum c).

3.5. Solid-state CP/MAS ^{13}C NMR spectra

CP/MAS ^{13}C NMR spectroscopy is one of the most effective methods to characterize lignocellulosic materials directly and is able to provide detailed information on solid samples. Fig. 3 shows the CP/MAS ^{13}C-NMR spectra of native cellulose (spectrum a) and regenerated cellulose sample 23 (spectrum b) and sample 38 (spectrum c). The region between 50 and 105 ppm is characterized for the carbons of carbohydrate. The signals from cellulose at 105.7 (C-1), 89.6 and 84.7 (C-4 of crystal cellulose and amorphous cellulose, respectively), 75.8 and 73.5 (C-2, C-3 and C-5), 65.8 and 63.8 (C-6 of crystal cellulose and amorphous cellulose, respectively) are all observed (Focher et al., 2001; Maunu, 2002; Bardet et al., 2002; Liu et al., 2007a; Liu et al., 2009). However, the spectra also exhibited differences at C-4 and C-6 signals. In spectrum b (sample 23), signals for C-4 and C-6 of crystal cellulose disappeared, indicating that the regenerated sample from ultrasonic assisted dissolution cellulose contained mainly amorphous structure. Nonetheless, signal specific of C-4 of amorphous cellulose was degenerated and that of crystal cellulose was reappeared in spectrum c (sample 38). The two signals contributing to C-6 of crystal and amorphous cellulose were merged into one. This spectrum hinted that cellulose macromolecular formed crystalline structure in the regeneration process after dissolve in [C_4mim]Cl with relative high power of microwave heating.

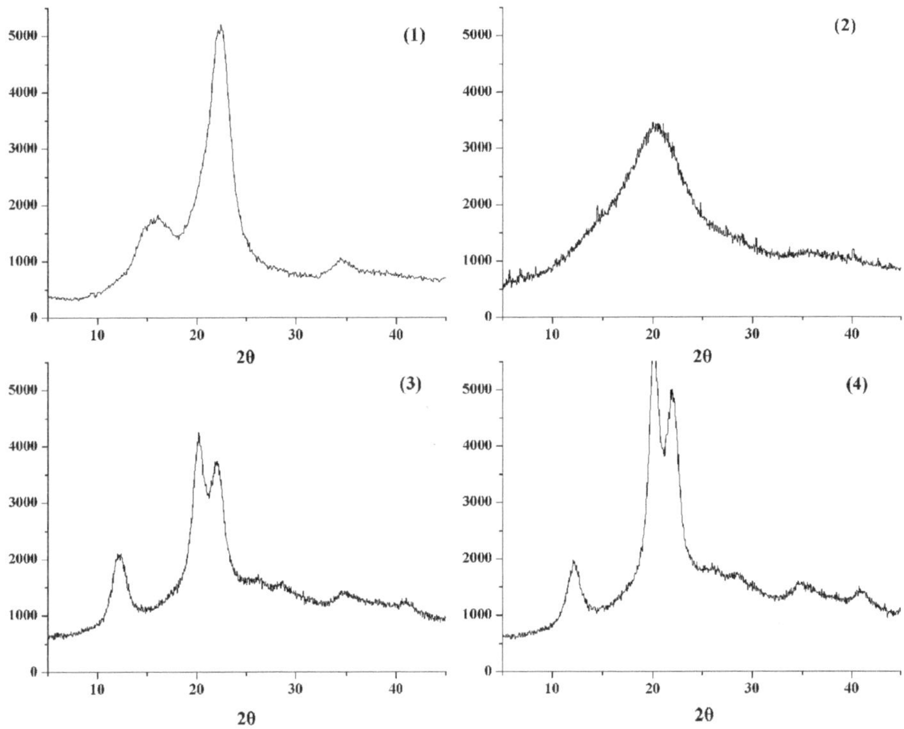

Figure 4. Wide-angle X-ray diffraction curves of native (spectrum 1) and regenerated cellulose sample 20 (spectrum 2), 31 (spectrum 3), and 34 (spectrum4).

3.6. Wide-angle X-ray diffraction

Wide-angle X-ray diffraction is the one of the most valuable technique to measure the crystalline structure of material. In this article wide-angle X-ray diffraction was used to characterize the influence of different pretreatment methods, ultrasonic irradiation and microwave heating on crystalline structure of regenerated cellulose. Fig. 4 exhibits the wide-angle X-ray diffractogram of native (spectrum 1) and regenerated cellulose sample 20 (spectrum 2), 31 (spectrum 3), and 34 (spectrum4). Spectrum 1 gave strong crystalline peak at 22.3° for (002) crystal plane and at 16.1° merged from peaks of (110) and ($1\bar{1}0$) crystal plane, indicating a cellulose I crystalline structure for native cellulose (Focher et al., 2001; Oh et al., 2005). After dissolution with ultrasonic irradiation and regeneration, cellulose sample comprised mainly amorphous structure and little cellulose II crystalline structure, suggested by a broad peak at 19.8° in the diffratogram (Isogai et al., 1989), as shown in spectrum 2. However, spectrum of sample 31 and 34 displayed three obvious crystalline peaks at 12.2°, 20.2° and 22.0°. They were corresponding to (110), ($1\bar{1}0$) and (002) planes of cellulose II crytal (Zuluaga et al., 2009), respectively. Evidently, samples with microwave heating after regeneration possessed more cellulose II crystalline structure and higher crystallinity, which was in well accordance with the results of NMR spectra (Fig. 3).

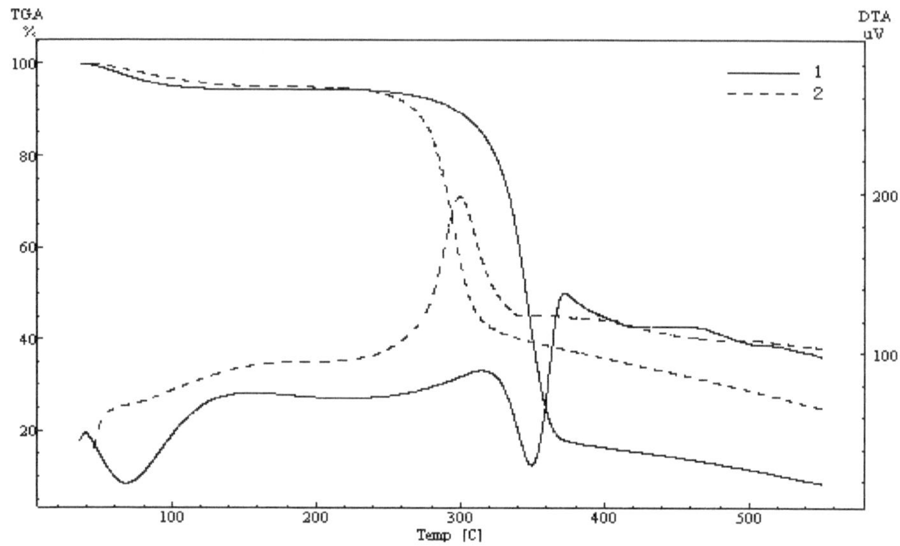

Figure 5. TGA/DTA curves of native cellulose (spectrum 1) and regenerated cellulose sample 17 (spectrum 2).

3.7. Thermal analysis

Fig. 5 shows the TGA and DTG curves of native cellulose (spectrum 1) and regenerated cellulose sample 17 (spectrum 2). The small initial drops occurring near 100 °C in both cases are due to the evaporation of retained moisture. The native cellulose starts to decompose at 228 °C, whereas the regenerated cellulose samples 17 begin to decompose at 211 °C. The decomposition temperature at 50% weight loss occurs at 342 °C for native cellulose and 330 °C for regenerated cellulose sample 17. These decreasing trends of decomposition temperature implied that the thermal stability of regenerated cellulose was lower than that of native cellulose, which was probably due to partial hydrolysis and degradation of macromolecular cellulose during pretreatment with microwave irradiation and dissolution. Similar results have been reported during the dissolution and modification of cellulose in ionic liquid (Swatloski et al., 2002; Liu et al., 2007b; Li et al., 2011). Pyrolysis residues are primarily indecomposable inorganic salts. As seen in Fig. 5, the residual char yield of native cellulose was lower than that of regenerated cellulose sample 17. The higher pyrolysis residues of regenerated cellulose suggested that some impurities from ILs probably involved in cellulose in the dissolution process, which might be the mind reason contributed to the over 100% yield of some samples.

4. Conclusion

Ethanol treatment, mechanical milling, ultrasonic and microwave irradiation were demonstrated to be effective pretreatment methods to improve cellulose dissolution in [C_4mim]Cl. Cellulose sample with 75 W ultrasonic pretreated for 40 min could be dissolved within 0.8 h. Moreover, all the pretreatment methods except mechanical milling did not seriously degrade the cellulose macromolecular. DP of the native cellulose was 1263 and DP of the pretreated samples ranged from about 900 to 1150. DP of regenerated samples were between 800 and 1110. In addition, ultrasonic irradiation and microwave heating were tested to be more effective when applied during the dissolution process. The total dissolution time of the cellulose sample irradiated with 30W ultrasonic for 20 min was 60 min. Under 400 W microwave heating at 115 °C, it took only 8.5 min to dissolve cellulose sample in [C_4mim]Cl. The application of ultrasonic and microwave heating, especially the microwave heating, broke the cellulose macromolecular at certain degree. Regenerated samples from ultrasonic irradiated cellulose were major amorphous cellulose whereas samples from microwave heating cellulose comprised cellulose II crystalline structure.

Author details

Wu Lan, Chuan-Fu Liu* and Feng-Xia Yue
State Key Laboratory of Pulp and Paper Engineering, South China University of Technology, Guangzhou, PR China

Run-Cang Sun
State Key Laboratory of Pulp and Paper Engineering, South China University of Technology, Guangzhou, PR China

College of Material Science and Technology, Beijing Forestry University, Beijing, PR China

Acknowledgement

The authors are grateful for the financial support of this research from the National Natural Science Foundation of China (30972325 and 31170550), Program for New Century Excellent Talents in University (NCET-11-0154), National Basic Research Program of China (2010CB732201), and the 111 Project.

* Corresponding Author

5. References

Adewuyi, Y.G. (2001). Sonochemistry: Environmental science and engineering applications, *Industrial & Engineering Chemistry Research* 40(22): 4681-4715.

Aliyu, M. & Hepher, M.J. (2000). Effects of ultrasound energy on degradation of cellulose material, *Ultrasonics Sonochemistry* 7(4): 265-268.

Bardet, M., Foray, M.F. & Tran, Q.K. (2002). High-resolution solid-state cpmas nmr study of archaeological woods, *Analytical Chemistry* 74(17): 4386-4390.

Cao, Y., Wu, J., Zhang, J., Li, H.Q., Zhang, Y. & He, J.S. (2009). Room temperature ionic liquids (rtils): A new and versatile platform for cellulose processing and derivatization, *Chemical Engineering Journal* 147(1): 13-21.

Cao, Y., Zhang, J., He, J.S., Li, H.Q. & Zhang, Y. (2010). Homogeneous acetylation of cellulose at relatively high concentrations in an ionic liquid, *Chinese Journal of Chemical Engineering* 18(3): 515-522.

Evtuguin, D.V., Neto, C.P., Silva, A.M.S., Domingues, P.M., Amado, F.M.L., Robert, D. & Faix, O. (2001). Comprehensive study on the chemical structure of dioxane lignin from plantation eucalyptus globulus wood, *Journal of Agricultural and Food Chemistry* 49(9): 4252-4261.

Feng, L. & Chen, Z.I. (2008). Research progress on dissolution and functional modification of cellulose in ionic liquids, *Journal of Molecular Liquids* 142(1-3): 1-5.

Focher, B., Palma, M.T., Canetti, M., Torri, G., Cosentino, C. & Gastaldi, G. (2001). Structural differences between non-wood plant celluloses: Evidence from solid state nmr, vibrational spectroscopy and x-ray diffractometry, *Industrial Crops and Products* 13(3): 193-208.

Forsyth, S.A., MacFarlane, D.R., Thomson, R.J. & von Itzstein, M. (2002). Rapid, clean, and mild o-acetylation of alcohols and carbohydrates in an ionic liquid, *Chemical Communications* (7): 714-715.

Fujimoto, A., Matsumoto, Y., Chang, H.M. & Meshitsuka, G. (2005). Quantitative evaluation of milling effects on lignin structure during the isolation process of milled wood lignin, *Journal of Wood Science* 51(1): 89-91.

Gadhe, J.B., Gupta, R.B. & Elder, T. (2006). Surface modification of lignocellulosic fibers using high-frequency ultrasound, *Cellulose* 13(1): 9-22.

Heinze, T., Schwikal, K. & Barthel, S. (2005). Ionic liquids as reaction medium in cellulose functionalization, *Macromolecular Bioscience* 5(6): 520-525.

Hermanutz, F., Gaehr, F., Uerdingen, E., Meister, F. & Kosan, B. (2008). New developments in dissolving and processing of cellulose in ionic liquids, *Macromolecular Symposia* 262: 23-27.

Hou, A.Q., Wang, X.J. & Wu, L.H. (2008). Effect of microwave irradiation on the physical properties and morphological structures of cotton cellulose, *Carbohydrate Polymers* 74(4): 934-937.

Isogai, A., Usuda, M., Kato, T., Uryu, T. & Atalla, R.H. (1989). Solid-state cp/mas carbon-13 nmr study of cellulose polymorphs, *Macromolecules* 22(7): 3168-3172.

Kardos, N. & Luche, J.L. (2001). Sonochemistry of carbohydrate compounds, *Carbohydrate Research* 332(2): 115-131.

Li, W.Y., Wu, L., Chen, D., Liu, C.F. & Sun, R.C. (2011). DMAP-catalyzed phthalylation of cellulose with phthalic anhydride in bmim cl, *BioResources* 6(3): 2375-2385.

Liu, C.F., Sun, R.C., Zhang, A.P., Ren, J.L. & Geng, Z.C. (2006). Structural and thermal characterization of sugarcane bagasse cellulose succinates prepared in ionic liquid, *Polymer Degradation and Stability* 91(12): 3040-3047.

Liu, C.F., Sun, R.C., Qin, M.H., Zhang, A.P., Ren, J.L., Xu, F., Ye, J. & Wu, S.B. (2007a). Chemical modification of ultrasound-pretreated sugarcane bagasse with maleic anhydride, *Industrial Crops and Products* 26(2): 212-219.

Liu, C.F., Sun, R.C., Zhang, A.P., Ren, J.L., Wang, X.A., Qin, M.H., Chao, Z.N. & Luo, W. (2007b). Homogeneous modification of sugarcane bagasse cellulose with succinic anhydride using a ionic liquid as reaction medium, *Carbohydrate Research* 342(7): 919-926.

Liu, C.F., Zhang, A.P., Li, W.Y., Yue, F.X. & Sun, R.C. (2009). Homogeneous modification of cellulose in ionic liquid with succinic anhydride using N-bromosuccinimide as a catalyst, *Journal of Agricultural and Food Chemistry* 57(5): 1814-1820.

Liu, C.F. & Sun, R.C. (2010). Chapter 5. Cellulose. *in*: Sun, R.-C. (Ed.) *Cereal straw as a resource for sustainable biomaterials and biofuels: Chemistry, extractives, lignins, hemicelluloses and cellulose.* Elsevier, Amsterdam, the Netherlands, pp. 131-167.

Ma, H., Liu, W.W., Chen, X., Wu, Y.J. & Yu, Z.L. (2009). Enhanced enzymatic saccharification of rice straw by microwave pretreatment, *Bioresource Technology* 100(3): 1279-1284.

Maunu, S.L. (2002). Nmr studies of wood and wood products, *Progress in Nuclear Magnetic Resonance Spectroscopy* 40(2): 151-174.

Mikkola, J.P., Kirilin, A., Tuuf, J.C., Pranovich, A., Holmbom, B., Kustov, L.M., Murzin, D.Y. & Salmi, T. (2007). Ultrasound enhancement of cellulose processing in ionic liquids: From dissolution towards functionalization, *Green Chemistry* 9(11): 1229-1237.

Oh, S.Y., Yoo, D.I., Shin, Y., Kim, H.C., Kim, H.Y., Chung, Y.S., Park, W.H. & Youk, J.H. (2005). Crystalline structure analysis of cellulose treated with sodium hydroxide and carbon dioxide by means of x-ray diffraction and ftir spectroscopy, *Carbohydrate Research* 340(15): 2376-2391.

Potthast, A., Rosenau, T. & Kosma, P. (2006). Analysis of oxidized functionalities in cellulose, *Polysaccharides II* 205: 1-48.

Ramier, J., Renard, E. & Grande, D. (2012). Microwave-assisted synthesis and characterization of biodegradable block copolyesters based on poly(3-hydroxyalkanoate)s and poly(d,l-lactide), *Journal of Polymer Science Part A-polymer Chemistry* 50(7): 1445-1455.

Satge, C., Verneuil, B., Branland, P., Granet, R., Krausz, P., Rozier, J. & Petit, C. (2002). Rapid homogeneous esterification of cellulose induced by microwave irradiation, *Carbohydrate Polymers* 49(3): 373-376.

Swatloski, R.P., Spear, S.K., Holbrey, J.D. & Rogers, R.D. (2002). Dissolution of cellose with ionic liquids, *Journal of the American Chemical Society* 124(18): 4974-4975.

Tang, A.M., Zhang, H.W., Chen, G., Xie, G.H. & Liang, W.Z. (2005). Influence of ultrasound treatment on accessibility and regioselective oxidation reactivity of cellulose, *Ultrasonics Sonochemistry* 12(6): 467-472.

Turner, M.B., Spear, S.K., Huddleston, J.G., Holbrey, J.D. & Rogers, R.D. (2003). Ionic liquid salt-induced inactivation and unfolding of cellulase from trichoderma reesei, *Green Chemistry* 5(4): 443-447.

Zhang, H., Wu, J., Zhang, J. & He, J.S. (2005). 1-allyl-3-methylimidazolium chloride room temperature ionic liquid: A new and powerful nonderivatizing solvent for cellulose, *Macromolecules* 38(20): 8272-8277.

Zuluaga, R., Putaux, J.L., Cruz, J., Velez, J., Mondragon, I. & Ganan, P. (2009). Cellulose microfibrils from banana rachis: Effect of alkaline treatments on structural and morphological features, *Carbohydrate Polymers* 76(1): 51-59.

Chapter 8

Electric Properties of Carboxymethyl Cellulose

Alexandar Metodiev Zhivkov

Additional information is available at the end of the chapter

1. Introduction

The linear polymer carboxymethyl cellulose (CMC) as a polyelectrolyte is an object of consideration in this review. The emphasis is on the electric properties of CMC both as a free chain in solution and adsorbed on the solid surface. A special attention is paid to the mobility of counterions, electrostatically associated with the CMC polyelectrolyte chain.

2. Chemical structure

2.1. Molecular mass

Carboxymethyl cellulose $[C_6H_7O(OH)_{3-x}(OCH_2COOH)_x]_n$ is a derivative of the regenerated cellulose $[C_6H_{10}O_5]_n$ with hydroxy-acetic acid (hydroxy ethanoic acid) $CH_2(OH)COOH$ or sodium monochloroacetate $ClCH_2COONa$. The CMC backbone consist of D-glucose residues linked by β-1,4-linkage. The molecular mass of one glucose unit in CMC chain is $m_{CMC} = 146.14 + 75.04{\cdot}x$, where $x \leq 3$ is the degree of substitution (DS). As a rule CMC is produced as sodium salt $[C_6H_7O(OH)_{3-x}(OCH_2COONa)_x]_n$ with DS = 0.4–1.2, then the molecular mass per unit is $m_{NaCMC} = 146.14 + 97.03{\cdot}x \approx 185–263$ g/mol.

2.2. Structure of the chain backbone

As derivative of the cellulose (poly-β-D-glucose) [1,2,3] CMC has inherited its main structural peculiarities: (a) rigidity of the glucose units (6-atoms rings in "armchair" conformation); (b) an almost fully extended conformation owing to impossibility for rotation round the C–O–C bonds between the glucose residues because of strong steric limitations (a consequence of the β-configuration at C-1 atom in 1,4-linkage); and (c) orientation of the bigger constituent groups (–OH, $–CH_2OH$ and $–OCH_2COOH$) in the equatorial plane out of the saccharic ring. The cross-sectional dimension of the CMC chain is a sum of the diameter of the glucose ring (0.5 nm) and the size of the carboxymethyl group (0.4 nm, situated on the both side of the chain); the effective diameter of the chain is rather higher because the glucose residues are oriented not line in line but somewhat tilted along the backbone line [4]. The replacement of the H-atom of –OH-group with $-CH_2COO^-$ causes a loss of the possibility of parallel association of CMC chains in sheet-like structures (a characteristic of the cellulose structure leading to supramolecular organization in microfibrils [5]) because of both steric hindrance and electrostatic repulsion but that does not exclude the possibility of intra- and intermolecular interaction by hydrogen bonds between the unsubstituted –OH groups.

In the literature the polymer unit of CMC chain is denoted as consisting of one or two glucose rings following the manner applying to the cellulose. The one unit way is chemically well-founded according the cellulose contents because all the glucose units are chemically identical (except those on the two ends of the chain). According to the conformation of the cellulose chain a more thought-out manner is the two glucose residues unit because this way allows distinguishing the cellulose (poly-β-D-glucose) and the amylose (poly-α-D-glucose), both linear chains of chemically identical glucose residues. The conformation of the polymer chains of these two polysaccharides differs drastically owing to the configuration at the anomeric C-1 atom: the β-1,4-linkage in the cellulose leads to inversed orientation of every second glucose ring according the previous one and to sterically conditioned full extended conformation because of impossible rotation round the C–O–C bonds between C-1 and C-4 atoms of the neighbour rings. In the case of the amylose the α-configuration at C-1 atom leads to uniform orientation of the glucose rings and to impossibility of extended conformation; due to the α-1,4-linkage amylose chain has a tendency to spiral conformation [5]. Since the linear charge density of the CMC chain is determined by the number of $-COO^-$ groups per glucose unit independently of its orientation, in this chapter the one glucose residue is accepted to denote the monomeric unit.

2.3. Distribution the carboxylic groups along the chain

The $-CH_2COOH$ group can be attached to every of the three hydroxyl groups of the cellulose monomer unit, so theoretically the degree of substitution (DS) can reach 3, but usually DS does not exceed 2. The distribution of $-CH_2COOH$ groups in the glucose unit and along the CMC chain is accepted to be random in the model [4] proposed to give the monomer composition of a sample with known DS. The model is based on the supposition that the substitute reaction occurs at random, i.e. each of the three –OH groups (at C-2, C-3 and C-6 position) of every monomer unit can be substituted in equal probability independently on the presence of another constituent in the same glucose unit. The experimental investigations [6,7,8] of the mole fraction of substituted monomers of CMC confirm the hypothesis for random distribution of the constituents. For example, at DS = 1.3 the mole fractions are about 0.2, 0.4, 0.3 and 0.1 for unsubstituted, mono-, di- and trisubstituted, respectively. On the base of these results it can be concluded that the $-CH_2COO^-$ groups are also randomly distributed along the CMC chain and practically there are no regions containing more than one or two uncharged monomer units when DS is higher than 1–2. The absence of long uncharged segments allows accepting that the negative charges are almost evenly distributed along the CMC chain at high linear charge density (high DS and high degree of dissociation); that is important condition for applying the model of uniformly charged cylinder to CMC chain.

2.4. Contour length

The distance between two glucose units in CMC chain is 0.515 nm [4]. The contour length L_c (the length of the chain backbone if it is fully stretched but without deformation of the valence angles and bonds) is determined as a product of the number of monomers n and the length of one unit: $L_c = 0.515 \cdot n$ (nm).

For example, at DS = 1.20 CMC chain with M_{NaCMC} = 250 kDa (sodium salt) has 952 glucose residues and counter length L_c = 490 nm.

3. Electric properties of CMC in aqueous medium

In this section the characteristics of CMC as a free polyelectrolyte chain are overlooked, i.e. the chain has conformational freedom that does not or depends on the presence of other macromolecules, respectively in diluted solution or semi-concentrated ones. The chain is immersed in salt-free medium or in electrolyte with valency z. In the first case the counterions originate from the ionizable group of the chain: H^+ or Na^+ (in the case of CMC or its sodium salt NaCMC). A part of counterions can be electrostatically adsorbed losing their freedom; the rest are scattered by the thermal energy kT and are randomly distributed around the chain forming an ions cloud.

3.1. Dissociation of the carboxylic groups in CMC chain

3.1.1. Apparent dissociation constant

The effective dissociation constant K_a (respectively $pK_a = -\log K_a$) of a weak polyelectrolyte depends on the neighbour electric charges in the chain and in the medium. The presence of other negative charges in the vicinity of a given COO^- group leads to increasing of the electric field intensity and the local concentration of H_3O^+ ions. As a result the probability of protonation of the COO^- group increases; respectively the equilibrium of the reaction $COOH \leftrightarrow COO^- + H^+$ is shifted to the non-dissociated form. That is why the dissociation capability of the ionizable groups in the polyelectrolyte chain can not be characterized by a constant like in the case of a simple (monomeric) acid (pK_a = pH at $\alpha = 1/2$). The apparent (effective) dissociation "constant" $pK_a = f(x,\alpha,\mu)$ of COOH groups in CMC chain is a function of the degree of substitution x, the degree of dissociation α (determined by pH of the medium), and the ionic strength μ; the first two factors determine the linear charge density of the chain.

3.1.2. Degree of dissociation

The degree of ionization (the fraction of the dissociated acid groups) α of a weak polyelectrolyte pK_a depends on the apparent dissociation constant $K_a(\alpha)$ at given α and the H^+ concentration:

$$\log[\alpha / (1-\alpha)] = \mathrm{pH} - \mathrm{pK}_a. \quad (1)$$

K_a can be represented as a sum of the intrinsic dissociation K_{a0} and the electrostatic term ΔK_a (owing to the influence of the neighbour electric charges):

$$\mathrm{pK}_a = \mathrm{pK}_{a0} + \Delta\mathrm{pK}_a, \quad (2)$$

At equilibrium the shift ΔpK_a is due to an alteration of the electrostatic free energy G_n of the polyacid at variation of n negatively charged groups [9]:

$$\Delta\mathrm{pK}_a = \left(0.434 / kT\right)(dG_n / dn). \quad (3)$$

3.1.3. *Experimental data*

Potentiometric titration of CMC aqueous solutions shows that the degree of acid dissociation (ionization) α increases in the range of pH 2–6. The fact is determined by the moderate value of the dissociation constant K_a of the –COOH groups and allows defining CMC as a weak polyelectrolyte.

An addition of a low molecular salt to the CMC solution leads to increasing of the degree of dissociation α because of the screening of the electric field around the polyelectrolyte chain; the influence of the salt increases with α and the degree of substitution x, i.e. with the linear charge density. For CMC with DS $x = 0.75$ the degree of dissociation increases from $\alpha = 0.23$ to 0.85 in the range pH 3.0 – 5.0 in 0.01 mol/L NaCl; at increasing of NaCl concentration up to 1 mol/L the dissociation increases from $\alpha = 0.43$ to 0.98 in the same pH diapason [8]. The increasing of the ionic strength from 0.01 to 1 mol/L shifts pK_a form 3.8 to 3.2; i.e. the dissociation at $\alpha = 0.5$ is quadruplicately increased due to the electrostatic screening by Na^+ ions.

3.1.4. *Intrinsic dissociation constant*

For aqueous solutions of CMC the experimental dependence of pK_a on the degree of dissociation α is linear with a slope $\Delta pK_a/\Delta\alpha$:

$$\mathrm{pK}_a = \mathrm{pK}_{a0} + (\Delta\mathrm{pK}_a / \Delta\alpha)\alpha. \quad (4)$$

That allows obtaining the intrinsic dissociation constant pK_{a0} by extrapolation to $\alpha = 0$ on the plot $pK_a = f(\alpha)$. Thus the extrapolated values are: $pK_{a0} \approx 3.3$ in salt-free solution of CMC [10]; $K_{a0} \approx 3.2$ in NaCl solution with concentration 0.01, 0.05 and 0.1 mol/L; $K_{a0} \approx 3.0$ in 0.5 mol/L and $K_{a0} \approx 2.95$ in 1 mol/L NaCl [8].

It is interesting to compare the K_{a0} of CMC with K_a of the acetic acid in diluted aqueous solution (when the electrostatic interactions between the next $-COO^-$ groups are negligible because of the big distance between acid molecules just in the opposite to the case of the polyelectrolyte chain). The thermodynamic acidity constant K_a of CH_3COOH in aqueous medium is equal to $1.754 \cdot 10^{-5}$ (calculated accounting the activity instead the concentration), thus $pK_a = 4.756$ (the negative logarithm of K_a). So, the intrinsic dissociation constant K_{a0} of the carboxylic groups of CMC (at $pK_{a0} \approx 3.0$ in 0.5 mol/L NaCl) is about 60 times higher than K_a of the acetic acid –COOH group. This observation is opposite to the expectation ($K_{a0} = K_a$ at $\alpha \to 0$ and $K_{a0} < K_a$ when $\alpha > 0$) if taking into account only the electrostatic interactions between the neighbour $-COO^-$ groups in the CMC chain. The difference in K_{a0} and K_a can be explained with the different electronic structure of the two chemical substances. That suggest that the intrinsic dissociation constant of the three –COOH groups are different, especially of that directly attached to the glucose rings at C-2 and C-3 atoms and that at C-6 position.

3.1.5. Theoretical approach

The pH-induced shift of the apparent dissociation constant ΔK_a determined by the change of the electrostatic free energy G_n of the polyion can be calculated by the work needed to display the elementary positive charge (proton) q_e from the polyacid surface (where the electric potential is ψ_0) to a infinite distance in the bulk (where $\psi_\infty = 0$ by definition). Then (dG_n/dn) in Eq. (3) effectively equals to $q_e\psi_0$.

One of the models used for this calculations is the that of uniformly charged cylinder with radii $a<b$ ($2a$ is the mean diameter of the polyelectrolyte chain and $2(b-a)$ is the size of the adsorbed counterions). The ΔK_a is proportional to the alteration of the electric potential owing to both the adsorbed (condensed) and diffuse counterions of all kinds (Na^+ in the case of NaCMC), represented by the potentials ψ_0 of the inner and ψ_1 of the external cylinder. The potential ψ_0 of the inner cylinder is determined by the degree of substitution x and the degree of dissociation α; the ψ_1 potential can be calculated from the surface charge density of the inner cylinder and the capacity of a cylinder capacitor with radii $b>a$ and interelectrode distance $b-a$. Then the shift ΔpK_a can be represented as a sum of two terms determined by the charge density of the inner cylinder and the potential on the external one [11]:

$$\Delta \mathrm{pK_a} = (0.217e^2 / \pi\varepsilon_0\varepsilon_r kT)\ln(b/a)x\alpha + \left(0.434e/kT\right)\psi_1, \tag{5}$$

where e is the elementary charge (proton); $\varepsilon_0\varepsilon_r$ is the absolute dielectric permittivity (electric constant ε_0 and the relative dielectric permittivity ε_r); kT is the thermal energy at absolute temperature T (k is the Boltzmann constant).

In the case of NaCMC the Eq. (5) takes into account the two kinds of counterions: H^+ ions determining the potential ψ_0 of the inner cylinder and adsorbed Na^+ cations determining ψ_1 of the external one. By fitting the experimental data for the degree of dissociation $\alpha = f(\mathrm{pH})$ (obtained by titration of CMC solution with NaOH) with the parameters of the model of uniformly charged cylinder in Ref. [8] are found the intrinsic dissociation constant $K_{a0} = 3.2$ (at 0.01–0.1 mol/L NaCl) and radius of the inner cylinder $a \approx 1$ nm; the last quantity increases from $a = 0.95$ to 1.15 nm when the degree of substitution grows from $x = 0.75$ to 1.25. So, the calculated diameter $2a \approx 2$ nm is two times higher than that of the molecular cross-section of CMC chain if taken as a sum of the size the glucose ring (0.5 nm) and one attached COOH group (0.4 nm).

3.2. Linear charge density of CMC chain

The dissociation of carboxylic groups in aqueous medium determines the electric properties of CMC as polyelectrolyte. The number of COO^- groups is determined by the degree of substitution (DS) and the degree of dissociation α. The linear charge density b^{-1} is equal to the ratio of ionized groups' number to the contour length L_c. For example, CMC chain with DS = 1.20 and M = 250 kg/mol (sodium salt) has 952 residues, L_c = 490 nm and mean distance between two neighbour COO^- groups b = 0.429 nm (b^{-1} = 2.33 charges/nm) at α = 1. For CMC with DS = 1.00 (b = 0.515 nm) the charge density is b^{-1} = 1.94 nm^{-1} at full dissociation or nearly one charge per nanometer at α = 1/2.

So defined charge density is determined by the chemical structure of CMC chain and could be designate as intrinsic linear charge density; it takes into account only chain's own charges (COO^- groups). The presence of adsorbed (condensed) counterions (Na^+ in the case of sodium salt of CMC) decreases the effective charge density. The value of α obtained by potentiometric titration reflects all kinds of electrostatic interactions influencing on ΔK_a including that of the condensed Na^+ counterions and the alteration of H^+ concentration in the vicinity of the polyelectrolyte chain. So, the using of experimental values of α gives the intrinsic linear charge density.

3.3. Counterions condensation on CMC chain

3.3.1. *Condensation phenomenon and theoretical approaches*

When a polyelectrolyte chain is immersed in aqueous solution of electrolyte the electrostatic field around the chain increases the local concentration of the diffuse counterions and decreases the one of the co-ions; the difference between the two concentrations is maximal at the chain surface and diminishes quasi-exponentially down to zero in the bulk. The distribution of the local concentration of the free ions is determined by the ratio between the electrostatic and the thermal energies. When this ratio increases above a critical value then part of counterions are electrostatically adsorbed on the chain forming temporary ion pairs with the electric charges of the polymer; the phenomenon is known as counterion condensation.

The distribution of the diffuse and condensed counterion considered as loosely and strongly associated with the polyion is a subject of several polyelectrolyte theories developed on the base of a few different models [12]; the most popular are: Katchalsky's cylindrical cell model [13] and the Oosawa-Manning two state model [14]. In the both models the polyelectrolyte chain is considered as an infinite impenetrable cylinder on which surface the charges are smeared as a non-discrete quantity and uniformly distributed along the cylinder axis.

The main imperfections of the cylindrical models are: (1) the real polyelectrolyte chain can be assumed as strait only for a rigid fragment shorter than the persistent length l_p and when the l_p is much longer than the chain diameter (the semi-flexible chains like CMC have a form of curved rod when their contour length is commensurable with l_p but the longer chains takes conformation of random coil); (2) the end-effects are neglected; that assumption is defensible only for enough long chains having big number of charges; (3) the charged groups are non-uniformly distributed along the chain backbone at incomplete degree of dissociation ($\alpha<1$) and/or at irregular distribution of the dissociable groups as in the case of CMC (see paragraph 2.3); and (4) the chain's charges can be considered as non-discrete quantity uniformly smeared on the cylinder only at a some distance from the surface (where the electric field is less inhomogeneous); at a shorter distance the charges should be considered as discrete and distributed according to the chain structure.

3.3.2. *Cylinder cell model*

In this model [15,16,17] a polyelectrolyte cylinder with radius a is localized in the center of an infinitely long cylinder cell with radius R; the distance ($R-a$) is divided into two zones. In the inner zone the centers of the adsorbed (condensed) counterions are uniformly distributed on a coaxial cylinder with radius b; the difference in radii ($b-a$) of the two cylinders is equal to the ion radius. The electric potential ψ_0 (on the inner cylinder surface) is determined by the electric charges of the polyelectrolyte chain themselves. The potential in the inner zone (at distance $b-a$) follows lineally as in a homogeneous solid state body; that allows to consider the polyelectrolyte chain as a coaxial cylindrical condenser. The presence of two kinds of counterions with different radius and valence can be considered by introducing an additional cylinder in the inner zone [18,19]. The potential ψ_1 of the cylinder with radius b is determined by the total chain's charge including the adsorbed ions. In the external zone ($R-b$) the diffuse counterions are uniformly scattered according the balance between the electrostatic $e\psi_1$ and thermal energy kT; their size is assumed to be negligibly small (point charges). The potential in the external zone follows quasi-exponentially (in the case of symmetrical electrolyte in the medium) according the non-liner Poisson-Boltzmann equation in its cylinder form [20,21].

So, the cylinder cell model is a one-dimensional analog of the two-dimensional electric double layer (EDL) used in the colloid chemistry to describe the electric properties of a flat solid/liquid interface: the two inner cylinders are analogs of the solid surface and Stern layer (the EDL dense part). The difference is that at low potential ($\psi_1 < 25$ mV in 1:1 electrolyte) $\psi(x)$ follows exponentially with the distance x in the case on flat surface while in the case of cylinder the potential distribution $\psi(x)$ is quasi-exponential. The main advantage of the cylindrical cell model is that it gives the radial counterions distribution around the polyelectrolyte chain; that is especially important in salt solutions. But its application is difficult because in the most cases it requires numerical solution of the non-liner Poisson-Boltzmann equation; the exact analytical solution is possible only for an infinite cylinder.

The application of the cylindrical cell model to NaCMC takes into account the two kinds of counterions (H^+ and Na^+) in their two states (adsorbed and diffuse): H^+ ions determine the potentials of the inner (ψ_0) cylinder and Na^+ cations on the external (ψ_1) one; and the both cations are in competition in the diffuse cloud determining the potential $\psi(x)$ together with the OH^- anions. The difference between the two counterions is principal at the surface: H^+ are potential determining for COO^- groups but Na^+ are adsorbed (condensed); in the diffuse cloud this difference disappear: H_3O^+ and Na^+ are in competition determined by their concentration in the bulk.

3.3.3. *Bjerrum length*

The dependence of electrostatic energy on the distance between two ions (a negative and positive) leads to condition for equality of the electrostatic and thermal energies; the critical distance is called Bjerrum length:

$$l_B = e^2 / \ 4\pi\varepsilon_0\varepsilon kT \quad (6)$$

where e is the elementary charge (electron); $\varepsilon_0\varepsilon$ is dielectric permittivity of the medium (the relative permittivity ε and electric constant ε_0 in SI); kT is the thermal energy (the Boltzmann constant k and the absolute temperature T). In aqueous medium: $l_B = 0.712$ nm at 20°C (ε= 80.08) and $l_B = 0.700$ nm at 25°C (ε= 78.25). The Bjerrum length is about 3.8 times higher than the crystal radius of Na atom and commensurable with the diameter of hydrated Na^+ ion (the counterion in salt-free solution of NaCMC).

Although Bjerrum had formulated the length l_B as a criterion for ion pair formation in a concentrated electrolyte solution, Eq. 6 is used in the Oosawa-Manning two-state polyelectrolyte theory for estimation of the electrostatic interactions between the polymer charges and small counterions.

3.3.4. *Two-state model*

In this model the counterions (electrostatically associated to the polyelectrolyte chain) are separated into two categories according their physical state: condensed (strongly bound and located inside a narrow potential valley along the polymer backbone) and free counterions (loosely bound and freely moving in the volume surrounding the chain). On the base of this model Manning has developed a simple theory for counterion condensation on isolated infinitively long cylinder immersed in a liquid medium with a vanishingly low ion concentration [22]. Such conditions appear in aqueous solutions of rigid or semi-flexible chains as DNA and CMC at low polymer concentration in salt free solution (counterions originate from the polymer sample: Na^+ in the case of NaCMC).

The Manning's model consider the polyelectrolyte chain as a liner polyion with average distance b_0 between its charges surrounded by a cylinder with radius a. The main parameter is the dimensionless charge parameter ξ determined by the linear charge density b_0^{-1} and the Bjerrum length $l_B \approx 0.7$ nm (at room temperature):

$$\xi \equiv l_B \ / b_0 = e^2 \ / \ 4\pi\varepsilon_0\varepsilon kTb_0. \quad (7)$$

A counterions condensation appears when ξ>1, i.e. when the mean distance between the two neighbour charges of the polyelectrolyte chain is smaller than 0.7 nm. When the degree of dissociation α increases until ξ<1, the linear charge density $b^{-1} = N/L_c$ (determined by the total number N of unit charges per unit contour length L_c) increases linearly up to a limit where $b_0 = l_B$; in this region $b = b_0$. At the point $\xi = 1$, the liner dependence $b^{-1} = f(\alpha)$ is broken and goes as a plateau at ξ>1, then $N = const$. I.e. at ξ>1 the effective charge density b^{-1} has its maximal value for a given linear polyelectrolyte and does not increase more although the chain's own charges b_0 continue to increase with α; this constancy of the total charge N is due to counterion condensation [23].

The fraction φ of the chain's charges neutralized by the condensed counterions with valency z is equal to:

$$\varphi = 1 - 1/z\xi = 1 - b_0/zl_B. \tag{8}$$

This equation allows determining the quantity of the condensed counterions when the quantity of the chain's own charges (per unit contour length) is known. Like so the condensed ions fraction is determined by the distance between the ionizable groups along the chain and the degree of dissociation α.

By that way the Manning's theory connect the counterions condensation with the chemical structure of the polyelectrolyte chain (the length of the polymer units and the nature of the ionizable groups) and the conditions of the medium (pH and temperature) together determining the quantity of chain's own charges (equal to the total counterions quantity). The theory allows defining two fractions of counterions having different mobile capability: the condensed ions are immobile (during the existence of the Bjerrum pairs) while the 'free' ones are mobile undergoing impacts with the solvent molecules.

The Manning's theory does not give exact value of the condensed ions fraction even when its two limits of validity (an infinitely long cylinder and vanishingly small concentration of small ions) are satisfied because Eq.(8) is derived on the base of analytical solution of the cylindrical Poisson-Boltzmann equation in its linearized form (assuming $\sinh \psi \approx \psi$) [24]. The numerical calculation of the non-linear cylindrical Poisson-Boltzmann equation shows that at small ξ (up to $\xi \approx 1/2$) the liner dependence $b^{-1} \equiv N/L_c = f(\alpha)$ coincides with the Manning's line but it is not broken at $\xi = 1$ and does not go on a plateau after this point. At ξ>1 the curve continue to increase although its slope decreases with ξ; this discrepancy increases with the ionic strength [8]. That means that the condensed ions fraction is smaller than the predicted by the Manning's theory. Nevertheless its experimental verification has shown an excellent agreement between the theoretical and experimental values in the case of double stranded DNA: at predicted $\varphi = 0.76$ the obtained by NMR condensed counterions fraction is 0.75 [25].

3.3.5. Application of Manning's theory to CMC

The distance between two glucose units in CMC chain (0.515 nm) and the degree of dissociation α determine the liner charge density b_0^{-1}. For CMC with degree of substitution DS = 1.00 the limit (ξ = 1) for the ions condensation appears at α = 0.72 according to Manning's theory. At full dissociation (α = 1) the charge density is b_0^{-1} = 1.94 nm^{-1} and the dimensionless parameter is ξ = 1.38 (at 20°C), so condensation of Na^+ counterions on COO^-

groups appears even in salt-free solution of NaCMC; the fraction of the condensed counterions is $\varphi = 1–1/\xi = 0.28$.

For CMC with DS = 1.20 the mean distance between two COOH groups is equal to 0.43 nm, so the charge density parameter is $\xi = 1.66$ at $\alpha = 1$. At pH 6.0 the degree of dissociation is $\alpha \approx 0.95$ [8], $\xi = 1.58$ and the condensed counterions fraction is equal to $\varphi = 0.37$. So, the CMC chain remains negatively charged when it is immersed in salt-free aqueous medium at pH 6.0; about 63% of the dissociated groups are compensated by the diffuse counterions in the ionic atmosphere.

3.3.6. *Relation between condensation and dissociation*

The apparent dissociation constant pK_a of the acid groups is shifted with ΔpK_a relatively to its intrinsic value pK_{a0} due to the increased local concentration of H_3O^+ ions because of the strong electric field around the polyelectrolyte chain; the value of ΔpK_a is proportional to the charge density $b_0^{-1}(\alpha)$ determined by the degree of dissociation α (Eq.3). It is reasonable to suppose that ΔpK_a depends on counterions condensation both indirectly (by the effective value of b^{-1}) and directly (forming ions pair [COO^-Na^+] instead COOH and influencing on the polarization of O–H electron-pair link in COOH groups). This means that the dependence of ΔpK_a on α should have different course at $\xi<1$ and at $\xi>1$. This hypotheses can be experimentally tested by determining $\alpha = f(\text{pH})$ and $pK_a = f(\alpha)$ by potentiometric titration of aqueous solution of CMC.

For CMC with DS = 0.80 and 0.90 the charge parameter ξ is equal to 1 at $\alpha = 0.90$ and 0.80; so, at these values of α the dependence $pK_a(\alpha)$ should change its slope. This effect must be stronger at high ionic strength where the counterion condensation increases. Although in the last case the Manning's theory is inapplicable, the limiting value $\xi=1$ also can be used as the dependence $b^{-1}(\alpha)$ begins to deflect from line in the region of $\xi \approx 1$ as the numerical solution of the non-linear Poisson-Boltzmann equation shows [8].

The potentiometric titration $\alpha = f(\text{pH})$ [8] shows that in salt-free solution the dependence $pK_a = f(\alpha)$ is almost linear in the range $\alpha = 0.1–0.9$ for CMC with DS = 0.81. The addition of NaCl up to 1 mol/L decreases the slope of the dependences $pK_a(\alpha)$ but they are absolutely liner in the studded interval $\alpha = 0.25–0.97$ for CMC with DS = 0.91.

The absence of a curvature of the experimental dependence $pK_a(\alpha)$ at $\xi \approx 1$ shows out that counterions condensation does not influence the degree of dissociation (in the frame of the experimental accuracy). This result can be explained with the different nature of H^+ and Na^+ binding to COO^- groups: the proton is involved in covalent linkage (coupled electron pair between H and O atoms) while the sodium ion is electrostatically bound. The different energy of these two kinds of banding leads to longer life of the COO–H links in comparison to the COO^-Na^+ bonds and respectively to much higher affinity of COO^- groups to H^+ than to Na^+ ion.

Nevertheless, the presence of Na^+ ions in the medium influences the $COOH \leftrightarrow COO^- + H^+$ equilibrium by the local concentration of H^+ in the diffuse cloud where H_3O^+ and Na^+ ions are in competition being electrostatically equivalent; the evidence for that is the decreasing slope of the dependence $pK_a(\alpha)$ at increasing concentration of NaCl [8]. This fact together with the absence of visible influence of Na^+ condensation on α shows out that the physical state of the counterions (immobile or free moving) is not as important as its concentration in the vicinity of the polyelectrolyte chain. This conclusion demonstrates the advantages of the cylindrical cell model (distinguishing the counterions by their distance from the surface and considering their local concentration) in comparison with the two-state model (separating the counterions into two categories according their motility and disregarding the 'free' counterions distribution).

4. Conformation of free CMC chain in aqueous medium

As a polyelectrolyte the conformation of CMC chain is determined by its chemical structure and electric properties determining the chain rigidity and linear charge density; the both quantities also depend on the medium parameters (pH, ionic strength and temperature). The electrostatic interactions could be neglected at low pH and/or high salt concentration; then the chain could be considered as a non-charged.

The main structural peculiarity of the cellulose and its derivatives are the rigidity of the glucose rings and the impossibility of rotation around the C–O–C bonds between the neighbour rings because of β-1,4-linkage; the both factors determine the rigidity of the chain backbone. Although very restricted oscillations round the C–O–C bonds occur and that decrease orientational correlation of the monomer units (regarding the line defined by the first init). That means that the orientation of every glucose unit strongly dependents on the orientation of significant number its neighbours but the orientational correlation decreases slowly along the chain. So, the CMC chain can be defined as a semi-flexible; that allows applying the model of worm-like chain.

4.1. Theoretical definitions

The conformation of an individual polymer chain is described by its end-to-end distance h (the distance between the first and the last polymer units) or by radius of gyration R_g (averaged mean-square distance between the units and the center of mass); these quantities are averaged in time for one chain or for an ensemble of chains at a given moment: $\langle h\rangle$ and $\langle R_g\rangle$. The CMC chain with contour length L_c has $n = L_c/l$ units with length l = 0.515 nm. The value of l is taken equal to the size of the glucose residue (instead the length of C–C bound as in the case of flexible polymers) because of the rigidity of the glucose rings.

4.1.1. Unperturbed random coil

In the model of free-articulated chain (Kuhn's model) the real chain is replaced by a hypothetical chain with length $L_c = A_K N_K$ consisting $N_K>1$ segments with length $A_K = sl$ (every segment has s monomeric units with length l); the rotation between the neighbour segments is completely free. For an unperturbed random coil the Kuhn's theory gives for the mean-square end-to-end distance $\langle h^2\rangle$:

$$\langle h^2 \rangle_0 = A_K L_c \tag{9}$$

and for the statistical segment length:

$$N_K = L_c^2 / \langle h^2 \rangle_0 . \tag{10}$$

This equation allows determining A_K and N_K from the experimentally measured molecular mass $M \sim L_c$ and $\langle h^2 \rangle_0$ or the mean-square radius of gyration $\langle R_g^2 \rangle$. For an unperturbed random coil at $N_K >> 1$:

$$\langle R_g^2 \rangle_0 = \langle h^2 \rangle_0 / 6 = (A_K / 6) L_c . \tag{11}$$

The distribution of the segments of an unperturbed random coil is Gaussian at (a) $L_c/A_K > 10$ and (b) when the polymer is immersed in "ideal solvent" (θ-solution or θ-temperature when the interactions polymer-polymer, polymer-solvent and solvent-solvent are compensated); for such coil $\langle h^2 \rangle_0$ and $\langle R_g^2 \rangle_0$ are proportional to the couture length L_c according the Eqs. (9) and (11) [26].

4.1.2. Expanded random coil

Out of the θ-conditions the proportionality between $\langle h^2 \rangle$, $\langle R_g^2 \rangle$ and L_c is infringed but the chain can be return to conformation of unperturbed random coil changing the thermodynamic properties of the solvent by the temperature and addition of salts; an indication for θ-point is an absence of dependence of $\langle h^2 \rangle$ or $\langle R_g^2 \rangle$ on the polymer concentration.

The linear expansion factor α_e is determined by the ratio between the random coil size (averaged end-to-end distance $\langle h^2 \rangle^{1/2}$ or radius of gyration $\langle R_g^2 \rangle^{1/2}$) and its unperturbed value:

$$\alpha_e = \langle h^2 \rangle^{1/2} / \langle h^2 \rangle_0^{1/2} = \langle R_g^2 \rangle^{1/2} / \langle R_g^2 \rangle_0^{1/2} . \tag{12}$$

In "good solvent" ($\alpha_e > 1$) or "bad solvent" ($\alpha_e < 1$) the chain has more expended or compact conformation, respectively. The dependence of α_e on the medium conditions is explained by the term "excluded volume" (equal to zero in "ideal solvent") which includes the physical volume of the chain and the difference of repulsion and attraction forces between polymer segments when distant chain units encounter owing their thermal motion; the probability of such collisions increases with the chain length and decreases with the rigidity of the chain: $\alpha_e \sim L_c/A_K$. The excluded volume additionally increases in the case of polyelectrolytes (at low polymer concentration) because of the electrostatic repulsion between the distant intermolecular chain units.

4.1.3. Persistent length

At $L_c/A_K < 10$ the proportionality between $\langle h^2 \rangle$, $\langle R_g^2 \rangle$ and L_c is infringed (the segment distribution is non-Gaussian even in θ-solvent), so the Eqs. 9-12 are inapplicable. To describe the conformation of such short and rigid chains the worm-like chain model is introduced; its peculiarity is the continuously changing chain curvature.

The averaged angle $\langle\varphi\rangle_L$ between the tangents to two points at distance L_1 along a worm-like chain is given by the expression:

$$\cos\langle\varphi\rangle_L = \exp(-L_1 / l_p) \tag{13}$$

where the persistent length l_p describes the e fold decreasing of the orientational correlation. In such way l_p is a measure for the rigidity (local stiffness) of the worm-like chain; at $L_c \leq l_p$ the chain can be described as a more or less curved rod. Comparison with the Kuhn model (at N_K>1) gives the ratio: $l_p = (1/2)\, A_K$.

The unperturbed radius of gyration $\langle R_g^2\rangle_0^{1/2}$ can be expressed by the ratio between the persistent and contour lengths $y = l_p/L_c$:

$$\langle R_g^2\rangle_0 = L_c^2\{(1/3)y - y^2 + 2y^3 - 2y^4[1-\exp(-1/y)]\}. \tag{14}$$

This equation allows describing the size of chins of all length and rigidity including the both limits: rigid rod (very shot and rigid chains, then $\langle R_g^2\rangle_0 = (1/12)\, L_c^2$) and Gaussian coil (long and flexible chains at L_c/l_p >20 in θ-solvent).

The random coil expansion in good solvent leads to increase of the effective persistent length $l_p = l_{p0} + l_{pe}$, where the two terms represent the intrinsic persistent length and the electrostatic component, respectively. The electrostatic term decreases with the ionic strength of the medium; that manifest itself as decreasing of the random coil size ($\langle h^2\rangle^{1/2}$ or $\langle R_g^2\rangle^{1/2}$) when the salt concentration of the polymer solution increases.

4.2. Experimental methods

The classical techniques for experimental investigation of the polymer chain conformation in liquid medium are those of the viscometry and the static light scattering.

4.2.1. *Viscometry*

The viscometry is the technique most frequently employed in physical chemistry of polymer solutions. The method is based on the measure of the dynamic viscosity of the solvent η_0 and of polymer solution η with concentration c and extrapolation of the relative viscosity η/η_0 to $c = 0$. The characteristic viscosity (the limiting viscosity number) defined as:

$$[\eta] = \lim_{c\to 0}[(\eta/\eta_0 - 1)/c] \tag{15}$$

is connected with the molecular mass M and the end-to-end size $\langle h^2\rangle_0^{1/2}$ of the macromolecules:

$$[\eta] = KM^a, \text{ and } \quad [\eta] = \Phi\langle h^2\rangle^{3/2} / M \tag{16}$$

where K and a are constants specific for a given polymer/solvent solution; $\Phi \approx 2.1 \cdot 10^{23}$ (when the dimensions of $[\eta]$ and $\langle h^2 \rangle_0^{1/2}$ are cm^3/g and cm, respectively) is a unspecific constant for linear chains with conformation of random coil in good solvent.

The constant a gives information about the form of the macromolecules: $a = 1.8$ for rigid thin rod with axes ratio ≥100; for unperturbed random coil $a = 0.5$ and increases up to 0.8 for an expanded coil. The method is widely used due to its experimental simplicity but it is not absolute requiring calibration with monodisperse fractions with independently determined molecular mass M. Once the constant K and a are known the equations (16) allows determining the mass and size of the random coil by measuring $[\eta]$.

4.2.2. *Small angle static light scattering*

The second widely used method for determination of the polymer chain mass and conformation is the static light scattering [27]. Its main advantage is that it is absolute: for determination of the squared radius of gyration $\langle R_g^2 \rangle$ and molecular mass M a calibration with polymer fractions is not required. Because the relative size $(\langle R_g^2 \rangle^{1/2}/\lambda$ is small relatively to the wavelength λ (in the medium), the small angle approximation (equation of Guinier) is applicable for the light scattering of polymer solutions with weight concentration c:

$$cH / R_\theta = 1/M + (16\pi^2/3)(n_0/\lambda_0)^2 \langle R_g^{\ 2} \rangle \tilde{n} \sin^2(\theta/2), \tag{17}$$

where $H \sim [(n_0^2/\lambda_0^4)(dn/dc)^2]$ is an optical constant (specific for a given pair polymer/solvent but not depending on c in diluted solution) determined by the refractive indexes of the polymer solution n and the medium n_0 at wavelength λ_0 (in vacuum); the increment of the refractive index $(dn/dc) = (n-n_0)/c$ (at $c \to 0$) is measured by refractometry. The scattering coefficient (Rayleigh ratio) $R_\theta = I/I_0$ is determined by the quantities measured in the experiment: intensities of the incident light I_0 and the scattered I at angle θ. For determination of $\langle R_g^2 \rangle$ it is not required to know the absolute values of c and H: it is enough to measure the dependence $1/I_\theta \sim \sin^2(\theta/2)$ at a few c and extrapolate $1/I_\theta$ to $c = 0$. The extrapolated to $\theta = 0$ concentration dependence $c/R_{0^\circ} \sim A_2 c$ (not considered in Eq. 17) gives the second virial coefficient A_2 which allows determination of the random coil expansion factor α_e [28].

4.3. Factors determining CMC chain conformation

4.3.1. *Rigidity and molecular mass*

The impossibility of rotation between the neighbour glucose units reflects in high persistent length $l_p = 5$–20 nm of the cellulose and its derivatives [29]; this value is one-two orders higher than for polymers with flexible chains ($l_p = 0.2$ nm for poly(ethylene glycol) [26]). The relatively high value of the l_p allows to define CMC chain as worm-like and suggest that (depending on the contour length L_c) its conformation is rod-like at $L_c < l_p$ or random coil at $L_c > l_p$. As a rule CMC samples have enough high M and than $L_c >> l_p$; that predicts random coil conformation. This expectation is conformed experimentally by viscometry and light scattering.

The viscometry measurements of aqueous solutions of cellulose and its derivatives showed out that the factor a in Eq. (16) lies near to $a \approx 0.8$ [29]; this value is characteristic for expanded random coil. The light dependence on the presence of charged or non-charged lateral groups conform the supposition that the CMC rigidity is determined mainly by the β-1,4-linkage in the chain backbone; for comparison: $a \approx 0.5$ (defined for non-expanded random coil) for amylose acetate due to α-1,4-linkage [26]).

CMC conformation is studied by small angle static light scattering (at eleven fixed angles θ = 44°–151°) in combination with size exclusion (gel permutation) chromatography and refractive index detector [8]. That technique allows continuous measuring of R_θ and c while the given fraction is eluted and on-line determining of $\langle R_g^2 \rangle$ and M from the initial slope of $cR_\theta^{-1} = f\,[M^{-1}, \sin^2(\theta/2)]$ according to Eq. (17). The measurements of CMC samples with degree of substitution DS = 0.75–1.25 have given $\langle R_g^2 \rangle^{1/2}$ = 30–140 nm and mean-weight mass M_w = 120–1100 kg/mol (these masses correspond to contour length $L_c \approx$ 260–2800 nm). On Fig. 1 is shown the experimental dependence of R_g on mean-weight M_w or mean-number M_n molecular mass; the difference between M_w and M_n is due to the polydispersity of the polymer samples (the polydispersity index is $M_w/M_n \approx$ 1.6–2.7).

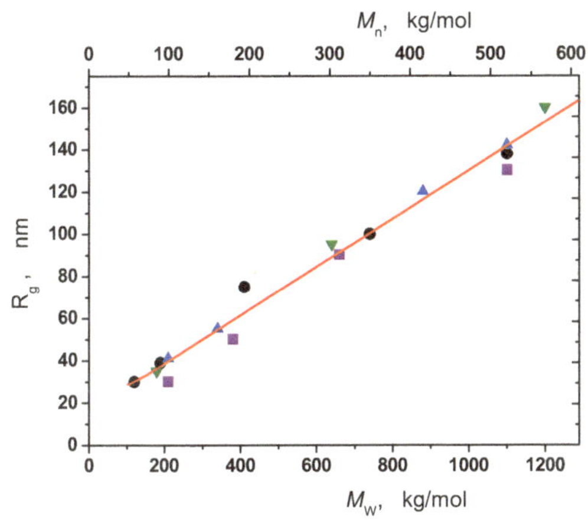

Figure 1. Dependence of the radius of gyration R_g on the mean-weight molecular mass M_w (the bottom abscissa), or mean-number molecular mass M_n (the top abscissa) of CMC with DS = 0.75 (,), 0.91 (7) 0.99 (B) and 1.25 (!). The values of R_g are measured by static light scattering; the data are taken from Ref. [8].

The above cited experimental results allow describing the conformation of CMC chains with $l_p \approx$ 15–17 nm at different molecular mass M. At $L_c \leq l_p$ the CMC chain has a conformation of bend rod which curvature changes in time due to the thermal motion; the chain with $L_c \approx 15 - 17$ nm has 29–33 glucose units and M = 3648–4134 g/mol (in the case of NaCMC with DS = 1.0). At $L_c > 10A_K$ the CMC chain has a conformation of random coil; a chain 10 times longer than the Kuhn segment ($A_K = 2l_p \approx$ 30–34 nm) has 582–660 glucose units and M = 142–160 kg/mol. So, NaCMC with M higher than 150 kg/mol has a conformation of some expanded random coil.

4.3.2. Linear charge density and ionic strength

The viscometry data [29] for high molecular CMC show out that the expansion factor α_e in aqueous medium (good solvent for CMC, $\alpha_e>1$) is additionally increased due to the electrostatic repulsion between COO^- groups; the effect increases in the range pH 2–6 ($\alpha \approx 1$ at pH 7) due to the liner charge density $b^{-1}(\alpha)$ increasing with the degree of dissociation α. The expansion of the random coil reflects in increased value of the effective persistent length $l_p = l_{p0} + l_{pe}$ and its electrostatic component l_{pe}; the effect is influenced by the ionic strength of the medium. The value of the intrinsic (structural) component l_{p0} of the persistent length is estimated to be 5.0 nm in high salt concentration (where the electrostatic interactions are negligible) [30]. In water/cadoxen (Cd^{2+}-ethylendiamine complex) mixtures $l_{p0} = 8.5$ nm was obtained [31].

Static light scattering [8] also show out that the random coil size of CMC depends on the ionic strength; the effect is proportional to the number of Kuhn segments $N_K = L_c^2 / 6\langle R_g^2\rangle_0$ for chain with contour length L_c. At $N_K = 100$ the radius of gyration decreases from $\langle R_g^2\rangle^{1/2} \approx 190$ nm (in 0.02 mol/L $NaNO_3$) to 140 nm (in 0.1 mol/L). That indicates for diminishing of α_e and l_{pe} with the ionic strength owing to decrease of the electrostatic component of the excluded volume (the origin for the random coil expansion).

It seems that the decreasing of $\langle R_g^2\rangle^{1/2}$ with salt concentration is caused not entirely by the electrostatic shielding of COO^- groups; an additional factor is the structural rigidity (l_{p0}) diminished with dehydration of CMC chain. This effect is probably one of the reasons for high difference (reaching 3 times) of the values of l_{p0} obtained at low and high ionic strength; another reason is the imperfections of the theories employing to estimate l_{p0} from data obtained by hydrodynamical (viscometry, dynamic light scattering) and statical (static light scattering, potentiometric titration) methods.

At pH 7.0 ($\alpha \approx 1$) the light scattering data show out a linear dependence of $\langle R_g^2\rangle^{1/2}$ on M in the range $M_w = 120$–1100 kg/mol (Fig. 1); linear is also the dependence of $\langle R_g^2\rangle^{1/2}$ on the number of statistical segments N_K [8]. The both dependences do not depend measurably on the degree of substitution DS = 0.75–1.25 (in this range the charge parameter ξ increases from 1.04 to 1.73). This independence on DS could be considered as evidence for counterion condensation; i.e. the effective linear charge density b^{-1} does not increase (at $\xi>1$) with chain's charges (the number of COO^- groups per unit length of CMC backbone) due to adsorption of Na^+ ions. This conclusion is done in Ref. [8] but it is not enough reliable because $\langle R_g^2\rangle^{1/2}$ of CMC chain is determined mainly by its structural parameters (l_{p0} and L_c) but lightly depends on its electric properties (b^{-1}) because $l_{p0} >> l_{pe}$.

4.4. CMC according to the wormlike chain models

For CMC the salt concentration needed to reach the state of unperturbed random coil (θ-solvent) has been estimated to be about 5 mol/L NaCl [32]. Because this concentration is practically unreachable, the CMC random coil is always expanded. That is why an electrostatic worm-chain theory is needed to evaluate the unperturbed random coil parameters, in particular to distinguish the structural and electrostatic components of the excluded volume and the persistent length.

In Ref. [8] the experimental data (obtained by static light scattering) have been used to calculate the CMC molecular parameters according to the models of Odijk [33] and Davis [32]; the last is better validated. The Davis' model gives the following parameters (for CMC with DS = 1.25 in 0.02 mol/L $NaNO_3$): (*a*) expansion factor increasing with M in the range $\alpha_e \approx 1.16$–1.40 at $M_w = 120$–1100 kg/mol ($L_c = 254$–2330 nm); (*b*) electrostatic component of the persistent length $l_{pe} = 2.9$ nm; and (*c*) intrinsic persistent length (estimated as a fitting parameter) $l_{p0} = 17$ nm. In 0.10 mol/L $NaNO_3$ these quantities are: $\alpha_e \approx 1.07$–1.14; $l_{pe} = 0.8$ nm; and $l_{p0} = 15$ nm [8].

In the same work $l_{p0} \approx 6$ nm (for CMC with DS = 0.9 in salt-free and in 0.01 mol/L NaCl) and $l_{p0} \approx 3.5$ nm (in 0.1 mol/L NaCl) was estimated from dependence pK_a (α) (obtained from titration curves α(pH)); the values of l_{p0} are calculated using the theory of Katchalsky and Lifson [34]. This values are 3–4 fold decreased according to the obtained by Davis's theory (from scattering data for $R_g(M)$); the discrepancy reflects rather the imperfections of the theoretical models for wormlike chain than the inaccuracy of the employed experimental methods. It seems that the Davis model gives more realistic values because the Katchalsky-Lifson model overestimates the electrostatic energy [8].

4.5. Peculiarity of CMC chain

The comparison between the intrinsic $l_{p0} \approx 15$–17 nm and electrostatic $l_{pe} \approx 1$–3 nm components of the persistent length (both obtained by Davis model) shows that the high rigidity of CMC chain is predominantly due to its chemical structure but the electrostatic repulsion plays a minor role although the linear charged density is high at neutral pH (about 2 charges per nm); that leads to relatively low expansion factor ($\alpha_e \approx 1.2$–1.4) even for very long chains ($L_c = 0.3$–$2{,}3$ μm) at low ionic strength.

For comparison: the uncharged chain of poly(ethylene glycol) (PEG) with contour length L_c= 26 nm has $N_K = 35$ statistical segments with length $A_K = 0.73$ nm, unperturbed radius of gyration $\langle R_g^2 \rangle_0^{1/2} \approx 1.8$ nm and expansion factor $\alpha_e = 1.31$ [35]. I.e. due to the flexibility of the PEG chain its Kuhn segment is about 40 times shorter than that of CMC.

5. Electric polarizability of CMC

5.1. Counterions polarizability of free CMC chains

5.1.1. Electric polarizability of polyelectrolytes in solution

When an external electric field with strength E is applied to polyelectrolyte solution an electric polarization γE appears (the coefficient γ is named polarizability) [36]. The polarization of polyelectrolytes in aqueous medium is determined mainly by migration of counterions along the chain; this assertion follows from the strong dependence of γ on the counterions quantity (determined by the linear charge density b^{-1} and pH-dependent by the degree of dissociation

α). The contribution of other kinds of polarization (electronic, atomic and orientation of polar groups) is negligible in fields with moderate strength; a measurable orientation of short rigid chains as DNA fragments is possible only in high voltage electric field [37].

In sinusoidal electric field the frequency ν determines the time for which the counterions move in one direction and results in the γ magnitude when the half-period is not enough long to reach equilibrium allocation; the critical frequency $\nu_{1/2}$ is defined as ν where γ decreases twofold. The value of $\nu_{1/2} = 1/2\pi\tau_i$ is determined by the relaxation time τ_i of the polarization which depends on the distance of migration and the diffusion coefficient of the counterions. So, the polarizability dispersion (the frequency dependence $\gamma = f(\nu)$) gives information for the counterions mobility.

The theories describing the polyelectrolyte polarization (the magnitude of γ and its frequency behaviour $\gamma(\nu)$) are based on different physical models considering migration of counterions with diffusion coefficient D_i on distance L along a linear polyion. Schwarz's equation [38,39] give $\nu_{1/2} = 4D_i/\pi L^2$ for strongly elongated particles with length L. According Mandel the polarization γ_i and the relaxation time τ_i of one-dimensional lattice with length L in the case of one type of counterions with valency z, mobility u and total number N are [40]:

$$\gamma_i = (z^2e^2 / 12kT)NL^2 \quad \text{and} \quad \tau_i = \left(1/12ukT\right)L^2 \qquad (18)$$

where e, k and T are the elementary charge, Boltzmann constant and the absolute temperature.

According Oosawa fluctuation theory the counterions polarization appear in both presence and absence of external electric field; in the last case a spontaneous dipole moments appears due to fluctuation of counterions concentration along the polyelectrolyte chain [41,42]. In the case of N number monovalent counterions the main component of the polarizability is:

$$\gamma_i = (e^2 / 2\pi kT)NL^2 \quad \text{and} \quad \tau_i = (1/4\pi^2ukT)L^2 \qquad (19)$$

The most of the theories consider polyelectrolyte polarization as displacement of one-state counterions or does not distinguish the contribution of the diffuse and condensed counterions. One of the exclusion is Minakata's stochastic theory [43] considering an infinite linear chain of discrete charges which counterions have different transitional mobility depending on the distance from the polyion. The theory shows that the counterions contribution depends on the potential barrier which they have to overcome jumping from one lattice site to the next, so the loosely bound (diffuse) counterions give the main electric polarization at higher frequency; the contribution of strongly bound (condensed) counterions should appear at lower ν. Thus, the two kinds of counterions are distinguished by their increment to the magnitude of γ and by the relaxation frequency $\nu_{1/2}$.

5.1.2. *Polarizability of CMC at random coil conformation*

The counterions polarizability is usually studied by the dielectric permittivity [44] or conductivity [45,46] of polyelectrolyte solutions; the first method is well working in salt-free solution and the second – in small-ions electrolyte. The measuring signal is electrical: the capacitance or active current between two electrodes. At these techniques the applied field does not influence the conformation of the polyelectrolyte chain because the field strength is very low and the polarization is due to counterions replacement only. An imperfection of the both methods is that they require high polymer volume fraction because the both polyelectrolyte and the bulk contribute to the measured signal (dielectric or conductivity current); that complicates the interpretation because of the intermolecular interactions at increased polymer concentration. Another problem is the electrode polarization; that limits the ion strength at permittivity measuring.

Dielectric spectroscopy of aqueous solutions of NaCMC shows two dispersions: high-frequency [47] and low-frequency [48]; the presence of two dispersions is typical for polyelectrolytes. The high-frequency dispersion is independent on the contour length but the low-frequency is strongly dependent on it; on this base the two frequency regions are attributed to counterions moving along chain's parts or along the hole polyelectrolyte chain, respectively [49,50].

5.1.3. *Polarizability of CMC at orientation in electric field*

At high field strength the counterions polarization is accompanied with orientation of the polyelectrolyte chain backbone; as a result optical anisotropy of the solution appears. Electro-optical methods are based on registration of the anisotropy of the refractive index Δn (birefringence) or the absorption coefficient (dichroism) [51]. A principal advantage of these techniques is that the bulk does not contribute to the measuring signal; that allows multiplying the sensitivity be increasing the optical path and work at lower polymer concentrations. Another important advantage is the possibility to obtain information for both electrical and hydrodynamical properties of the polyelectrolyte chain by measuring the steady-state Δn_s and transient Δn_t anisotropy. These properties can be separately studded by the measuring $\Delta n_s \sim \gamma E^2$ and $\Delta n_t \sim \Delta n_s \exp(-t/\tau_0)$ with time t after switching off the electric field (τ_0 is the relaxation time determined by the coefficient of friction of macromolecules with given size and form).

The capabilities of the electric birefringence to study aqueous solution of NaCMC are demonstrated in Ref. [52]. In this work the orientation of CMC with M = 30–250 kg/mol is done by applying of a single high voltage rectangular impulse; the field strength reaches $E \approx 2$ MV/m. The results show that the polarizability increases with M but the slope of $\gamma = f(L_c)$ decreases in the range $M \approx 5$–100 kg/mol and reaches plateau at $M \approx 250$ kg/mol. The found dependence of γ on L_c is different from predicted by Eqs. (17,18) ($\gamma \sim L^2$); the reason for that is revealed comparing the measured disorientation time τ_0 with calculated ones for rigid rod, bending rod and worm-like chain. The comparison show out that the experimental values of τ_0 corresponds to the rigid rod model ($\tau_0 \sim L^3$) only for very short chains with mass less than $M \approx 5$ kg/mol; this M corresponds to 20 glucose residues and $L_c \approx 10$ nm. This value of L_c coincides with the persistent length l_p = 5–15 nm obtained by viscometry and static light scattering (see section 4.3).

From the result in Ref. [52] it follows that the counterions are moving along the CMC chain on a distance commensurable with l_p. At higher distance the counterions displacement is limited by the local curvature of the chain backbone (which determine the field component along the chain); when the curvature is higher than right angle the counterions are moving in opposite directions along two neighbour chain segments. That means that l_p instead L_c should be used when estimating γ and τ_i. So, concerning counterions polarizability the long CMC chain can be presented as a consequence of rods likewise the free-articulated chain in the Kuhn model. That allows explaining the experimental fact that $\gamma(L_c)$ is almost constant at $M \approx 250$ kg/mol [52]; at these M the NaCMC chain with DS = 1 has $L_c \approx 530$ nm and about 16 statistical segments with length $A_K = 2l_p \approx 34$ nm.

Hydrodynamic behaviour of CMC in electric field suggests that in the case of long chain the orientation of its segments is accompanied by deformation of the random coil; the degree of deformation is determined by both polarizability and rigidity of the chain. The contribution of both electrical and mechanical properties complicates the interpretation of the experimental results about the polarizability of chains with $L_c > l_p$.

Demonstration of that are dependences $\Delta n_s(\nu)$ of CMC with 230, 1100 and 2000 glucose units ($L_c \approx 120$, 570 and 1030 nm) described in Ref. [53]. In this work the sinusoidal field with low strength in combination with very sensitive detection (based on an amplitude modulation of the birefringence signal) is used. The results show out that the dispersion curves $\Delta n_s(\nu)$ are quite different at frequencies under and up a few kilohertz. In the hertz diapason Δn_s has negative values indicating predominant orientation across the electric field; the absolute value $|-\Delta n_s|$ decreases with diminishing of the degree of dissociation α and with increasing of M, C_{CMC} and C_{NaCl} (concentrations of CMC and the added NaCl, respectively); Δn_s becomes positive (orientation along the field) at high values of M, C_{CMC} and C_{NaCl} even at $\alpha = 1$. The dependence on α, M, C_{CMC} and C_{NaCl} in kilohertz diapason is analogical although there Δn_s is positive in all conditions. The dependence of $\Delta n_s(\nu)$ on α and C_{NaCl} in both low and high frequency diapasons confirms that the orientation-deformation of CMC chain is due to counterion polarization. In the same time the dependency of $\Delta n_s(\nu)$ on M and C_{CMC} suggest a complicated relation between of the electrically induced conformation of CMC chain and its electrical (γ) and mechanical (l_p) properties; an manifestation of that is the unexpected negative value of Δn_s. This complexity does not allow application of the theories developed for counterions polarizability $\gamma(L)$ of straight rigid polyions and do very difficult the interpretation of the experimental dependence $\Delta n_s(\nu)$ in order obtaining information about $\gamma(L_c)$ at $L_c >> l_p$.

5.2. Polarizability of CMC adsorbed on colloid particles

5.2.1. Electrophoretic mobility of CMC-alumina particles

A polyelectrolyte adsorption on solid/liquid interface allows overcoming the difficulties connected with the electrically induced conformation of long CMC chains. In this case the parts of the adsorbed chain are immobilized on the hard surface and that hinders chain deformation. An additional advantage has the adsorption on colloid particles with submicron size; then an electric field with moderate strength is enough to orientate the particles due to the their big size (in comparison with the polymer random coil). Then the electro-optical method employed allows registering reliably the orientational degree averaged over the particles ensemble; due to high sensitivity of this technique it is possible to work at low particle concentration.

The adsorption of polyelectrolytes alters the electric and hydrodynamic properties of the solid/liquid interface; an indication for that is the change of the electrophoretic mobility u_{el} of the particles. On Fig. 2 is shown the dependence of u_{el} on the concentration C_{CMC} of NaCMC at adsorption on alumina (γ-Al_2O_3) particles [54]; $u_{el} = v/E$ is measured by the technique of micro-electrophoresis (microscopic measuring of the velocity v in electric field with strength E). The dependence $u_{el}(C_{CMC})$ is studded at pH 6 where the alumina surface is positive (the isoelectric point is at pH 8.5) and the COO^- groups of CMC are almost fully dissociated; the opposite charge sign is a precondition for electrostatic adsorption. The change of u_{el} sign from positive to negative indicates over-equivalent adsorption of CMC and predominant exposition of the polyelectrolyte charges. The steep slope of the curve before the isoelectric point reflects the well known fact that all added polyelectrolyte is adsorbed on the surface at low polymer concentration; that is a characteristic of polyelectrolyte adsorption on an oppositely charged surface [55,56]. The plateau at high C_{CMC} reflects the saturated adsorption because electrostatic repulsion of already recharged surface and the CMC macromolecules.

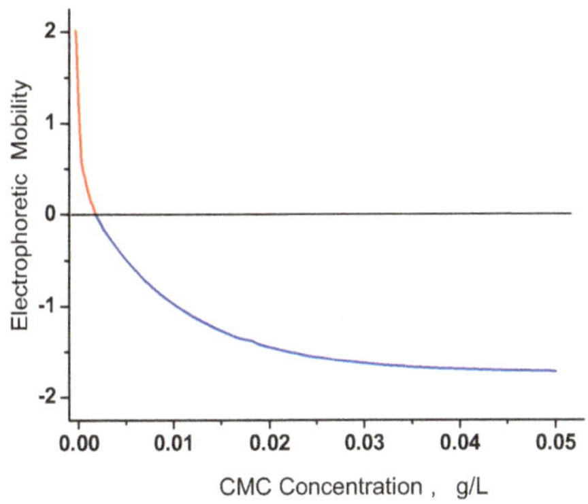

Figure 2. Dependence of electrophoretic mobility u_{el} [10^{-8} (m/s)/(V/m)] of γ-Al_2O_3 particles on the concentration C_{CMC} [g/L] of NaCMC in water-polymer suspension.

The concentration dependence $u_{el}(C_{CMC})$ shows out that the surface electric properties of the CMC covered alumina particles are determined predominantly by the charge of the adsorbed layer; the particle surface charge does not manifest itself apparently because of electrostatic and hydrodynamic shielding by the polyelectrolyte layer. This gives opportunity to study electro-optically the electric polarizability of CMC being adsorbed on the colloid particles; in the case the hard core of coated particles determines their optical properties but the orientation depends on the counterions polarization in the polyelectrolyte layer.

5.2.2. *Electric light scattering in suspension*

The theory of this method [57] is developed in the Rayleigh-Debye-Gans approximation [58] which restricts: (a) the relative refractive index $n = n_1/n_0$ (n_1 and n_0 are the refractive indexes of the particles and the medium, respectively); and (b) the relative size L/λ (where L is the particle length and λ is the wavelength in the medium; $\lambda = \lambda_0/n_0$ when λ_0 is the wavelength in the vacuum). The light scattering intensity I_0 at random orientation of the particles is determined by the function of internal interference $P(\theta)$ (form-factor) at scattering angle θ [27]:

$$I_0 = kcHMP(\theta), \tag{20}$$

where: k – the apparatus constant determined by the scattering volume and the solid angle of the photoreceiver; c – the particle weight concentration; H – the optical constant defined by λ_0, n_0 and n_1; M – the particle mass.

When an electric field is applied to the suspension, the light scattering intensity I_E is changed due to particles orientation [59]. The absolute electro-optical effect (EOE) $\Delta I = I_E - I_0$ can be determined by the functions of internal interference at certain degree of orientation $P(\theta, E)$ and at random orientation $P(\theta)$ [60]:

$$\Delta I = kcHM[P(\theta,E) \; - P(\theta)], \tag{21}$$

At an orientation degree Φ (varying from 0 at random orientation to 1 at full orientation):

$$P(\theta,E) \; = P(0) + A\left(KL\right) \times \Phi(\gamma,E,T), \tag{22}$$

where: $KL = 2\pi(L/\lambda)sin(\theta/2)$. The optical function $A(KL)$ is determined by the form and relative size L/λ of the particles; the expressions for $A(KL)$ are different at low and high degree of orientation. The value of Φ at steady-state EOE is a function of the electric polarizability γ, the electric field strength E and the temperature T.

The relative EOE $\Delta I/I_0$ does not depend on k, c, H, M and it is defined at a moment t only by $P(\theta,E)$ and $P(\theta)$, which are functions of the particles' form, size and optical anisotropy:

$$\Delta I_t \; / I_0 = \; [P(\theta,E) \; / \; P(\theta)] \; - \; 1 = \; [A\left(KL\right) \; / \; P(\theta)] \times \Phi(\gamma,E,T,t). \tag{23}$$

In the steady-state $\Phi(\gamma, E, T)$ depends only on the ratio between the orientation energy γE^2 and the energy of random motion kT. Then EOE at low degrees of orientation ($\gamma E^2 << kT$) is:

$$\Delta I_s \; / \; I_0 = \; [A\left(KL\right) \; / \; P(\theta)] \times (\gamma E^2 \; / \; 15kT). \tag{24}$$

The electro-optical results described below are obtained at $\theta = 90°$ by cell with platinum electrodes at interelectrode distance of 2.6 mm. The light scattering intensities I_0 (in absence of electric field) and ΔI_s (in sinusoidal field with strength $E \leq 54$ kV/m) were measured after reaching steady-state polymer adsorption.

5.2.3. Kilohertz polarizability of CMC-alumina particles

Figures 3 and 4 represent the field dependence $\Delta I/I_0 = f(E^2)$ at several CMC concentrations. The linearity of the graphics means that the degree of orientation is low, i.e. $\gamma E^2 << kT$. Therefore γ is proportional to the slope of the lines $[(\Delta I/I_0)/E^2]$ according to eq. (23).

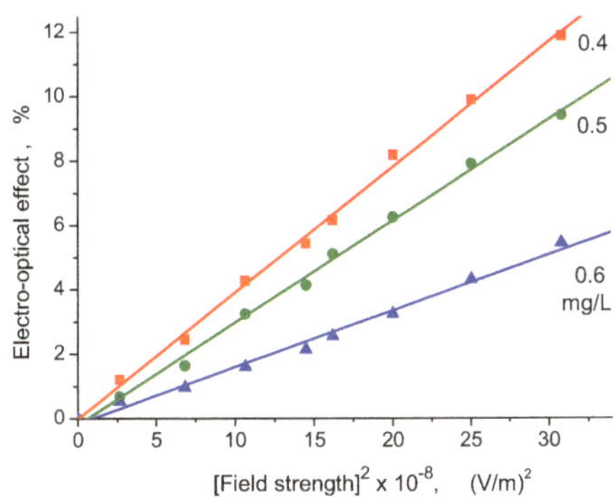

Figure 3. Dependence of steady-state EOE $(\Delta I_s/I_0)\times 100$ in water-polymer suspension of γ-Al_2O_3 particles on the squared field strength E^2 [10^8 V^2m^{-2}] at 1 kHz and NaCMC concentrations $C_{CMC} = (4-6)\times 10^{-4}$ g/L NaCMC (under the recharging point).

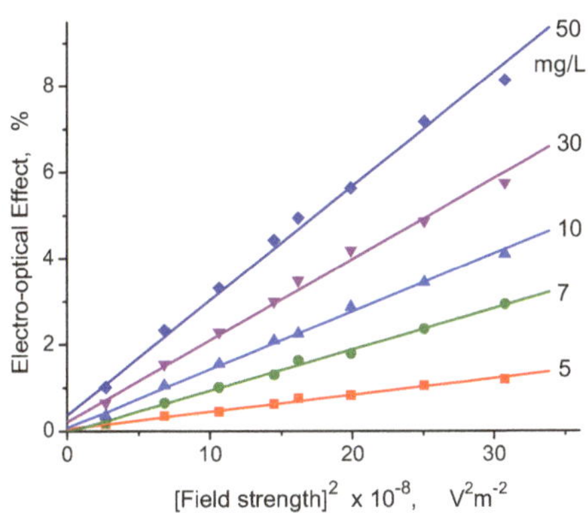

Figure 4. Field strength dependence (the same as in Fig. 3) at $C_{CMC} = (5-50)\times 10^{-3}$ g/L NaCMC (above the recharging point).

Fig. 3 shows out that the polarizability decreases with CMC concentration but Fig. 4 demonstrates just the opposite: γ increases with C_{CMC}. The both trends correlate with the concentration dependence of the electrophoretic mobility $u_{el}(C_{CMC})$ (Fig. 2) which reflects the alteration in the total surface charge of the particles with adsorbed CMC. That means that the magnitude of γ depends mainly on the total charge of the diffuse counterions (which decreases before the recharging point and increases after it). So, the kilohertz polarizability is due predominatingly to the diffuse counterion but that does not exclude participation of condensed ions.

5.2.4. High frequency dispersion of CMC-alumina polarization

The change of the half-period $1/2\nu$ of the applied sinusoidal electric field allows altering the ions migration distance and by that the contribution to γ of the counterions being in two physical states: adsorbed and diffuse. According the theory (see paragraph *5.1.1*) the condensed cations have diminished mobility along the CMC chain due to the strong electrostatic interaction with the charged COO^- groups. The critical frequency $\nu_{1/2} \sim D_i/L^2$ depends on the polyion length L and the diffusion coefficient D_i of the counterions electrically associated to the polyelectrolyte chain (Eqs. 17, 18); so, at constant L (absence of particles aggregation, see Ref. [54]) the value of $\nu_{1/2}$ gives information about the counterions mobility. The dispersion dependence $\gamma(\nu)$ shows the magnitude and $\nu_{1/2}$ of the polarizability components with different D_i/L; that allow distinguishing the contribution of the condensed and diffuse counterions.

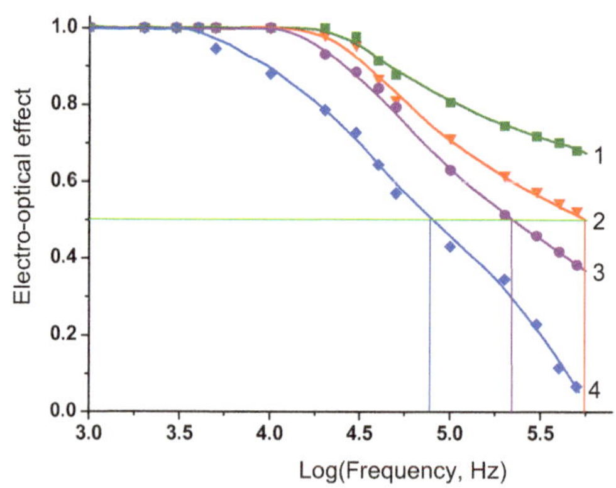

Figure 5. Normalized high frequency dependence of EOE ($\Delta I/I_0$) of alumina particles in water (curve 1) and in aqueous solution NaCMC (M_w = 250 kg/mol) at concentrations: $4{\cdot}10^{-4}$ g/L (curve 2), $6{\cdot}10^{-4}$ g/L (curve 3) and $1{\cdot}10^{-2}$ g/L (curve 4). The EOE $\Delta I_s/I_0$ at 1 kHz is taken to be one for all curves.

The frequency dependences of EOE of alumina particles with size 0.3 µm in CMC solution is shown in Fig. 5. The total particle charge is positive for curves 1–3 and negative for curve 4 (recharged surface). The critical frequency decreases with the adsorbed CMC amount: $\nu_{1/2} \approx 550$ kHz (curve 2), $\nu_{1/2} \approx 220$ kHz (curve 3) and $\nu_{1/2} \approx 80$ kHz (curve 4).

The form of the dispersion curves $\gamma(\nu)$ indicates that the polarization is due to one kind of counterions; in the case of different D_i/L the curves should have a shoulder with fixed $\nu_{1/2}$ but increasing increment while the CMC adsorbed amount grows. The value of $\nu_{1/2}$ at saturated adsorption (curve 4) is not very different from that at small CMC surface concentration (curves 3 and 2) and bare particle (curve 1) determined by the diffuse counterions polarization; this fact together with the absence of noticeable shoulder allows concluding that the polarizability of the CMC covered surface (curve 4) is due to the diffuse counterions around the particles. The displacement of the $\nu_{1/2}$ to lower frequencies could be explained with their decreased mobility because of the presence of polymer layer, i.e. the effective diffusion coefficient D_i is smaller than that of free ions in solution.

Nevertheless, the results in Fig. 5 do not exclude unambiguously the condensed counterions participation in the polarization. The absence of the increasing shoulder at CMC adsorption could be owing to equal ratio D_i/L for the condensed and diffuse counterions. In this case D_{Na} of the condensed Na^+ ions should be 400 times less than that of the diffuse ones assuming $L = l_p = 15$ nm for bound counterions and $L = 0.3$ μm (the particle size) for the free ones.

To exclude the possibility that the values of $\nu_{1/2} \sim D_i/L^2$ of the diffuse and condensed counterions polarization coincide due to unique combination of L and D_i we have studied the concentration dependence $\gamma(C_{CMC})$ which reflects the magnitude of γ as function of the quantity of both kinds of counterions [54]. The measurements are performed at $\nu = 1$ kHz where the half-period $1/2\nu$ is long enough to reach the equilibrium distribution of counterions in any moment t in a field with instantaneous strength $E_t = E_0\sin(2\pi\nu t)$; indication for that is the plateau at all C_{CMC} (Fig. 5).

5.2.5. Concentration dependence of CMC-alumina polarizability

The contribution of the diffuse and condensed counterions can be estimated by comparing the concentration dependences of the polarizability $\gamma(C_{CMC})$ and mobility $u_{el}(C_{CMC})$. This approach is based on the fact that the condensed counterions move together with the polyelectrolyte chain in a permanent electric field, that is why they are not seen in methods based on the translational motion in a d.c. electric field [61]. I.e. the u_{el} gives information about the quantity of the diffuse counterions only while the polarizability in a sinusoidal field reflects participation of the all movable counterions.

When the amount of adsorbed CMC increases with C_{CMC} then the ratio condensed/diffuse counterions alters but in a different way before and after the isoelectric point. The quantity of condensed ions increases for all C_{CMC} concentrations where adsorption appears but the quantity of the diffuse counterions decreases before the recharging point and increases after it. The comparison between $\gamma(C_{CMC})$ and $u_{el}(C_{CMC})$ at lower C_{CMC} concentrations can give unambiguous information about the relative contribution of the diffuse and condensed counterions due to the opposite change of their quantity. In this concentration range the slope of the dependence $\gamma(C_{CMC})$ should be negative but smaller than that of $u_{el}(C_{CMC})$ taking into account that the adsorbed CMC introduces approximately one condensed and two diffuse Na^+ counterions on every three COO^- groups (see paragraph 3.3.5).

Fig. 6 refutes this expectation: both the curves have the same slope before the isoelectric point; this disproves the hypothesis that the condensed ions participate in the polarization. The difference at the higher CMC concentrations can be explained by the different influence of the polymer layer on the behaviour of the counterions in direct and sinusoidal electric field and by the way the figure is constructed: the two quantities are equated ($\gamma = u_{el}$) at $C_{CMC} = 0$.

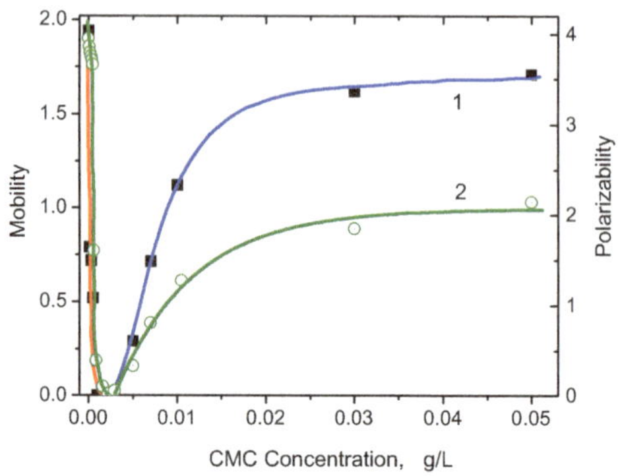

Figure 6. Fig. 6. Dependences of the absolute value $|u_{el}|$ of the electrophoretic mobility (curve 1, left ordinate) and the electric polarizability γ (curve 2, right ordinate) of γ-Al_2O_3 particles on the concentration C_{CMC} of NaCMC. The u_{el} and γ are given in relative units assumed to be equal in the absence of CMC. The sign of u_{el} is positive at C_{CMC} under 3 mg/L (the red part of the curve) and negative above this concentration (the blue part).

So, the correlation between the concentration dependences $\gamma(C_{CMC})$ and $u_{el}(C_{CMC})$ indicates that the kilohertz polarizability of the alumina particles with adsorbed CMC is determined by the diffuse counterions.

5.2.6. Comparing with the literature

A low frequency shift of the polarizability with $\nu_{1/2} \approx 3$ kHz is found at saturated CMC adsorption on ferrioxide particles [62]. That value of $\nu_{1/2}$ is 25 times lower than ours (curve 4 on Fig. 5) although in both experiments the same CMC specimen (M_w = 250 kg/mol, DS = 1.2) is studied. The difference in $\nu_{1/2}$ could be related to the different particles employed; this suggests that the value of $\nu_{1/2}$ is determined by L rather than by D_i. An additional argument is that the $\nu_{1/2} \approx 80$ kHz of CMC coated alumina particles lies in the frequency region specific for the diffuse counterions polarizability of elongated ellipsoids with $L \approx 0.3$ μm (taking into account that $\nu_{1/2} \sim 1/L^2$) [63]. That means that the interfacial polarization of CMC coated particles is determined by the diffuse counterions displacement around the whole particle but not along the CMC chains. If the bound counterions are the origin of the electric polarization then a shoulder on the dispersion curve should appear and its increment must be proportional to the surface CMC amount.

In Ref. [62] a principally different conclusion is drawn: the electric polarizability of CMC coated particles is due to the condensed counterions. The interpretation is based on the low-frequency shift of the dispersion curves employing the Schwarz's equation $\nu_{1/2} = 4D_i/\pi L^2$ and using the literature value $D_i = 6 \cdot 10^{-11}$ m²/s [64] for bound Na^+ in a CMC solution; the CMC chain is assumed to be a straight polyion with length L = 505 nm (equal to the contour length L_c of CMC with M = 250 kg/mol). Then, accepting that the estimated $\nu_{1/2}$ = 300 Hz is close to the experimentally obtained $\nu_{1/2} \approx 3$ kHz, the authors conclude that the orientation of CMC coated particles in the electric field is due to condensed counterions polarization along the adsorbed polyelectrolyte chains.

The interpretation in Ref. [62] is wrong because the CMC chain with a contour length L_c = 0.5 μm cannot be considered as rod-like neither free in the solution nor adsorbed on the particle surface. It is well known that parts of adsorbed chain take a quasi-planar (two-dimensional) conformation and the rest of the chain retains a three-dimensional conformation [65], so the form of adsorbed CMC is predetermined by its conformation in solution. A free CMC chain with M = 250 kg/mol has the conformation of a random coil with $N_K \approx 15$ statistical segments (see section 4.3) and a radius of gyration R_g = 44 nm (Fig. 1). So, for a CMC chain with L_c = 0.5 μm the magnitude of γ and the critical frequency $\nu_{1/2}$ cannot be estimated using the contour length L_c in the equations derived for the polarizability of straight polyions. This conclusion is confirmed by results for CMC in aqueous solution: the experimental dependence $\gamma(L_c)$ does not correspond to $\gamma \sim L_c^2$ predicted by Eq. (17, 18) but has a plateau at $M \geq 200$ kg/mol [52]; so, the dependence $\nu_{1/2} \sim L^{-2}$ (according Schwarz's equation) is also not applicable for long CMC chains.

My opinion is that the polarizability of the polymer covered particles is due to the diffuse counterions but their mobility is decreased because of the presence of the polymer layer, i.e. the effective diffusion coefficient D_i is smaller than that of the free ions in solution. The reason for such interpretation is the fact that the dispersion curve displacement (towards lower $\nu_{1/2}$) increases with the CMC amount on the particle surface.

Author details

Alexandar Metodiev Zhivkov
Institute of Physical Chemistry, Bulgarian Academy of Sciences, Acad. G. Bonchev Str., Sofia, Bulgaria

6. References

[1] D.A. Rees, *Adv. Carbohydr. Chem. Biochem.* 24 (1969) 267-332.
[2] V.S.R. Rao, N. Yathindra, P.R. Sundararajan, *Biopolymers* 8 (1969) 325-333.
[3] B.K. Sathyanarayana, V.S.R. Rao, *Biopolymers* 11 (1972) 1379-1394.
[4] E. Ott, H.M. Spurlin, M.W. Grafflin, *Cellulose and Cellulose Derivatives*, part II, *High Polymers*, vol. V, Interscience publishers, New York, 1954.

[5] D.E. Metzler, *Biochemistry*, Academic press, New York, 1977.
[6] J. Reuben, H.T. Connor, *Carbohydr. Res.* 115 (1983) 1-13.
[7] Z. Ma, Z. Wiebang, Z. Li, *Chinese J. Polym. Sci.* 7 (1989) 45.
[8] C.W. Hoogendam, A. de Keizer, M.A. Cohen Stuart, B.H. Bijsterbosch, J.A.M. Smit, J.A.P.P. van Dijk, P.M. van der Horst, J.G. Batellaan, *Macromolecules* 31 (1998) 6297-6309.
[9] A. Katchalsky, N. Shavit, H. Eisenberg, *J. Polym. Sci.* 13 (1954) 69-84.
[10] F.H. Crowdhury, S.M. Neale, *J. Polym. Sci.* 1 (1963) 2881.
[11] L. Zhang, T. Takematsu, T. Norisuye, *Macromolecules* 20 (1987) 2882.
[12] A.V. Dobrynin, M. Rubinstain, *Prog. Polym. Sci.* 30 (2005) 1049-1118.
[13] A. Katchalsky, Z. Alexandrowicz, In: B.E. Conway, R.G. Baradas (Eds.), *Chemical physics of ionic solutions*, London, Wiley, 1966. p.295-346.
[14] F. Oosawa, *Polyelectrolytes*, New York, Marcel Dekker, 1971.
[15] R.M. Fuoss, A. Katchalsky, S. Lifson, *Proc. Natl. Acad. Sci. USA* 37 (1951) 579-586.
[16] Z. Alexandrowicz, A. Katchalsky, *J. Polym. Sci. Part A* 1 (1963) 3231-3260.
[17] A. Katchalsky, *Pure Appl. Chem.* 26 (1971) 327-374.
[18] T. Nishio, A. Minakata, *J. Chem. Phys.* 113 (2000) 10784-10792.
[19] T. Nishio, A. Minakata, *J. Phys. Chem. B*, 107 (2003) 8140-8145.
[20] H. Ohshima, K. Furusawa (eds.), *Electrical Phenomena at Interfaces: Fundamentals, Measurements and Applications*, 2ed edition, Dekker, New York, 1998.
[21] H. Ohshima, *Theory of Colloid and Interfacial Electric Phenomena*, Elsevier/Academic Press, Amsterdam, 2006.
[22] G.S. Manning, *J. Chem. Phys.* 51 (1969) 924-933.
[23] G.S. Manning, *Polyelectrolyte binding*, 12 (1979) 443-449.
[24] H. Ohshima, *Biophysical Chemistry of Biointerfaces*, Wiley, 2010.
[25] C.R. Cantor, P.R. Schimmel, *Biophysical Chemistry*, W.H. Freeman and Company, San Francisco, 1980.
[26] S.R. Rafikov, V.P. Budtov, Y.B. Monakov, *Introduction in physico-chemistry of polymer solutions*, Nauka, Moscow, 1978.
[27] B.E. Eskin, *Light Scattering by Polymer Solutions and Macromolecule Properties*, Nauka, Leningrad, 1986.
[28] B.E. Eskin, *Scattering of Light by Polymer Solutions*, Nauka, Moscow, 1973.
[29] M. Yalpani, *Polysaccharides*, Elsevier, New York, 1988.
[30] M. Rinaudo, in: J.F. Kennedy, G.O. Phillips, P.O. Williams, L. Piculell (eds), *Cellulose and cellulose derivates: physicochemical aspects and industrial applications*, Woodhead Publishing Limited, Cambrige, U.K., 1995.
[31] P.N. Lavrenko, O.V. Okatova, V.N. Tsvetkov, H. Dautzenberg, B. Philipp, *Polymer* 31 (1990) 348-352.
[32] R.M. Davis, *Macromolecules* 24 (1991) 1149-1155.
[33] T. Odijk, *Biopolymers* 18 (1979) 3111-3113.
[34] A. Katchalsky, S. Lifson, *J. Polym. Sci.* 13 (1954) 43-55.
[35] A.M. Zhivkov, *J. Colloid Interface Sci.* 113 (2007) 122-127.

[36] S.S. Dukhin, V.N. Shilov *Dielectric Phenomena and Double Layer in Disperse Systems and Polyelectrolytes*, Wiley, New York, 1974.
[37] D. Porschke, M. Antosiewicz, in: S.P. Stoylov and M.V. Stoimenova (eds.), *Molecular and Colloid Electro-optics*, Taylor & Francis, New York, 2007.
[38] G. Schwarz, *J. Phys. Chem.* 66 (1962) 2636-2641.
[39] M. Mandel, A. Jenard, *Transactions Faraday Soc.* 59 (1963) 2158-3169.
[40] M. Mandel, *Mol. Phys.* 4 (1961) 489-496.
[41] F. Oosawa, *Biopolymers* 9 (1970) 677-688.
[42] A. Minakata, N. Imai, F. Oosawa, *Biopolymers* 11 (1972) 347-354.
[43] A. Warashina, A. Minakata, *J. Chem. Phys.* 58 (1973) 4743-4749.
[44] M. Mandel, in: S. Krause (Ed.) *Molecular Electro-Optics*, Plenum Press, New York, 1981.
[45] A. Minakata, H. Takahashi, T. Nishio, J. Nagaya, A. Tanioka, *Colloids & Surfaces A*:209 (2002) 213-218.
[46] A. Minakata, T. Nishio, M. Kimura, S.-I. Yano, Y. Tanaka, T. Shimizu, *Colloids & Surfaces B: Biointerfaces* 56 (2007) 277-280.
[47] L.-G. Allgen, S. Roswall, *J. Polymer Sci.* 12 (1954) 229-236.
[48] A. Oga, A. Takimoto, R. Hayakawa, Y. Wada, *Rept. Progr. Polymer Phys. Japan*, 24 (1981) 77.
[49] M. Mandel, *Ann. New York Acad. Sci.* 303 (1977) 74-87.
[50] Th. Vreugdenhil, F. van der Touw, M. Mandel, *Biophys. Chem.* 10 (1979) 67-80.
[51] Fredericq E., Houssier C., *Electric Dichroism and Electric Birefringence*, Clarendon press, Oxford, 1973.
[52] A.R. Foweraker, B.R. Jennings, *Polymer* 16 (1975) 720-724.
[53] T. Yamamoto, Y. Mori, N. Ookubo, R. Hayakawa, Y. Wada, Colloid & Polymer Sci. 260 (1982) 20-26.
[54] A.M. Zhivkov, R.P. Hristov, *J. Physics: Condensed Matter* 22 (2010) 494112 (7 pp).
[55] J. Gregory, S. Barany, *Advances Colloid. Interface Sci.*, 169 (2011) 1-12.
[56] F. Bordi, S. Sennato, D. Truzzolillo, *J. Phys: Condens. Matter* 21 (2009) 203102 (26 pp.).
[57] S.P. Stoylov, *Colloid Electro-optics - Theory, Techniques and Applications*, Acad. Press, London, 1991.
[58] H.C. van de Hulst, *Light Scattering by Small Particles*, John Wiley, New York, 1957.
[59] A.A. Spartakov, A.A. Trusov, A.V. Voitylov, V.V. Vojtylov, in: S.P. Stoylov and M.V. Stoimenova (eds.), *Molecular and Colloid Electro-optics*, Taylor & Francis, New York, 2007.
[60] A.M. Zhivkov, in: S.P. Stoylov and M.V. Stoimenova (eds.), *Molecular and Colloid Electro-optics*, Taylor & Francis, New York, 2007.
[61] M.C. Stuart, R. de Vries, H. Lyklema, *Polyelectrolytes*, in: Lyclema H. (Ed.), *Fundamentals of interface and colloid science*, vol. V, Elsevier, London, 2005.
[62] Ts. Radeva, K. Kamburova, *J. Colloid Interface Sci.* 293 (2006) 290-295.
[63] V.N. Shilov, Yu.B. Borkovskaja, S.N. Budankova, in: S.P. Stoylov, M. Stoimenova (**eds.**), *Molecular and Colloid Electro-optics*, Taylor & Francis, New York, 2007.
[64] N. Oocubo, Y. Hirai, K. Ito, R. Hayakawa, *Macromolecules* 22 (1989) 1359-1366.

[65] G.J. Fleer, M.A. Cohen-Stuart, J.M.H.M. Scheutjens, T. Cosgove, B. Vincent, *Polymers at interfaces*, London, Chapman & Hall, 1993.

Chapter 9

Implications of Cellulose in Modeling the Behavior of Vegetal Additive Materials in Clay Based Ceramics: Technical and Archaeological Issues

Ferenc Kristály

Additional information is available at the end of the chapter

1. Introduction

The separate sections are structured so to present novel results of personal experiments and investigations combined with literature references on major domains where cellulose as a model material is involved in clay bricks and archaeometry research. Each section presents a step in materials characterization or testing, which requires the cellulose as model material to decipher the results and observation. Each section may represent a separate step of a chain of investigations, but the information from each is needed to understand the behavior of vegetal materials in clay mixtures, and especially the need for cellulose as a model material.

Working with vegetal waste materials in clay brick industry may seem complex at first. But using the traditional analytical techniques of mineralogy the properties and behavior of different types of vegetal materials may be deciphered. X-ray diffraction reveals that large amounts of vegetal materials components are crystalline (although on a nanometric scale) and are made up by cellulose. This is confirmed by optical microscopy in polarizing light also. Scanning electron microscopy is the best way to characterize the fabric of different vegetal materials. Beyond chemical and physical composition, the fabric – fibre dimensions, orientation and textural porosity – is also a basic property, regarding physical properties of vegetal material grains. Among physical properties is included elasticity, which was observed as main factor in behavior of clay mixtures with vegetal materials regarding deformation at high pressures. Another property of vegetal materials is to form remnants after thermal oxidation. The above mentioned analytical techniques help to demonstrate that these remnants are inorganic, mineral and not char (or carbon based) and that their chemical composition and fabric is characteristic for each vegetal material and plant part.

The path of remnants formation was followed by thermal analytical techniques and combined sequential firing and optical microscopy techniques. By thermal analysis also the decomposition reactions were determined and evolved gaseous products identified. Here, the role of cellulose was again crucial, since the thermal reactions allow quantification of crystalline and non-crystalline components, if one is aware of characteristic reactions. The heat contribution of vegetal material oxidation was observed to influence the behavior of clay minerals and composition of fired clay products also.

Abbreviations used for investigation techniques: X-ray Powder Diffraction – XPD, Optical Microscopy in Transmitted Polarized Light – OMTPL, Scanning Electron Microscopy – SEM, Energy Dispersive Spectrometry – EDS, Differential Thermal Analysis – DTA, Thermogravimetry – TG, Derivative Thermogravimetry – DTG, Evolved Gas Analysis – EGA, VM – vegetal materials.

2. Cellulose as crystalline component in vegetal materials

XPD was performed on <50μm grain size random powder specimens. Samples were obtained by grinding liquid N_2 dried materials of sawdust (SD), sunflower seeds hull (SSH) and rice husks (RH). Patterns were obtained in Bragg-Brentano geometry with Cu-Kα radiation, on top-loaded samples. Patterns were evaluated by Search/Match on ICDD PDF-2 (2005) and the best match was "native cellulose". All patterns display two intense broad peaks at ~15°(2θ) and ~ 22°(2θ) and a smaller one at ~35°(2θ) (Fig 1). These correspond, in shape and position with data reported by Thygesen et al. (2005). Peak broadening indicates poor crystallinity, <10nm (fiber length), as usually given in literature (Das et al. 2010, Thygesen et al. 2005, Zhao et al. 2007). As in the case of several plant materials (Ververis et al. 2004) the crystalline cellulose content is approximated to be ~20-40% of the investigated materials. Crystallite sizes (Table 1) were estimated from peak FWHM by the Scherrer-formula, structure-less solution, in TOPAS. Instrumental contribution was removed by the Fundamental Parameters Approach peak shape (profile) determination. The amorphous content was modeled by the broad peak method (hump centered at ~28°2θ), as the fraction of hump area from the total scattered intensity. However, results are to be treated as semi-quantitative, due to intense peak broadening, overlapping and grain size effects.

These crystallite sizes are not directly observed by microscopy techniques. Correction of texture (preferred orientation) and stress-strain effects might be necessary since the individual crystallites usually build up into fibers, rod-shaped quasi crystals. However, for the current applications these aspects might be disregarded and semi-quantitative results are used in the experiments. But, as a main conclusion, crystallinity determination of VM-s from XRD needs combined evaluation techniques and more importantly, combined analytical methods. For instance, cellulose content based on TA gave highest value for SSH, but a clear delimitation from lignin was not possible.

OMTPL of VM grains in ~30μm thin sections revealed a polycrystalline structure. The microscopy observations were conducted with 1 Nicol, crossed Nicols and crossed Nicols with gypsum compensator. The plant fibers in the grains are made up by fibrils, intergrown with crystallographic orientation. Sections were obtained in various directions to the fibers. On the SD grains (Fig 2) the fibers are cut perpendicular.

Position		Peak area (cps*°(2θ))			Crystallite sizes (nm)		
°(2θ)	(hkl)	SD	SSH	RH	SD	SSH	RH
~14.9	?	6.54	1.98	0.81	1.7	2.2	4.5
~16.6	(111)	3.84	2.23	1.91	2.1	2.6	2.6
~20.6	(021)	1.12	4.21	4.89	6.1	3	2.1
~22.3	(002)	20.34	10.74	13.12	3.3	3.6	2.5
29.44[a]	*(040)*	*1.59*			1566		
~34.7		1.48	0.84	0.70	7.4	11.9	16
Crystalline Σ_{area}	cps	33.32	20.00	21.43	area of cellulose peaks		
Amorphous Σ_{area}	cps	70	53	45	area of ~28°2θ hump		
crystallinity	%	32	24	32	relative percents		

Table 1. Crystallite sizes for cellulose (a – Ca-oxalate or oxalic acid peak)

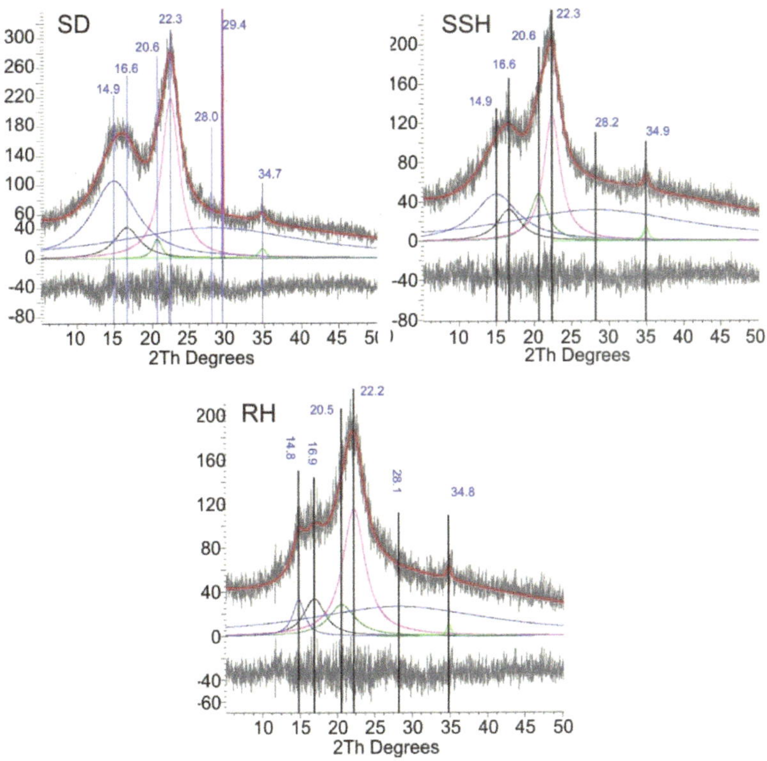

Figure 1. Cellulose peaks and amorphous hump (peak at ~28°(2Th)

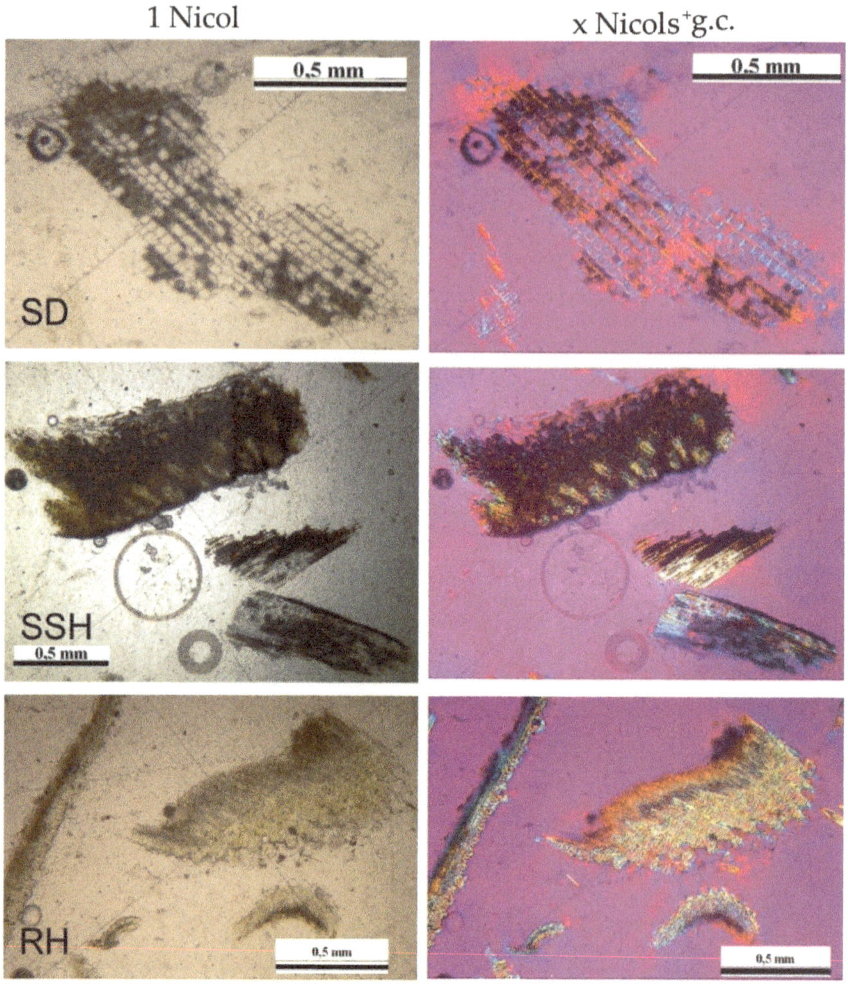

Figure 2. OMTPL images of vegetal grains (x Nichols=crossed Nichols,g.c.=gypsum compensator)

On the image with crossed Nicols and gypsum compensator, the different crystallographic orientation of fiber groups is observed. For SSH the largest grain shows the section perpendicular to the elongation of the hulls, displaying micrometric bunches of fibrils in similar crystallographic orientations. The fibrils are on the outer, harder part of the hulls. To the inner parts, fibrils disappear and an amorphous material is seen, probably non-crystalline cellulose and lignin. RH is shown in 3 orientations, the largest grain is oblique to the platy husks, the longest is perpendicular and along the elongation, while the small curved grains are perpendicular to the platy husks and elongation too. As compared to SD and SSH, the smallest fibrils are observed. A layered structure is seen: massive lignin on the outer part, long fibrils parallel to elongation and platy-ness in the middle and small fibrils building up the inner undulated surface.

3. Fabric (microstructure) of vegetal grains, plant-specific features, characteristic chemical elements in the cellulose and lignin

SEM and EDS on polished surface raw vegetal samples evidenced the chemical elements in the organic compounds and the fine-scale structure of fibers also. For sample preparation methodology and experimental details see Kristály & Gömze (2008). In the SD grains texture the section perpendicular to the fibers shows the polygonal structure (Fig 3, SD-a) with various dimensions of the individual polygons (Fig 3, SD-b). These in a natural state are empty on the inside and are sections of channels running parallel to the grains elongation (Fig 3, SD-c). Thus, the elongated grains of SD are due to the existence of long fibers in one direction only.

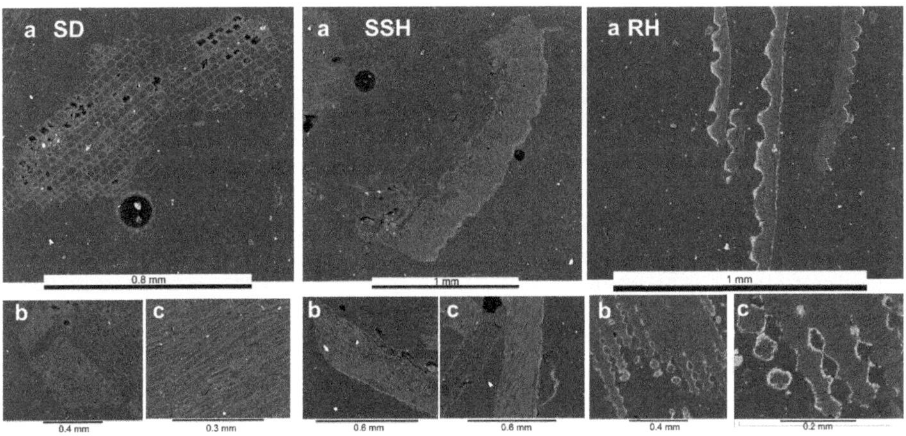

Figure 3. BSE images of vegetal material grains

For SSH grains, the differentiated texture is better observed, the light gray shell is the K-enriched organic material. The section parallel to the hulls elongation (Fig 3, SSH-a) displays a sponge-like structure, covered by the K-rich shell. On detailed oblique sections (Fig 3, SSH-b and c) the micrometric polygonal cavities of the sponge-like structure turn out to be fibers – channels running perpendicular to the elongation and in parallel with the K-rich shell. In RH grains the white strips on one side of the grains (Fig 3, RH-a) are the Si-rich (up to 40% Si) cellulose parts, on the inner side of the husks. Small amount of Si-enrichment is also observed on the outer shell of husks and in the fine-fibrous mass. Perpendicular to the elongation the uneven inner surface is visible. In section parallel with the elongation, the micrometric compact fibers perpendicular to the elongation are seen (Fig 3, RH-c). In the plane of elongation, the Si-rich uneven inner surface forms a hollow reticular planar texture (Fig 3, RH-b and RH-c). This structure may have pillaring role, when the structure is undertaken to mechanical deformations.

Using the routine SEM and EDS sample preparation and investigations of mineralogy, we could evidence basic features of vegetal materials and important differences in their fabric which determine their behaviour in various conditions/applications.

4. Thermal decomposition of vegetal materials as cellulose depolymerization, evolved heat and gases, fabric transformation

Simultaneous DTA, TG and DTG ("derivatography") was performed in order to observe the thermal decomposition reactions of vegetal materials. The most important observations are that 1) all investigated vegetal materials display very similar thermal reactions (Kristály et al. 2010a); 2) the type of reactions depend on the O_2 content of reaction media and 3) activation energies and heat of reaction vary depending on the inorganic content of the vegetal materials. The thermal reactions observed on SD, SSH and RH samples are summarized in Table 2 as DTA and DTG peak temperatures. The 2nd reaction is of depolymerization, while the 3rd one is combustion in the 5 mg samples and partial carbonization in the 200 mg samples. We could learn from these reactions that oxidation as decomposition of organic materials occurs only if the amount of investigated materials is low so that O_2 content of a static air atmosphere is sufficient. The carbonization of cellulose is supported by the black residue left at 700°C. The residue for 5 mg samples was white and glassy powder.

Nr.	SD 5mg		SD 200mg		SSH 5mg		SSH 200mg		RH 5 mg		RH 200mg	
	DTA	DTG	DTA	DTG	DTA	DTG	DTA	DTG	DTA	DTG	DTA	DTG
1	-	-	120 ↓	120	-	-	120 ↓	120	-	-	127 ↓	130
2	349 ↑	336	296 ↑	290	318 ↑	290	270 ↑	280	330 ↑	306	285 ↑	285
3	458 ↑	464	445 ↓	352	478 ↑	500	394 ↓	330	445 ↑	445	336 ↑	338
4			450 ↓	-			445 ↓				487 ↓	-
5											515 ↓	-

Table 2. Peak temperatures of DTA curves on VM, without peak deconvolution.

Deconvolution approach on DTA curves revealed several overlapping peaks, which were determined for 5 mg samples. Deconvolution was performed in Fityk 0.9.4. (Wojdyr 2010) using split-PearsonVII function on DTA curves with base-line correction in WinDer. The base-line shifts after correction are due to the heat diffusion effect of crucibles and loss of sample as related to the constant volume (mass) of inert material. At least 3 major exothermic peaks, followed by at least one endothermic peak are observed (Fig 4). Comparing values from Fig 4 to those of Table 2 the small differences are observed for the peaks observed without deconvolution (for the 5mg samples). The peak between 2nd and 3rd exothermic shows similar values for SD, SSH and RH, but its intensity varies largely. It represents the oxidation of functional groups liberated by breaking polymer chains. The endothermic peak following oxidation peaks is not a carbonization reaction. It is the reorganization into inorganic material of non-volatile (mineral) elements in the VM. This also explains the large differences in the peak T(°C). Additional reactions are observed on RH, which are assigned to be the SiO_2 crystallization – reorganization processes. The large endothermic peak >700°C are instrumental contributions, the asymmetry caused by the almost empty sample crucible vs. the α-Al_2O_3 containing crucible. Heat of reaction (A_{peak} [$°C^2$]/M_{sample} [mg]) is 53.97 for SD, 64.53 for SSH and 31.38 for RH (where A_{peak} was taken as the total exothermic peaks). The solid residues were 0.2 mg for SD, 0.6 for SSH and 1.1 for RH.

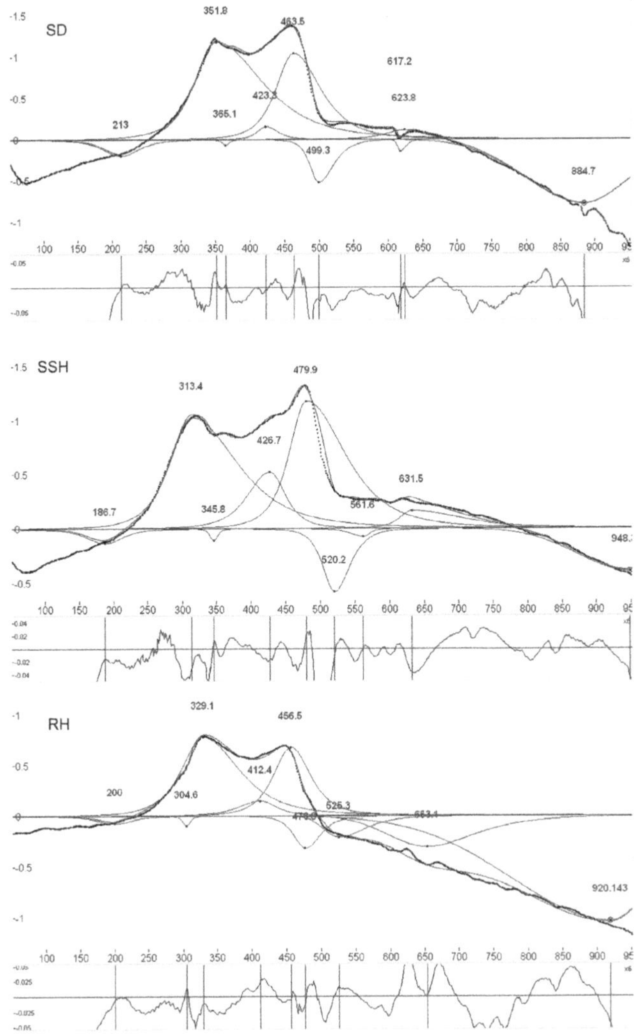

Figure 4. Deconvoluted DTA curves (5mg sample, 10°C/min linear heating)

EGA by means of quadrupole mass spectrometry (QMS) is a useful tool to observe volatile products for thermal reactions. In our experiments (Kristály et al. 2010) we have used equipment at University of Debrecen (Hungary) (Szöőr & Bohátka 1985). Simultaneous detection of various atomic/ionic or molecular mass compounds is possible. By evacuating the evolved gases trough a capillary system into a high vacuum drive, the gases from the derivatograph furnace are directly driven to the spectrometer. Thus, static-air atmosphere investigations are possible, on sample amounts ranging from 5mg to several hundreds of mg-s. Liberation of H_2O and OH^- are observed at similar reactions (Fig 5), indicating that OH-to-H_2O reactions support the oxidations. However, in the second intense reaction (3rd among deconvoluted peaks) less (or no) water and more CO_2 is produced. This observation supports the depolymerization-to-combustion route of thermal decomposition. The three peak system of decomposition in SSH is also observed in the QMS patterns. The reaction path is supported by data from literature (Table 3), but also the presence of other cellulose like polysaccharides is evidenced.

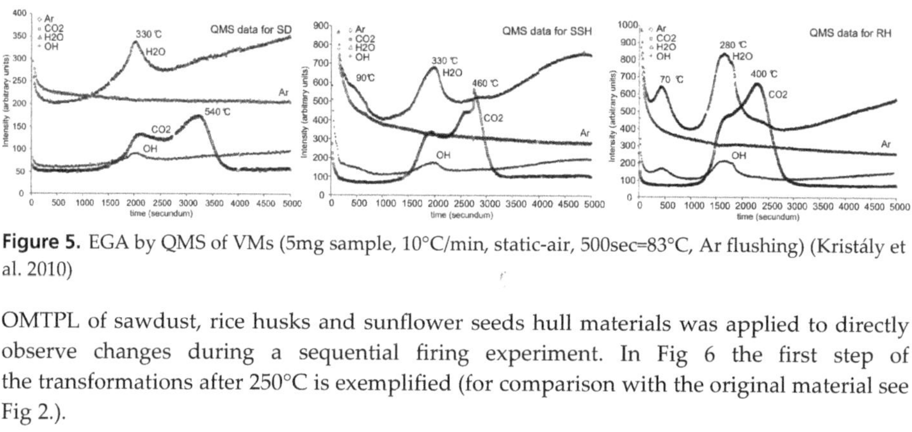

Figure 5. EGA by QMS of VMs (5mg sample, 10°C/min, static-air, 500sec=83°C, Ar flushing) (Kristály et al. 2010)

OMTPL of sawdust, rice husks and sunflower seeds hull materials was applied to directly observe changes during a sequential firing experiment. In Fig 6 the first step of the transformations after 250°C is exemplified (for comparison with the original material see Fig 2.).

nr.	Reaction type	Our results SD	SSH	RH	Szcześniak et.al. 2008	H. Yang et.al. (2007) cell.[1]	hemic.[2]	lign.[3]
1	exothermal	352	314	330	300	270	350	cont.
2	exothermal	467	480	457	440	*no combustion**		

Table 3. Exothermal peak temperatures, compared to experiments on cellulose and derivatives (*-inert atmosphere)

The images with crossed Nicols and gypsum compensator enabled to observe the vitrification/amorphization of VM textures. SD has preserved most of its crystalline fibrils as evidenced by the variation of interference colors upon specimen rotation. SSH already presents some degree of amorphization as black spots in the fibrillar fabric, and some degree of shrinkage is observed, as gap between clay matrix and SSH grains. RH was undertaken to total amorphization, indicating that H_2O liberated is part of the Si – cellulose polymer and individual crystallites are below detection limit of OMTPL. This fact is also sustained by the brownish dark color of the vegetal mass, which at higher temperatures is cleared and transformed to glass. At 450°C (Fig 7) the total combustion of organic components is observed without shrinkage of SD and RH grains and ~70% by volume shrinkage of SSH grains.

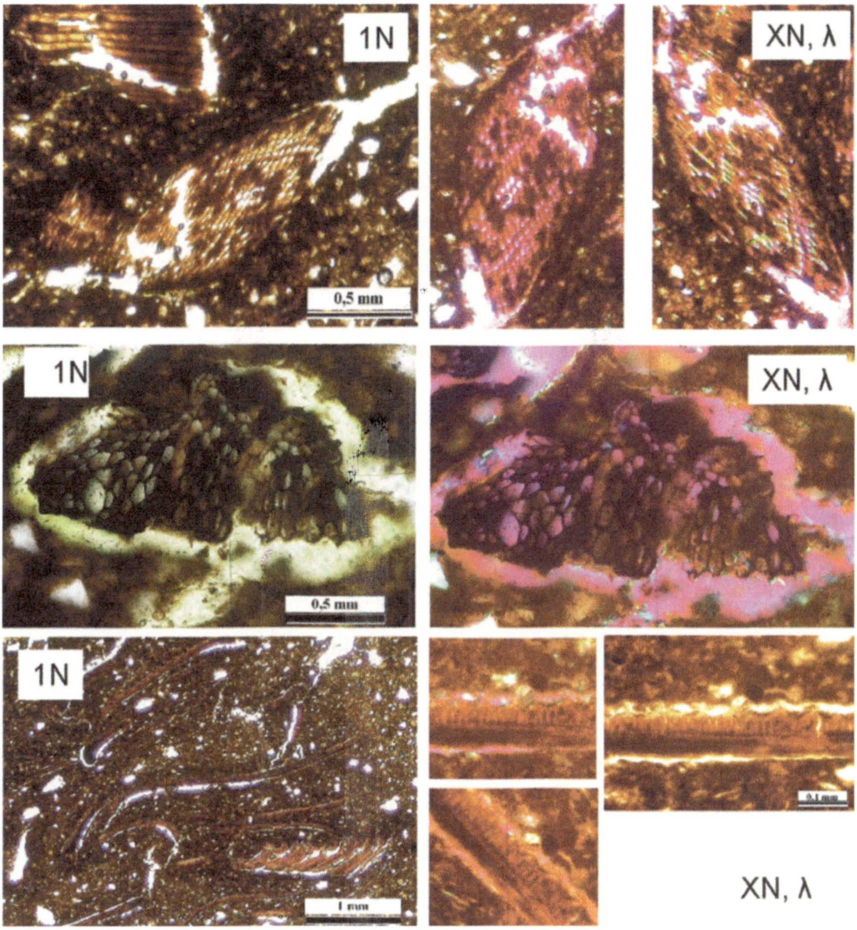

Figure 6. OMTPL images of VM at 250°C

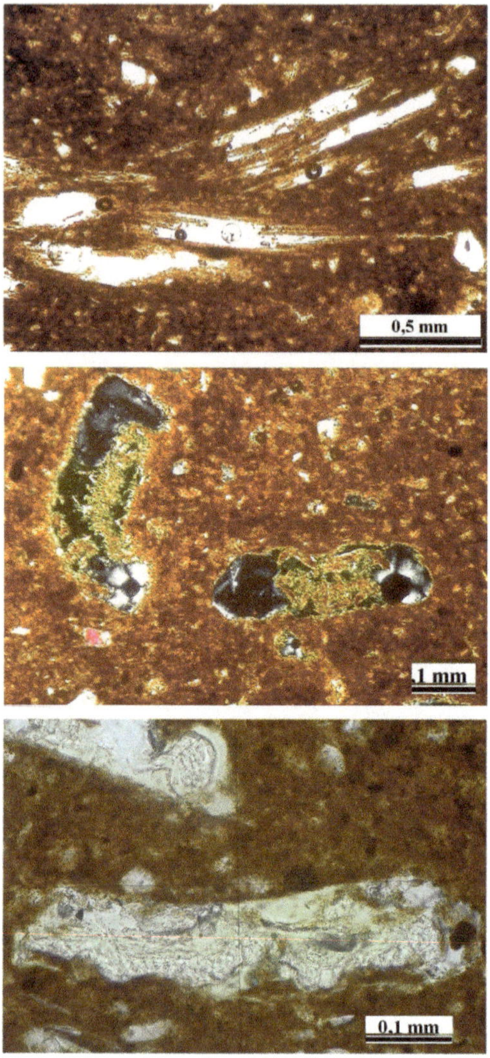

Figure 7. VM remnants at 450°C (top-SD, 1N; centre-SSH, xN; bottom-RH, 1N)

In comparison with the TA results, the second exothermic peak continues above this temperature due to the diffusion of heat in the clay material and sample holder crucible. The characterization of remnants in sequential firing experiments is made difficult trough the thin specimen preparation from metastable materials. This step of experiment leads eventually to further necessary investigations to understand what the products are and how they influence the properties and composition of the materials they are admixed in.

5. Solid-state remnants of thermally decomposed vegetal materials

As indicated by TG results, vegetal materials do not decompose totally to volatile compounds. The solid-state remnants are the oxidized form of alkalis, alkaline earths, metals and semimetals contained in the organic compounds (or as biominerals e.g. Ca-oxalate crystals). These materials must be differentiated from the char and ash type remnants, which form due to incomplete combustion of organic matter. As evidenced by XRD the remnants are different for each vegetal material.

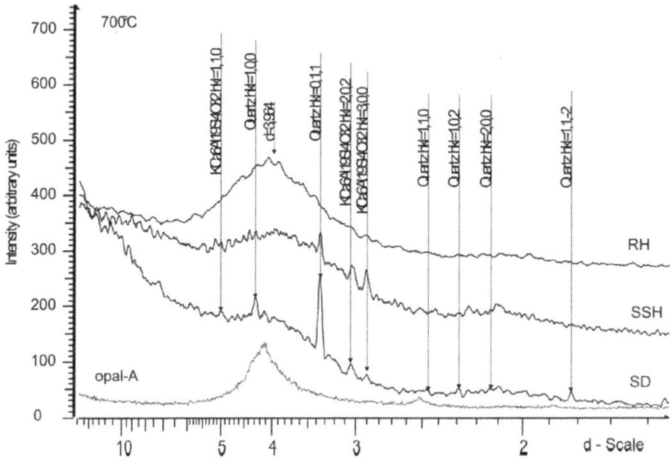

Figure 8. XRD patterns of VM remnants (smoothed, with original background, Y-scale translated)

Amorphous silica (perhaps nano-crystalline cristobalite) remains from RH, while SD and SSH remnants are more complex, with crystalline components too (Fig 8). For comparison, an opal XRD pattern is also shown. The most crystalline material as related to the amorphous hump is observed in the SD remnants and quartz is the dominant crystalline component. The SSH produces mainly silicates of Na, Mg and Ca with K as main cation, explained by the original chemistry of SSH grains. RH produced only amorphous SiO_2 as expected. In special treatment procedures, cristobalite and tridymite can be obtained from rice husks ash (Abou-Sekkina et al. 2010).

OMTPL investigations on materials fired at 900°C revealed the final state of remnants (Kristály & Gömze 2008), since the recrystallized matrix offered a good enough mechanical protection for remnants in the pores. Basically, the fabric observed in raw state is observed for SD, SSH displays the shrinkage affected material while RH has the hollow glassy network formed.

SEM on remnant materials was performed to observe their morphology. Peculiar details were revealed which enable us to link together the raw composition, thermal behavior and mechanical properties of VM (Fig 10).

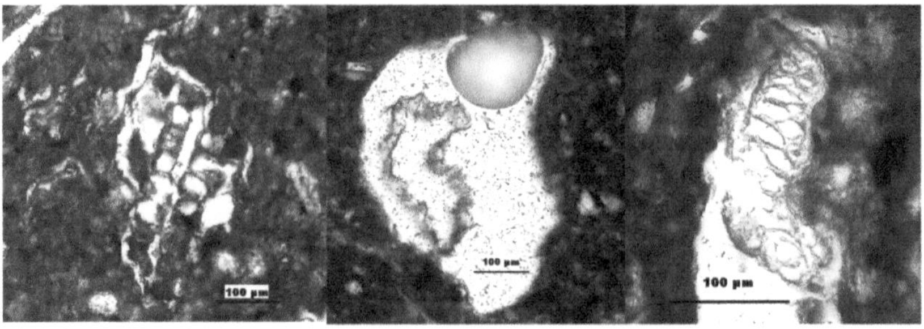

Figure 9. OMTPL images of VM remnants is thin section (left SD, center SSH, right RH).

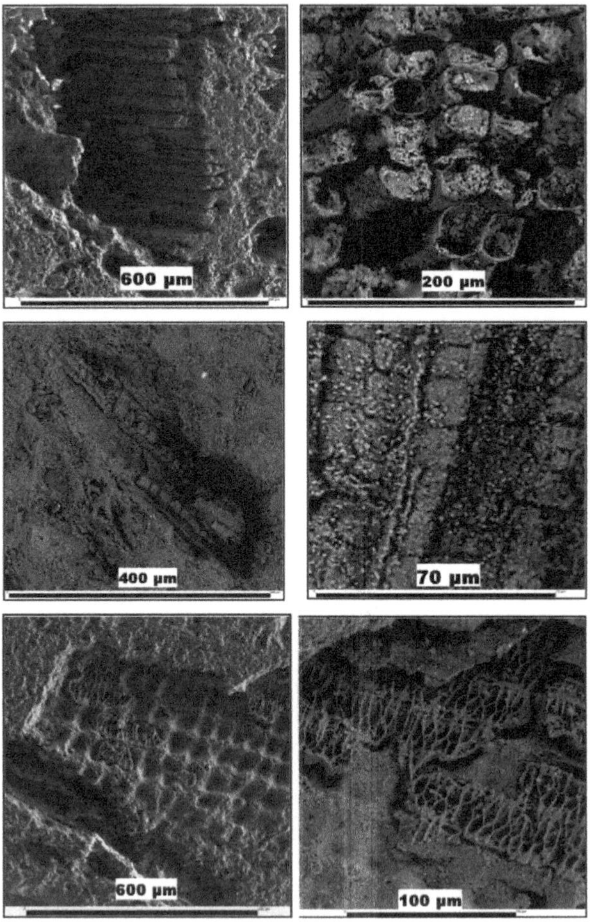

Figure 10. SEM images of VM remnants (top SD, center SSH, bottom RH).

The SD grains preserve the fibers in their raw form, forming empty microtubes. While between the grains and recrystallized clay matrix no voids can be observed, the individual tubes show distinct boundaries on fractured surfaces. Tube diameters vary in the 10-20μm range according to their dimensions in the raw state. SSH grains leave less material behind due to the significant shrinkage by firing. However, the polygonal aspect of remnants still is observed, according to fabric made up by bunches of fibrils observed by OMTPL. The most spectacular remnants are formed by RH grains, under the form of isolated microporous systems. A thin submicrometric layer is formed by the outer part of grains. In the inside a hollow honeycomb-like structure is revealed. The SiO_2 network is made up by the walls of original cells after the organic matter is decomposed. Additionally, characteristic impressions are formed by outer protuberances of the husks. The observations prove that in VM the most well preserved parts are those with higher cation content. Besides importance for archaeometry, these results were found to explain thermal properties of clay bricks in which these materials are used as pore forming additives. Thermal conductivity experiments revealed that SSH produces the highest drop in conductivity, while the smallest effect is observed for SD (Bánhidi et al. 2007, Bánhidi & Gömze 2008). The trend observed for density and porosity of fired clay bodies with these VM additives is similar: lowest density and highest porosity for mixtures with SSH, then SD and finally RH (Fig 11).

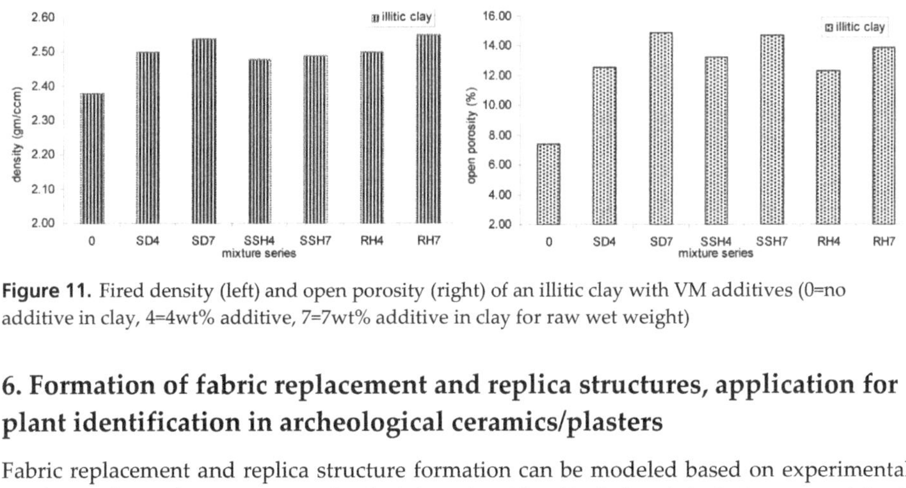

Figure 11. Fired density (left) and open porosity (right) of an illitic clay with VM additives (0=no additive in clay, 4=4wt% additive, 7=7wt% additive in clay for raw wet weight)

6. Formation of fabric replacement and replica structures, application for plant identification in archeological ceramics/plasters

Fabric replacement and replica structure formation can be modeled based on experimental results, and actual remnants can be deciphered to identify their original plant materials. A classification is needed not only for VM types, but for the different parts of plants. The hardest parts are grains and stems, material covering the grains (hulls, husks, etc.), stalks and leaves. As VM types, we may consider cereals, different wood materials and bean-like crops. Regarding cereals three major parts are often encountered in archaeological materials: grains, husks, stalk (straw) and leaves. In the case of bean-like crops situation is similar. However, wooden materials could be expected less frequent as tempers in ceramics. However, just like grass, they can be either intentionally added or accidentally enclosed in clay masses.

In archaeology, remnants of plant materials offer data about the lifestyle of extinct cultures. The remnants are mainly referred to as "phytoliths" describing the mineral nature of materials with vegetal origins. Their materials are usually SiO_2 or $CaCO_3$ as result of silicification or carbonation. By heat induced transformations the silicification is specific, and in lack of microbeam analysis, cations in the SiO_2-glass are often not determined. According to the previous sections of this paper, evidence turned up that inorganic remnants and their chemistry is plant specific. One of the direct applications of these results was in the case of medieval clay bricks (Kristály et al. 2011). The large number of pores was indicative for use of additives (or tempers, in archeological terms) and microscopic remnants were observed. A plant biological investigation of remnants showed that the species was *Triticum Monococcum* L. a wheat being cropped up to the 12th century, in the region from which samples originates. The historical dating of brick material based on their archeological context was estimated to be from years 1100 A.D. As shown in Fig 12 the characteristic morphology of husks is preserved, down to micrometric precision. One part of remnants is a glassy layer on the clay matrix, another part being the tiny nanoporous husk fibers. Other typical preserved details are found as the grain shells (Fig 13) which may seem as whole wheat grains by the naked eye but on a microscopic scale it is revealed that a shell of the form of grain is filled by the clay material and the shape is then preserved. Notice the thin white glassy film on the border of grain-replica and clay matrix.

Figure 12. Remnants of *T. Monococcum* L. husks(upper photo by István Nyilas, BSE images by Péter Pekker)

Phytolith research does not focus only on heat induced remnants, but on all plant remnants indifferent of transformation processes. In all cases, but mainly in ceramic related findings cellulose is the key component for the preservation of shape, morphology and size of the VM parts in replicas and replacement structures too.

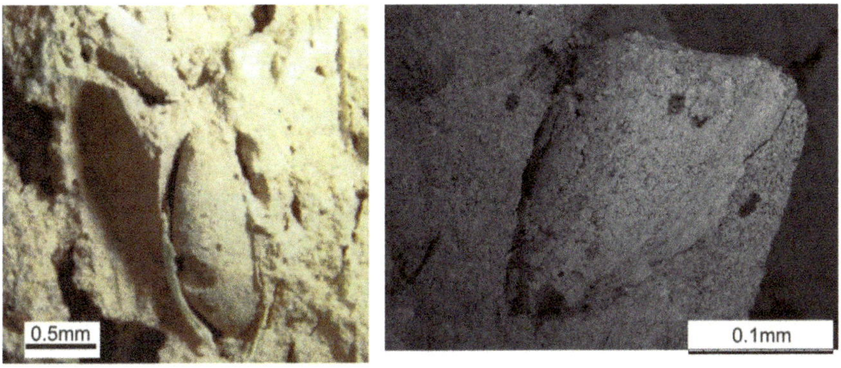

Figure 13. Remnants of *T. Monococcum* L. grains(left photo by István Nyilas, BSE images by Péter Pekker)

Other plant materials are also frequently recognized based on their remnants and the research field of *Phytoarchaeology* or *Archaeobotany* has become basic in archaeological research.

7. Elastic behavior of cellulose-rich vegetal materials

Shaping and deformation studies trough vacuum extruded masses of brick clays and vegetal materials showed that plant materials bear specific elasticity (Kristály & Kocserha 2010). Albeit a direct statistical correlation was not possible between the texture, composition and elasticity of VM samples, specific trends were recognized. First, the expansion of VM-free clay materials was tested to create a realistic baseline for the phenomena. Values <1% of the extruded diameter were observed, showing a good evacuation of air during extruding process. The further trends can be summarized as depending both on the quantity of cellulose in VM as the amount of VM in clay mixture but also the mineralogy of raw clays (Fig 14), expressed in muscovite content.

The results were obtained on experimental series of 10 pieces each being measured, for every clay type. In Fig 14 the expansion of four types of clays (LZ=carbonatic, LS=carbonatic muscovitic, S=illite-muscovitic, K=illitic-muscovitic) is shown with regard to VM type, admixed amount and extruding order for 10 test pieces. Besides the compacting effect of extruder head, the effect of VM and their quantity is evidenced. Comparing Fig 15 to Fig 14 it is clear that the main influence is to be searched in VM composition and DTA peak areas bear the information. An approach to cellulose content was made by relative ranking of DTA exothermal peaks. This way, the smallest amount of cellulose (and similar materials) is found in RH, also observed by the other methods of investigation. SD shows to bear more cellulose and SSH has the highest content. This ranking is also indicative for VM elasticity, since cellulose fibrils directly control it. On the other hand the fabric type observed by microscopy is indicative for the elasticity of VM type: the SD is tubular-fibrous, SSH is sponge-like while RH is hollow, massive. Based on the distribution of structural porosity and chemical elements, SSH is expected to bear the highest elasticity.

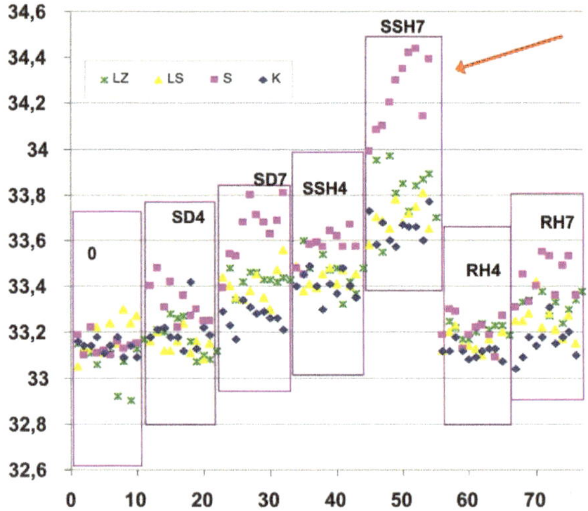

Figure 14. Expansion of different clay types with different VM additives

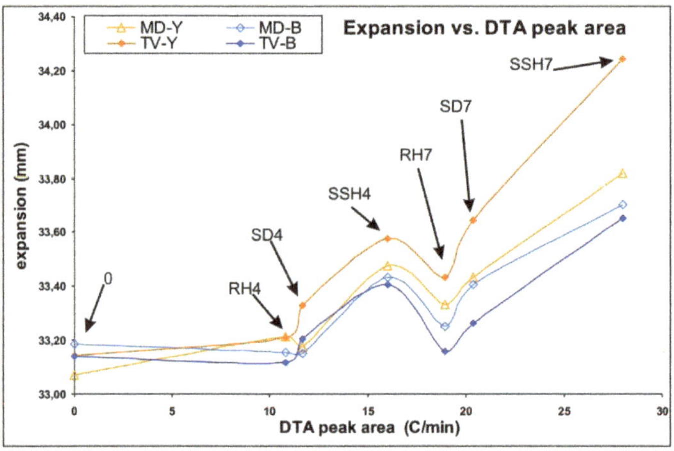

Figure 15. Expansion vs DTA peak area.

8. Effect of porous cellulose structures and thermal decomposition on clay mineral dehydration and dehydroxylation

On the TA measurements involving illitic clay with SSH a change in exothermal heat quantity and illite dehydroxylation was observed as compared to the raw clays and SSH material. DTA peaks indicate that heat evolved by cellulose combustion is decreased in the presence of illitic material for SSH containing mixture, while the high temperature (>750°C) OH^- loss of illite-sericite material is decreased in terms of heat absorption. This observation was validated by a second set of TA measurements also (Table 4). A reasonable explanation is the nature of SSH fabric, according to term "sponge-like" it has a suction effect over the

	SD	SSH	RH
raw VM	55.71	68.75	31.38
illtic-1	53.76	34.65	55.06
illitic-2	42.56	38.02	42.22
carbonatic-1	56.68	55.05	59.69
carbonatic-2	65.42	59.01	62.18

Table 4. Reation heats of VM in mixtures (J/mg)

OH^- groups becoming instable in mineral structures by rising T(°C). TG and DTG readings also confirmed that the weight loss in the high temperature illite dehydroxylation is less than expected according to measurements on raw clay and other mixture samples (Fig 16). Background shifts are due to the different porosity before oxidation and to recrystallization in different degree after illite dehydroxylation. XPD on mixtures fired at 700°C indicated also that clay mineral relict structures (like smectite anhydrate, chlorite anhydrate) have lesser amount in mixtures with SSH than other VM.

This process could also be responsible for the lowest amorphous material content of fired mixture samples (section 8.).

9. Effect of the vegetal materials presence on the amorphous material content of fired clay bricks

Quantitative Rietveld-refinement of XPD on fired clay+VM mixtures indicated that in some cases the interaction of VM with clay material can lead to the reduction of amorphous content as residual metastable phase at firing (Fig 17).

Firing induces changes are mainly the crystallization of Ca and Mg aluminum-silicates and hematite but Rietveld-refinement with internal standard for amorphous content determination also revealed changes less frequently mentioned in literature. Such are crystallization of quartz (denoted Quartz2) and microcline (denoted Microcline max.) or the reduction of amorphous content by the presence of VM materials. The quartz formed during firing is crystallized from the SiO_2 liberated on clay minerals recrystallization. Note that its amount is higher in VM mixture samples only for the illitic (TV-B) clay. The composition is the result of transformation illitic material into K-feldspar and with cations from chlorite breakdown into spinel (Fig 18), in lack of Ca and Mg for silicate formation. Presence of H_2O at <700°C from muscovite and sericite dehydration is a catalyzing agent for grain boundary diffusion, a key process for feldspar and other minerals crystallization. Akermanite is formed from the smectitic-chloritic clay minerals. Reduction of amorphous content is accompanied by increase in microcline and quartz2 content, the heat effect of VM during firing, aiding the recrystallization.

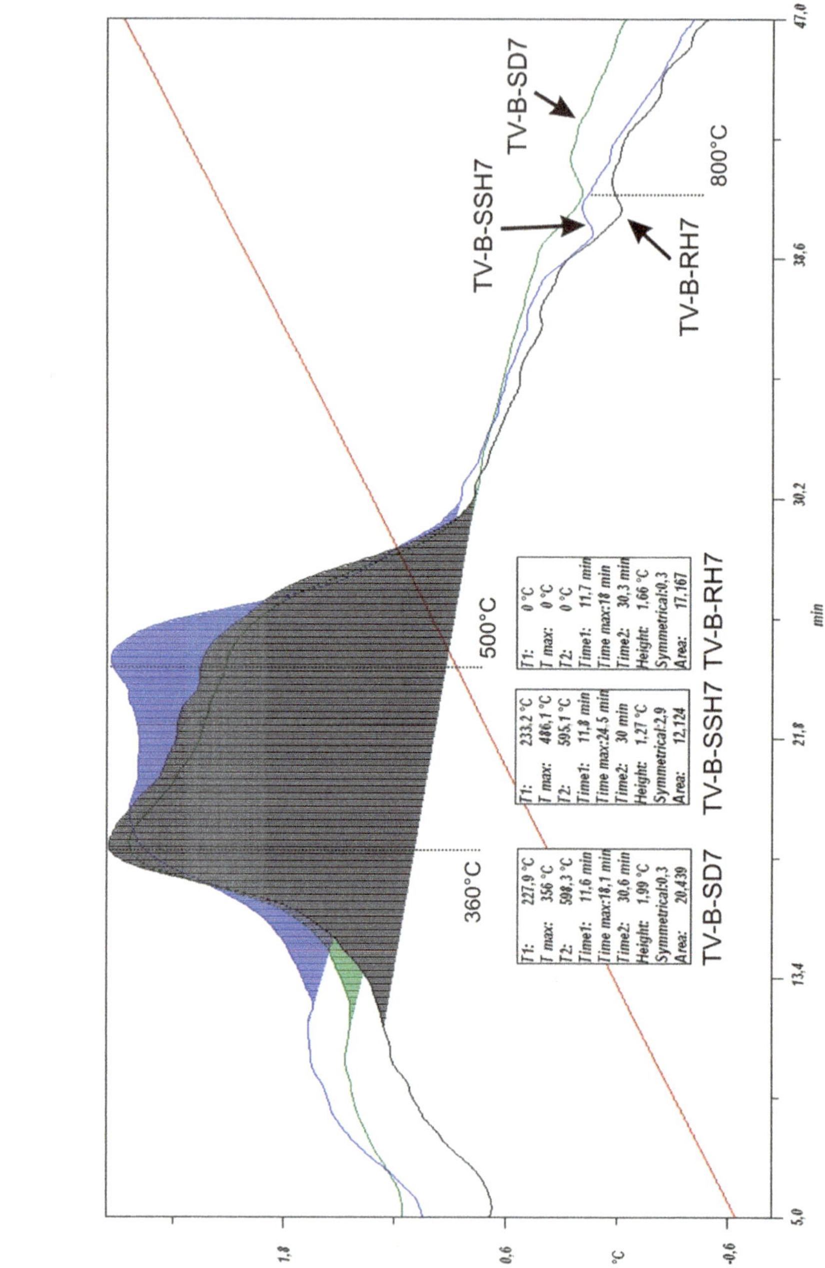

Figure 16. DTA curves for illitic clay (TV-B) with VM additives in 7wt%

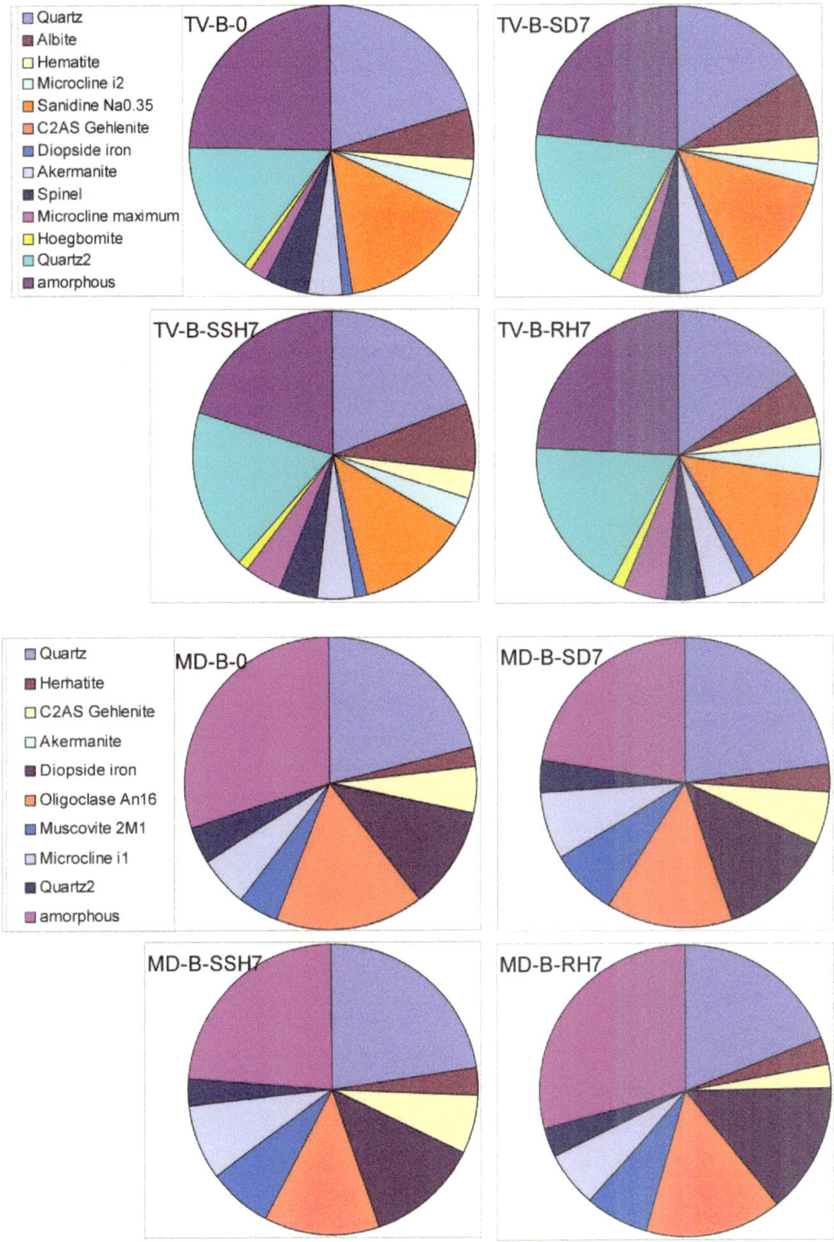

Figure 17. Quantitative mineralogical composition of fired blank and VM containing clays

	TV-B 0	TV-B SD7	TV-B SSH7	TV-B RH7	MD-B 0	MD-B SD7	MD-B SSH7	MD-B RH7
Akermanite	4	5	4	4				
Albite	6	7	7	5	2	3	3	3
Diopside (Fe)	1	1	1	1	11	13	12	14
Gehlenite					5	6	6	3
Hematite	2	3	3	3				
Hoegbomite	1	1	1	1				
Microcline	4	3	3	4				
Microcline (max.)	2	3	4	5	6	8	8	6
Muscovite 2M1					4	7	7	7
Oligoclase (An16)					16	14	13	15
Quartz	21	16	20	15	21	23	23	19
Quartz2	15	19	18	18	4	4	3	3
Sanidine (Na0.35)	15	14	13	13				
Spinel	5	4	4	4				
amorphous	25	23	20	24	30	22	24	29

Table 5. Absolute quantitative mineralogical composition of VM bearing fired clay mixtures

$$(K,Na)Al_2Si_3AlO_{10} \rightarrow (K,Na)AlSi_3O_8 + Al_2O_3 + H_2O$$

amorphous "meta-illite" → feldspar + amorphous? (↓ reagent) + from muscovite dehydroxilation

$$Mg_5Al_2Si_3O_{3.5} + \begin{matrix} 4Al_2O_3 \\ 4Fe_2O_3 \end{matrix} + 4.5O \rightarrow 5MgAl_2O_4 + 3SiO_2$$

amorphous "meta-chlorite" → spinel + amorphous? (↓ recrystallizes to quartz "Quartz2")

Figure 18. Two of the main reactions in illitic clay that determine the fired mineralogy (in the second reaction Fe may replace Al in spinels)

In the case of carbonatic clays the reduction of amorphous content is accompanied by increase in gehlenite, diopside and microline content, which were allowed to crystallize in higher amounts by the presence of VM.

10. Conclusions

To understand the behavior of VM and their interactions with clay matrix it's useful to approach them as cellulose based engineering materials. From the point of view of mineralogists, VM may not show too large differences as regarding the analytical techniques routinely used for geological materials characterization. The observations made on experiments with VM transformations are also useful in other research fields, like archaeometry, making it easier to recognize VM remnants and identify the original plants. As regards the interaction between vegetal materials and clay minerals, clay based fired products, DTA and XRD have shown that selective thermal interactions and recrystallization may take place.

Author details

Ferenc Kristály
University of Miskolc, Faculty of Earth Science and Engineering, Miskolc-Egyetemváros, Hungary

Acknowledgement

Help was provided by Dr. István Kocserha (Univesrity of Miskolc) at the preparation of mixtures and experimental set-up of clay-VM mixture firing. István Papp (Univesity of Debrecen) provided the measurements and helpful discussions on the EGA analysis. Preparation of this paper was supported by the European Union and the European Social Fund under the grant agreement no. TÁ MOP-4.2.1.B-10/2/KONV-2010-0001 in the framework of the New Hungary Development Plan.

11. References

Abou-Sekkina, M.M., Issa, R.M., Bastawysi, A.E-D.M. & El-Helece W.A. (2010) Characterization and Evaluation of Thermodynamic Parameters for Egyptian Heap Fired Rice Straw Ash (RSA). International Journal of Chemistry, Vol. 2/1, pp 81-88

Bánhidi V. & Gömze A.L. (2008) Improvement of insulation properties of conventional brick products. Materials Science Forum Vol. 589, pp 1-6

Bánhidi V., Gömze L.A. & Pázmándi P. (2007) Modification of heat conductivity in commercial brick products by recharging bio-waste material additives. MicroCAD 2007., Materials Science and Material Processing Technologies, 1-4, 2007

Das K., Ray D., Bandyopadhyay N. R., Sengupta S. (2010) Study of the Properties of Microcrystalline Cellulose Particles from Different Renewable Resources by XRD, FTIR, Nanoindentation, TGA and SEM. J. Polym. Environ. 18, pp 355–363

Kristály F. & Gömze A. L. (2008): Remnants of organic pore-forming additives in conventional clay brick materials: Optical Microscopy and Scanning Electron Microscopy study. Építőanyag, 2008/2, pp. 34-38

Kristály F., Kelemen É., Rózsa P., Nyilas I. & Papp I. (2011): Mineralogical investigations of medieval brick samples from Békés county (SE Hungary). Archeometry.

Kristály, F. & Kocserha, I. (2010): Correlations between combustion type additives and expansion after extrusion of clay bricks. Materials Science Forum Vol. 659, pp 43-48

Kristály, F., Gömze, A..L. & Papp, I. (2010): The transformation of added vegetal waste materials during clay brick firing. Materials Science Forum Vol. 659, pp 37-42

Szöőr Gy. & Bohátka S. (1985): Derivatograph-QMS system in geochemical research. Thermochimica Acta 92, pp. 395-398

Thygesen A., Oddershede J. , Lilholt H. , Thomsen A. B. & Stahl K. (2005) On the determination of crystallinity and cellulose content in plant fibers. Cellulose, 12, 563-576

Ververis C., Georghiou K., Christodoulakis N., Santas P. & Santas R. (2004) Fiber dimensions, lignin and cellulose content of various plant materials and their suitability for paper production. Industrial Crops and Products 19, pp 245–254

Wojdyr, M. (2010) Fityk: a general-purpose peak fitting program. Journal of Applied Crystallography, 43, 1126–1128.

Zhao H., Kwak J. H., Zhang Z. C., Brown H. M., Arey B. W. & Holladay J. E. (2007) Studying cellulose fiber structure by SEM, XRD, NMR and acid hydrolysis. Carbohydrate Polymers 68, pp 235–241

Chapter 10

Cellulose Langmuir-Blodgett Films for Moisture and Gaseous Molecular Sensing System

Hiroyuki Kusano, Shin-ichi Kimura and Masahiko Kitagawa

Additional information is available at the end of the chapter

1. Introduction

Cellulose is not only the main constituent of cell walls of a variety of plants but also one of the most important and versatile renewable resources in the world [1]. The cellulose derivative shows high sensitivity to moisture owing to its hydrophilic character on the glucose base unit and therefore a modification on side chains can act itself as ambipolar adsorber for ambient gas molecules. Thus, the well-defined and tailored cellulose film is a crucial step for the realization of gas sensitive polymer film dedicated for various applications, such as food gas sensitive unit based on the quartz crystal oscillator (QCO).

Langmuir - Blodgett (LB) method is well-established as the mono-molecular deposition technique in which the molecules exhibits well oriented and layered structures with a controllable manner, thus the LB based film is characterized as a film with the mono-layer designed function. Small molecular based LB films have disadvantages in terms of the stability and in-plane orientation, that is, molecular ordering in the plane of monolayer films or substrates, although they are known to form excellent films having periodic molecular arrays along the film deposition direction [2, 3].

Cellulose LB films, on the other hand, can overcome these problems, because of the inherent rigidity of the cellulose backbone, as well as the strictly designable molecular structure with reactive hydroxyl groups in glucose unit [4-7] and, therefore, the LB-cellulose film is expected to show the higher stability than that of the cast film because of the higher regularity and rigidity of the LB-stacked film [8, 9].

In this chapter, we review concisely the understanding and application of cellulose and thin film; especially LB thin film advancement, gas adsorption on cellulose molecular field with special interest in water molecule, merit and necessity of QCO measurement and stability and further applicability as an ethanol detecting technique. We also describe our results of characterization for the cellulose LB films, the QCO sensor systems and the characteristics of the cellulose LB QCO sensor.

2. History of cellulose LB films

In this section, the history of cellulose LB films is summarized. It is importance is to develop the effective highly tailored cellulose LB sensing film unit for future application.

In the 1980s, the successful application of the LB technique to polymeric systems including cellulose polymer was started. M. Roman has summarized the history of cellulose LB films in his recent extensive review [1]. Here, we have cited the heart of his review since the core importance is to develop the effective highly tailored cellulose LB sensing film unit for future application. He has classified the cellulose LB films into simple three model categories as follows;

1. Cellulose-Derivative Model Surfaces

T. Kawaguchi et al. and T. Itoh et al. have elucidated the molecular arrangement in the cellulose-derivative LB films [10,11]. F. Nakatsubo and co-workers have investigated the incorporation of chromophores into the films [12]. M.Tanaka et al. and we have developed the (bio)technological applications of the films [8,9,13]. K. Gotoh studied the detergency of a model aqueous detergent solution with respect to arachidic acid LB-layers on spin-coated cellulose acetate films using a quartz crystal microbalance [14].

2. Regenerated-Cellulose Model Surfaces

M. Schaub et al. investigated the use of the LB technique to deposit mono- and multilayers of the regenerated-cellulose films onto various substrates. By subsequently cleaving off the trimethylsilyl groups in humid HCl vapor, the cellulose molecular structure could be regenerated, yielding very smooth cellulose films. The thickness of these films was directly proportional to the number of monolayers [15,16].

F.F. Rossetti et al. elucidated and compared the nano-scale structures of regenerated cellulose surfaces prepared by the LB technique and by spin coating. The authors used non-contact AFM (NC- AFM) and grazing incidence small-angle x-ray scattering (GI-SAXS) to measure the characteristic in-plane length scales, as well as x-ray specular reflectometry to obtain information in the direction normal to the surface. Swelling experiments showed that significant swelling of the model surfaces occurred at relative humidities above 97% and that, especially in the thicker films, swelling was accompanied by an out-of-plane rearrangement of the cellulose bundles [17].

3. Native-Cellulose Model Surfaces

Regenerated cellulose LB films are prepared either cellulose solution in a suitable solvent system or by converting a thin film of a cellulose derivative through chemical reaction into a cellulose film. In either case, the native cellulose morphology is lost in the process and the thin cellulose film have a non-native morphology. Native-cellulose LB films are prepared from aqueous colloidal suspensions of nano-scale cellulose fibrils or fragment thereof.

We selected cellulose ester as the cellulose derivative for forming cellulose LB films, which is adopted to a humidity and/or alcoholic molecular sensor. The Cellulose ester has hydrophobic alkyl side chains and hydrophilic glucose unit in polymer back-bone structure with regularity. Highly oriented molecular closed-packing structure can be obtained in making mono-molecular film between air and water interface by using the cellulose ester. The well-oriented multi-layer of the cellulose ester is useful as the thin and tough membrane to prevent swelling of the film by intercalation of water molecules. Therefore, the cellulose ester LB film is suitable for the moisture and gas sensing systems in comparison with two other models.

Recently, Y. Habibi et al. studied the use of the LB technique to transfer monolayers of cellulose nano-crystals onto silicon wafer substrates and demonstrated that highly oriented cellulose nanocrystal films could also be prepared by solvent casting in an electric field [18].

3. Characterization of cellulose-derivative LB films

We illustrate our results of studies on the determination of the surface and bulk properties of the cellulose derivative (palmitoyl cellulose) LB films and confirm the formation of the well-defined solid-like film by using scanning probe microscopy (SPM), Fourier Transform-infrared (FT-IR) absorption spectroscopy, small-angle X-ray diffractometry(XRD) and high resolution transmission electron microscope (HR-TEM) [19-22].

3.1. Pressure-area isotherm

Figure 1 shows the surface pressure-area isotherm for the palmitoyl cellulose on the surface of double-distilled water at 290 K and inset in Figure 1 shows the molecular structure of the palmitoyl cellulose. Method for sample preparation has been described in detail elsewhere 8]. A monolayer of stearic acid was deposited at first on the indium tin oxide (ITO) substrate surface to enhance the contact and then well-defined X-type layers of the palmitoyl cellulose LB film were built up by using the horizontal lifting method at a trough temperature of 290 K. The limiting molecular area for the condensed region at the surface pressure from 5 to 30 mN/m of palmitoyl cellulose obtained from Fig. 1 was determined to be 60 Å^2 per glucose unit and the molecular area for condensation was about 44 Å^2 at 20 mN/m, which confirms the formation of the well-defined solid-like film with the area of about 50 Å^2. The value is less than that of cellulose esters, 54 to 66 Å^2 per glucose unit as reported by Kawaguchi et al. [10]. Therefore, it is emphasized that stable the palmitoyl cellulose Langmuir-monolayers were obtained by adjusting the optimum surface pressure during deposition at 20 mN/m.

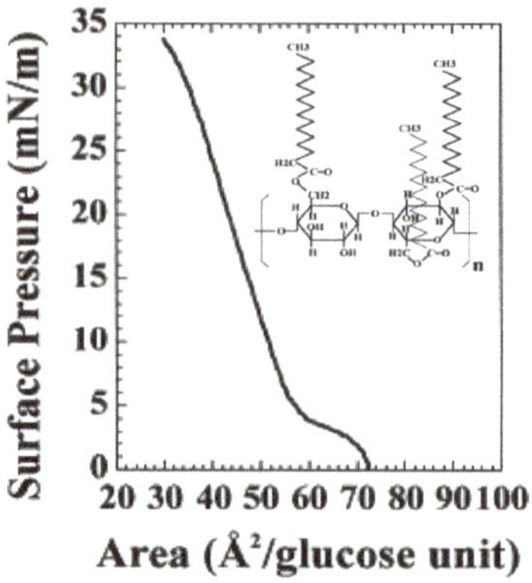

Figure 1. Surface pressure-area isotherm for palmitoyl cellulose at 290 K (inset: Molecular structure of the palmitoyl cellulose)

3.2. SPM

Figure 2 shows the SPM profile of the palmitoyl cellulose 5 layers LB film. SPM profile showed grains with average size of 200 nm in diameter. The observed height of the grain was about 11 nm for the five mono-layered LB film. Thus, the mono-layer thickness of 22 Å was estimated, which is consistent with the reported values between 20 and 26 Å [19].

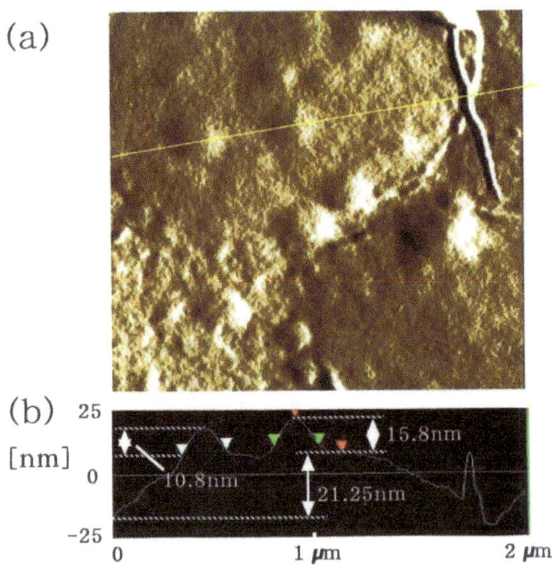

Figure 2. The SPM of palmitoyl cellulose 5 layered LB film. (a)SPM image and (b) SPM section analysis [19].

3.3. FT-IR

Figure 3(a) and (b) show expanded infrared spectra for the palmitoyl cellulose between 1500 cm^{-1} and 3000 cm^{-1}, for the LB films and the cast films [19]. In the FT-IR RAS measurement, p-polarized electric field was used, where the electric field vector of the incident light excites selectively the IR mode which has the molecular vibration perpendicular to the film surface, i.e., to the substrate surface. Hence, in the case of the well-defined palmitoyl cellulose LB film, the CH_2 asymmetric mode which has a stretching direction perpendicular to the substrate can be excited selectively compared with the CO mode which has a stretching direction parallel to the substrate. On the other hand, on the film that has been cast, each palmitoyl cellulose takes a random orientation with respect to the substrate surface. As a result, the orientation of CH_2 bonds and CO bonds are random. Therefore, intensities of CO vibration modes decrease in the LB films compared with the cast films. In this case, the ratio Im/Ik of LB film was 0.45, where Im is the intensity for the CH_2 asymmetric mode and Ik is the intensity for the C=O mode. The ratio for the cast film was 0.20. These results confirm the formation of the mono-layered LB film.

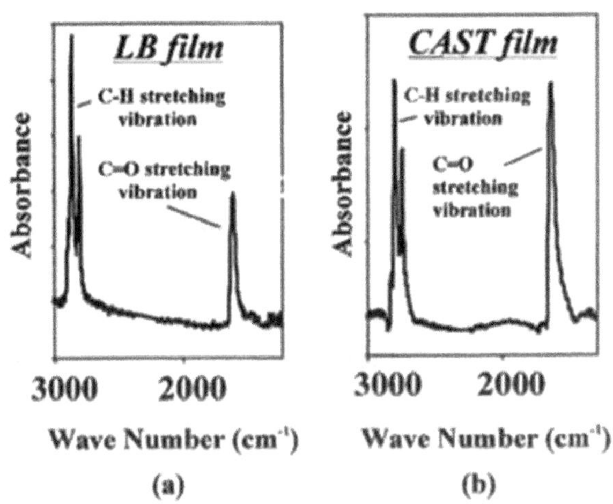

Figure 3. IR absorption between 1500cm^{-1} and 3000cm^{-1} for the palmitoyl cellulose films (a) LB film and (b) cast film.

3.4. XRD

Figure 4 shows In-plane X-ray diffraction pattern of the PC-LB film on the ITO substrate. The In-plane X-ray diffraction pattern were recorded on an X-ray diffractometer equipped with a 4-axes goniometer. Cuk α radiation from copper rotating anode was used for the experiment. Incident angle (ω) for the in-plane geometry was between 0.14° and 0.36°. Diffraction angle of In-plane geometry is defined as $2\theta\chi$. The peak intensity of $2\theta\chi$=21.1° (peak A) at ω= 0.14° was larger than that of at ω=0.20°. To the contrary, the peaks of

$2\theta\chi$=30.7°(peak B), 35.5°(peak C) appeared at ω=0.20° almost disappeared at ω= 0.14° [21]. The interplanar spacings of these two peaks (B, C) correspond to those of ITO(222), ITO(004).

The peak intensity of A decreased with the increase of intent angle. Contrary, the peak intensities of ITO substrate increased with increase incident angle. Consequently, the peak A is due to the diffraction peak of the PC-LB film, which appeared at the small incident angle of X-ray (ω= 0.14°). The interplanar spacing of the peak A of the PC-LB film was 4.20Å. This value was attributed to a distance between the palmitoyl chains in the PC molecules perpendicular to the glucose unit. Palmitoyl chains were long acyl group (16 carbon) and bonded to the hydroxyl group of the glucose unit as shown in Fig.1.

On the other hands, the size of apparent crystallite can be estimated from the peak broadening. Hence, the size of crystallite can be estimated by the Scherrer equation ($D= 0.9\lambda/(\beta\cos\theta)$), where D is the size of crystallite, λ is wave length of CuK α, β is the half width of the peak and θ is the Bragg angle. From the half width of the peak of $2\theta\chi$=21.1°, the size apparent of crystallite in the PC-LB film was about 15 Å.

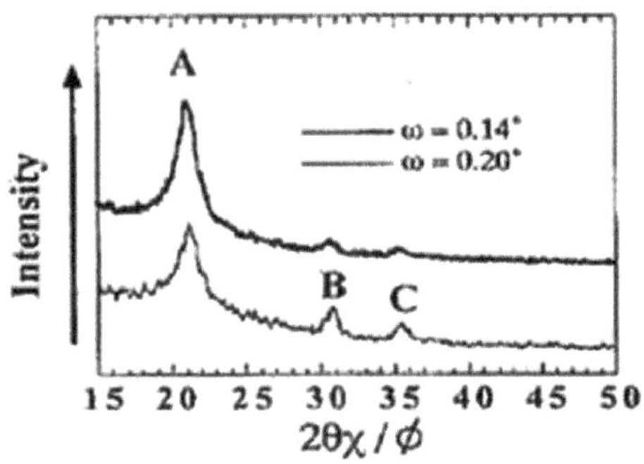

Figure 4. In-plane X-ray diffraction pattern of palmitoyl cellulose Langmuir-Blodgett film on the ITO substrate.

3.5. HR-TEM

Figure 5 is an example of a high resolution TEM image of the surface of Pt coated LB film condensed at 293 K under the surface pressure of 20 mNm^{-1}[20]. Grains with larger size than 200 nm in diameter have been observed.

Figure 6 shows a schematic model of the PC-LB film on the ITO substrate. The structure of the PC-LB film is regular in the direction parallel to the film substrate and the distance between the palmitoyl chains keep some same distance. The conformation of palmitoyl chains is most probably determined due to the cellulose main chains.

Figure 7 illustrates the model of grains in the PC-LB film. We already described the grain size was estimated 200nm by using SPM and TEM. Therefore, the structure of 200nm size grains in the PC-LB film is considered to be aggregation of many 15 Å size crystallites with 4.20 Å distance between the palmitoyl chains.

Figure 5. TEM image of the palmitoyl cellulose LB films deposited at 293K. Circle areas are palmitoyl cellulose grains.

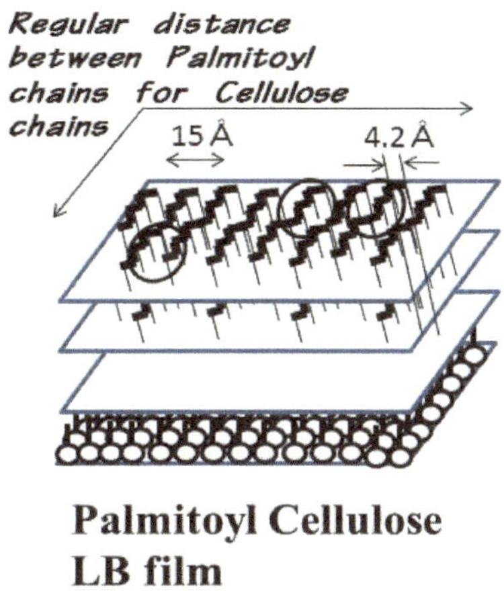

Figure 6. Schematic model of the PC-LB film on the ITO substrate.

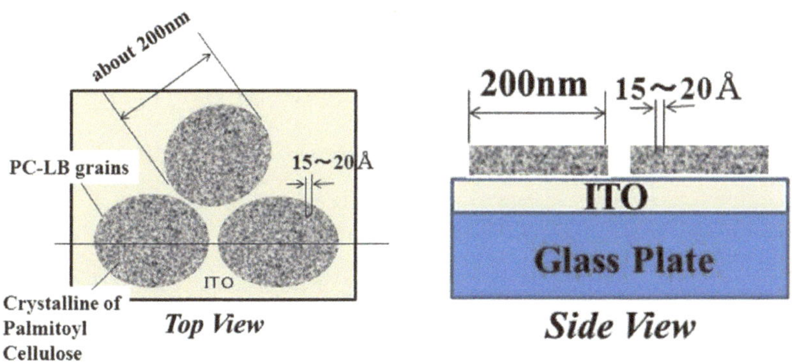

Figure 7. The model of grains in the PC-LB film.

4. Cellulose-derivative LB film based humidity sensing system

We show our study of the humidity measurement system using the QCO coated with the cellulose LB film [8,9]. We clarified the water absorption and desorption mechanism for the cellulose-derivative LB films. The dependence of the sorption properties on the number of layers clearly indicated that the two regions which hold in row and high humidity condition show different sorption characteristics, and thus the evidence showed that the modes of sorption must be different. Therefore, as the main conclusion of LB cellulose sensing study, the analysis based on the combination of the two sorption models was prefered. The stability test by soaking the sensors in water showed the superiority of LB based sensing-films with the structurally ordered cellulose molecules.

4.1. Quartz crystal oscillator measurement system

Figure 8 illustrates schematic diagram of the quartz crystal oscillator (QCO) coated with the cellulose LB film. A monolayer of stearic acid was deposited on the Au electrode to contact with the hydrophilic head of the stearic acid in order to build up good X-type layers of the palmitoyl cellulose LB films. The horizontal lifting method was used to prepare the X-type LB films on both sides of the QC.

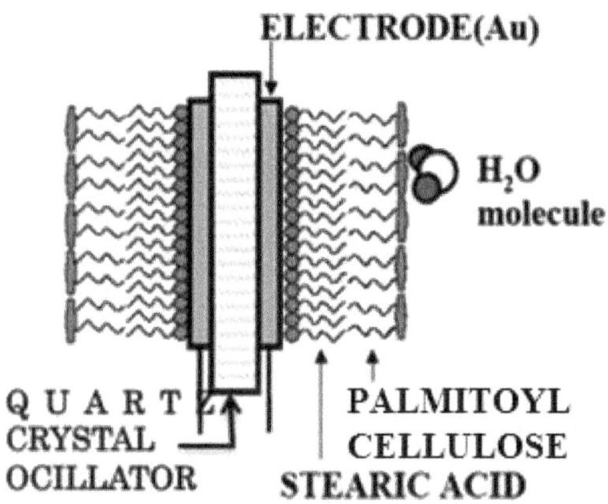

Figure 8. Schematic diagram of the quartz crystal oscillator (QCO) coated with the cellulose LB film.

The humidity measurement system using the quartz crystal oscillator (QCO) coated with the cellulose LB film was constructed with four portions; **A**:Quartz crystal oscillator (QCO) coated with cellulose LB films; **B**:Resonant circuit for conversion of QCO mechanical vibration to alternating electrical signal; **C**:Frequency counter for changing the alternating current signal to numerical data; **D**:Personal computer for stacking the numerical data.

The frequency of vibration in the QCO with the cellulose LB film changes by the amount of water molecules adsorbed on the surface. Resonating circuit converts the mechanical vibration of the QCO into electrical signal. Frequency of the electrical signals is counted and converted into numerical digital signal to input computer for stacking the data and calculation of the value of humidity. The Colpitts resonating electrical circuit was used for the QCO in this study. Inverted amplifier with CMOS IC was high speed enough for quick response of the high frequency (about 15MHz) of the QCO. The wave form of the output signals was monitored by the digital oscilloscope (model 54200A, Hewlett Packard Co.). The frequency was corrected with a digital counter (model SC-7202, Toyo Technica Co.). The humidity dependence of the frequency change of the quartz crystal with the cellulose LB films was measured in the desiccator held at a fixed relative humidity. Saturated salt solutions were used to control the humidity in the desiccators.

5. Adsorption characteristics

Figure 9(a) shows dependence of the number of sorbed water molecules on the relative humidity for a 4 monolayer LB films. A and B regions in which the temporal dependence of frequency variations with decreasing relative humidity were measured. Figure 9(b) shows desorption characteristics of sorbed water on the LB films, showing the temporal dependence of the frequency changes with decreasing relative humidity in regions A and B in Fig. 9, respectively.

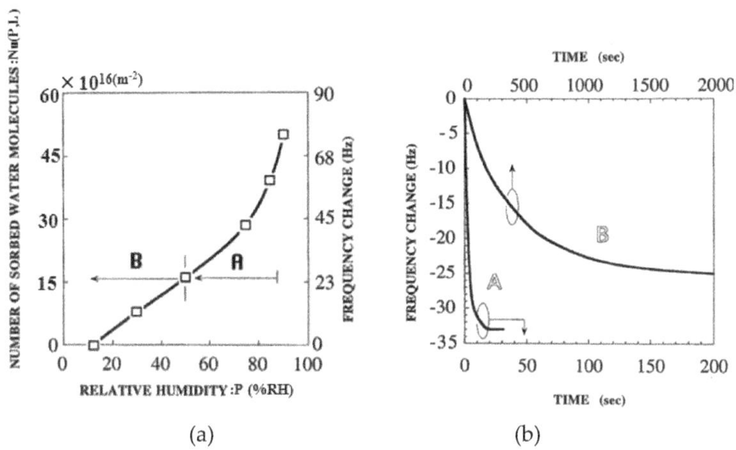

Figure 9. (a) Dependence of the number of sorbed water molecules on the relative humidity for a 4 monolayer LB films. A and B regions in which the temporal dependence of frequency variations with decreasing relative humidity were measured. (b) Desorption characteristics of sorbed water on LB films, showing the temporal dependence of the frequency changes with decreasing relative humidity in regions A and B in Fig. 9, respectively.

From the viewpoint of standard analysis of the sorption characteristics, Brunauer, Emmett and Teller (BET) analysis should be applied to each data set with an appropriate set of parameters. In the LB cellulose sensor case, the sorption process as an intercalation-like molecular process was inevitable. Therefore, the new model in the analysis of sorption properties of the LB monolayered humidity sensors was tested. First, the dependence of the sorption properties on the number of layers clearly indicates that the two regions, which hold in row and high humidity condition show different sorption characteristics, and thus the evidence showed that the modes of sorption must be different. Therefore, as the main conclusion of the LB cellulose sensing study, the analysis based on the combination of the two sorption models was prefered. Second, even if the perfection of the monolayer can be improved further, i.e. if the ordering of the monolayers is not so good, the number of monolayers, and thus the number of molecules, makes no difference to the results obtained in the experiment. This indicates the uniqueness of the existence of the observed two sorption processes in ordinary cellulose humidity sensors. Therefore, new formula could be regarded as providing a better description of sorption characteristics in the humidity and/or molecular sensing study.

5.1. Water adsorption model

The water molecule sorption characteristics of the cellulose LB monolayered films have been explained on the basis of the QCM method. These characteristics can be divided into two parts. In the low relative humidity region, desorption/sorption was observed to be slow, with a time constant of about 500 s, whereas in the high relative humidity region it was more rapid, with a time constant of less than 10 s. The dependence of the number of sorbed water molecules on the number of monolayers in the films reveals that two distinct mechanisms are involved in the relative humidity range examined, and the number of sorbed water molecules can be described by a combination of the Freundlich-type and the Polanyi-type functions, in the following form [8]:

$$N_u\,(p,L) = 1.5\times10^{16}\,(p - p_a)^{0.87}\,L + 6.0\times10^{20}(RT\,\mathrm{In}(p_0/p))^{-1} - 3.0\times10^{17}$$

where $N_u(p,L)$ is the number of sorbed water molecules per m^2, p is the relative humidity (%), p_a is the relative humidity at the start of the measurements (%), p_0 is the relative humidity at saturation (%), L is the number of layers in the LB film, R is the gas constant (8.314 J K^{-1} mol^{-1}), and T is the absolute temperature (K).

Figure 10 shows the sorption model (the structure of the palmitoyl cellulose LB film is shown in Fig. 1) including both Freundlich-type and Polanyi-type sorption.

In the low relative humidity region the characteristics can be described by the Freundlich isotherm, and show a linear dependence on the number of monolayers in the film. We interpret this as being due to the sorption of water into the cellulose LB monolayer plane, so that the Freundlich-type sorption mechanism determines the water sorption characteristics that predominate in this region. It is explained as sorption onto the hydrophilic sites in the glucose units in the monolayer. The experimentally determined number of sorbed water molecules is 4.7 per glucose unit at almost saturation, which is close to the average number of 4.4 hydrophilic sites per glucose unit in the sample.

In the high relative humidity region another mechanism dominates the sorption characteristics, as described by the potential theory of Polanyi. In this case, the amount of sorbed water corresponds to a layer thickness of about three water molecules at a fixed point of almost saturation, 90% RH. In this humidity region the sorption process is controlled by the free condensation of water molecules onto the film surface. The average number of sorbed water molecules per glucose unit calculated using the occupied area of the glucose unit and the water molecule is about 4.4, comparable to the experimentally determined value of about 4.7. These values coincide with the number of base sites not substituted by palmitoyl chloride. The number of hydrophilic sites, such as -OH and -C=O, per glucose unit in the sample is 4.3. Thus the number of sorbed water molecules is thought to be limited simultaneously by both steric occupancy and site occupancy factors. We have constructed the water sorption mechanism onto the glucose unit of the cellulose polymer LB film as a measure of gas molecular sorption phenomena [13].

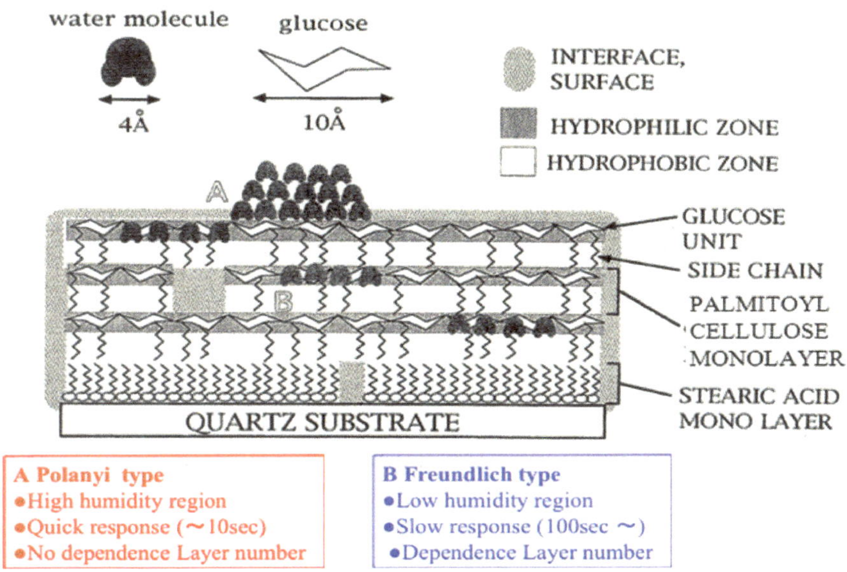

Figure 10. Schematic diagram of the proposed mechanism of water sorption in palmitoyl cellulose polymer LB film [13].

5.2. Durability test

Figure 11 shows durability of the LB-based QCO sensor against water. X is the amount of the cellulose molecules after soaking the QCO's into the water and then drying, X_0 is the initial amount reading at the measurement condition at 80%RH. We have then tested both the cellulose LB-QCO sensors and the cast cellulose bulk QCO sensors in the sequence of simulated high humidity condition by soaking sensors into the water bath at several cycles, as the durability test of the LB-cellulose based QCO sensor and the cast cellulose QCO sensor against water, since a major issue with gas sensors is their unstability to humidity because of today's requirement of stable measurement at high humidity for the use of agricultural factory in the near-saturation condition of water vapor.

The test condition of stability against water was defined to soak the sensors in water for 30 minutes at room temperature between the measurements. The stability was defined as the change rate, that is, the ratio of frequency change normalized by the initial value in the measurement of humidity at the condition of 80%RH and 30°C. The indicated value of humidity for the cast type cellulose sensor showed steep decrease in signal read out with increasing cycle of test measurement after each soaking. On the other hand, the LB-based QCO sensor did show no change in read out at all in the frequency change.

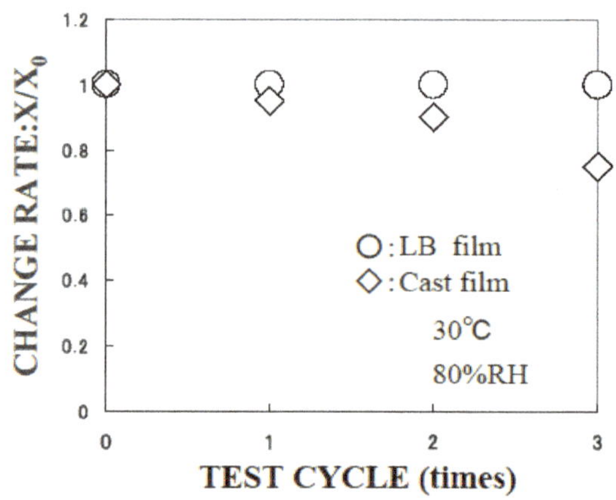

Figure 11. Durability of the LB-based QCO sensor against water. X is the amount of the cellulose molecules after soaking the QCO's into the water and then drying, X_0 is the initial amount reading at the measurement condition at 80%RH [9].

6. Cellulose-derivative LB film based alcohol sensing system

We describe our recent results of the adsorption characteristics of ethanol gas and the possibility of development of the sensitive and stable gaseous molecular-sensing system by the advanced QCO based LB cellulose sensing system [9].

6.1. Ethyl alcohol adsorption

Specific gaseous molecules such as ethanol have been tested. After a steady frequency was attained in a dry N_2 atmosphere and in an air atmosphere, the time-dependence of the frequency change of the LB-based QCO sensor was recorded while exposing to gaseous alcohols (Ethyl alcohol, Isopropyl alcohol and so on.).

Figure 12 shows temporal response of the frequency change of the LB-based QCO sensor exposing to gaseous Ethyl alcohol. The definite and reproducible sensor responses for these gaseous alcohols have been obtained in the temporal response of the frequency change of the LB-based QCO sensor exposed to evaporated Ethyl alcohol. Under the ethanol adsorption and desorption condition, the frequency changes with delay times of about 30s were observed and the response time was an order of magnitude shorter than that of water. Similar to the water adsorption, alcohol adsorption on molecular films seems to be able to be well explained and estimated on the basis of hydrophilicity of the LB cellulose films, as we have already established to explain the adsorption of water molecules on molecularly-well-oriented and structured LB film surface. Therefore, development of sensitive and stable moisture and gaseous molecular-sensing and discrimination system for alcoholic beverages gases is thus possible by the advanced QCO based LB cellulose sensing system.

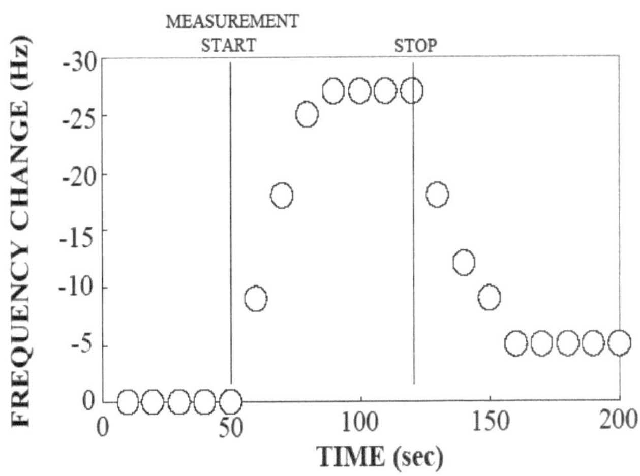

Figure 12. Temporal response of the frequency change of the LB-based QCO sensor exposing to gaseous Ethyl alcohol [9].

7. Perspectives

We have reviewed the studies and the advances of the cellulose Langmuir-Blodgett (LB) films and the cellulose LB based quartz crystal oscillator (QCO) sensors.

We described our results of characterization for the cellulose LB films by using scanning probe microscopy (SPM), Fourier Transform-infrared (FT-IR) absorption spectroscopy, small-angle X-ray diffractometry (XRD) and high resolution transmission electron microscope (HR-TEM).

We also described concisely the understanding and application of the cellulose and thin film; especially LB thin film advancement, gas adsorption on cellulose molecular field with special interest in water molecule, merit and necessity of QCO measurement and stability and further applicability as an ethanol detecting technique.

One of the most important points in developing alcoholic gas sensing is to realize simple discrimination between ethanol and water. We have examined the ethanol gas sensors by using the cellulose LB QCO. Similar to the water case, ethanol adsorption on molecular films seems to be able to be well defined on the basis of hydrophilicity of the LB cellulose films. Therefore, sensitive and stable moisture and gaseous molecular-sensing and discrimination system for alcoholic beverage gases are promising by the advanced QCO based LB cellulose sensing system in near future.

Author details

Hiroyuki Kusano and Shin-ichi Kimura
Tottori Institute of Industrial Technology, Wakabadai, Tottori, Japan

Masahiko Kitagawa
Department of Information and Electronics, Graduate School of Engineering, Tottori University, Koyama, Tottori, Japan

Acknowledgement

We would like to thank Takaaki Hishinuma of Tottori university for his help in the cause of this study.

8. References

[1] Roman M (2010) Model Cellulose Surface: History and Recent Advaces. ACS Symposium Series:3-53.

[2] Sugi M, Carter F. L, Siatkowski R. E, Wohltjen H, ed. (1988) Molecular Electronic Devices. Amsterdam, Elsevier.

[3] Petty M.C (1996) Langmuir-Blodgett films An introduction. Cambridge University Press, UK,.

[4] Schaub M, Fakirov C, Schmidt A, Lieser G, Wenz G, Wegner G, Albouy P.-A, Wu H, Foster M.D, Majrkzak C, Satija S, (1995) Macromolecules 28: 1221.

[5] Basque P, Gunzbourg Ad, Rondeau P, A, Ritcey M (1996) Langmuir 12:5614.

[6] Redl F.X, Lutz M, Daub J, (2001) Chem. Eur. J. :75350.

[7] Redl F.X, Kothe O, Rockl K, Bauer W, Daub J (2000) Macromol. Chem. Phys. 201:2091.

[8] Kusano H, Kimura S, Kitagawa M, Kobayashi H (1997) Thin Solid Films 295: 53-59.

[9] Kusano H, Kitagawa M (2008) IEICE TRANS. ELECTRON. E91-C, 12: 1876-1879.

[10] Kawaguchi T, Nakahara H, Fukuda K (1985) Monomolecular and Multimolecular Films of Cellulose Esters with Various Alkyl Chains. Thin Solid Films.133:29–38.

[11] Itoh T, Tsujii Y, Fukuda T, Miyamoto T, Ito S, Asada T, Yamamoto M (1991) Fluorescence Spectroscopy for a Cellulose Trioctadecanoate Monolayer at the Air-Water-Interface. Langmuir. 7:2803–2807.

[12] Sakakibara K, Ogawa Y, Nakatsubo (2007) F First cellulose Langmuir- Blodgett films towards photocurrent generation systems. Macromol. Rapid Commun. 28:1270–1275.

[13] Hillebrandt H, Tanaka M, Sackmann E (2002) A novel membrane charge sensor: Sensitive detection of surface charge at polymer/lipid composite films on indium tin oxide electrodes. J. Phys. Chem. B. 106:477–486.

[14] Gotoh K, Nakata Y, Tagawa M (2006) Evaluation of particle deposition in aqueous solutions by the quartz crystal microbalance method. Colloid Surf. A-Physicochem. Eng. Asp. 272:117–123.

[15] Schaub M, Wenz G, Wegner G, Stein A, Klemm D (1993) Ultrathin Films of Cellulose on Silicon-Wafers. Adv. Mater. 5:919-922.

[16] Buchholz V, Wegner G, Stemme S, Odberg L (1996) Regeneration, derivatization and utilization of cellulose in ultrathin films. Adv. Mater. 8399-402.

[17] Rossetti F. F, Panagiotou P, Schneck F, Dommach M, Funari S.S, Timmann A, Muller-Buschbaum P, Tanaka M (2008) Structures of regenerated cellulose films revealed by grazing incidence small-angle x-ray scattering. Biointerphases. 3:117-127.

[18] Habibi Y, Heim T, Douillard R (2008) AC electric field-assisted assembly and alignment of cellulose nanocrystals. J. Polym. Sci. Pt. B-Polym. Phys. 46:1430-1436.

[19] Kimura S, Kusano H, Kitagawa M, Kobayashi H (1999) Layer-by-layer characterization of cellulose Langmuir-Blodgett monolayer films. *Appl. Surf. Sci. 142*: 585–590.

[20] Kimura S, Kusano H, Kitagawa M, Kobayashi H (1999) TEM characterization of cellulose Langmuir-Blodgett monolayer films. *Appl. Surf. Sci., 142*: 579–585.

[21] Kimura S, Kitagawa M, Kusano H, Kobayashi H (2000) Electrical and dielectric properties of cellulose LB films. *Polym. Adv. Technol.* 11:723–726.

[22] Kimura S, Kitagawa M, Kusano H, Kobayashi H (2001) In and out plane X- ray diffraction of cellulose LB films. In *Novel Methods to Study Interfacial Layers;* Möbius, D., Miller, R., Eds.; Studies in Interface Science 11; Elsevier: Amsterdam, The Netherlands. 255–264.

Chapter 11

Removal of Excess Cellulose and Associated Polysaccharides in Fruit and Vegetable By-Products – Implication for Use in Feed for Monogastric Farm Animals

Annie King

Additional information is available at the end of the chapter

1. Introduction

As an abundant biomaterial, cellulose has many uses. Presently, many investigations are underway to explore its use as fuel. However, cellulose, properly treated, could be used as a source of nutrients for animals. Thus, in this chapter, a case for use of cellulose, accompanying polysaccharides and nutrients in fruit and vegetable by-products (waste) as feed ingredients in animal feed is developed. The production of fruits and vegetables throughout the world and the concomitant loss of by-products are discussed. Potentially, nutrients from by-products could be included in feed for monogastric animals, a source of protein for expanding populations, especially in food insecure parts of the world. The reasons for low use of by-products are delineated with particular attention focused on the indigestibility of lignocellulosic material. Selected treatments to reduce these compounds are examined.

2. Worldwide production of cellulose

As shown below in Figure 1, cellulose is the primary polysaccharide in plant cell walls and vegetable fibers. Wood, some clothing, food and feed (grain, stalk, straw) all contain cellulose.

In 1977, Ratledge estimated that about 9^{11} tonnes of cellulose were synthesized and replenished through photosynthesis annually. Fadel (1999) studied the status of selected by-products containing high quantities of cellulose and found that they represented almost two trillion Mcal of energy. The investigator also noted that the world's total by-products almost equaled one billion tonnes of dry matter. Since 1999, quantities of by-products containing high amounts of cellulose and associated compounds have increased in developed and developing countries. In their review, Das and Singh (2004) noted that, "Cellulose is the most abundant material in the universe and makes up more than 50% of the total organic carbon in the biosphere and one third of the vegetation." When considering the use of all cellulosic compounds for fuel, it has been noted that globally, they represent 670 billion barrels of oil; this is 20 times the present worldwide oil consumption (Anon, 2011a).

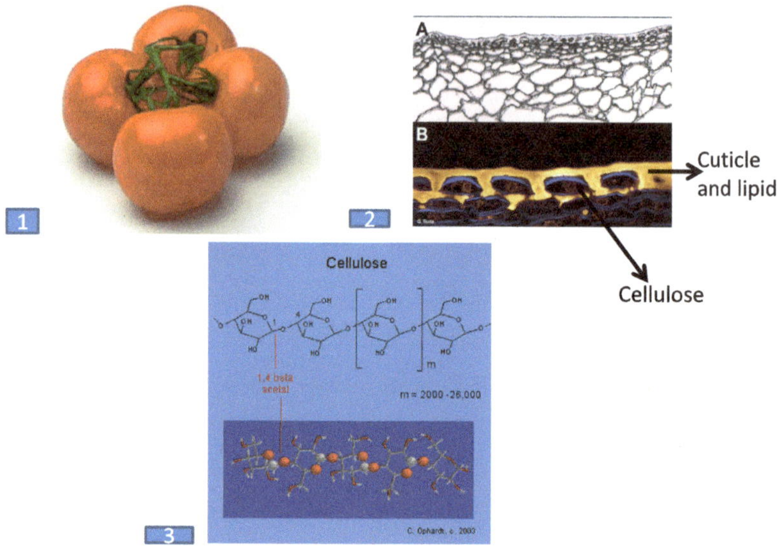

1. Biology-blog, 2011.
2 Rose Laboratory, 2012.
3. Carbohydrate – cellulose, 2012.

Figure 1. Cellulose is an abundant material. Shown: (1) tomatoes; (2) tomato epidermis and cuticle – A. light micrograph of an unstained section; B. confocal optical section showing the cuticle and lipids stained orange and cellulose of the polysaccharide cell wall stained blue and (3) chemical structure of cellulose.

3. Cellulose - Fruit and vegetable production and loss

China, India and Brazil produce the most fruits in the world and the first two countries along with the United States produce the most vegetables (Tables 1 and 2, Freshel, 2012).

4. Loss of nutrients in agricultural by-products

Some cellulose is used for food, feed and fuel while a large quantity, along with valuable nutrients, is wasted as agricultural by-products. For the purposes of discussion in this chapter, fruit and vegetable by-products are defined as nutritious plant material not harvested and/ or wasted during and after production of food products (Figure 2).

Country	2008 (in T)	Average 03-07 (in T)	Var ø 03-07/08 (%)
China	184,560,585	160,912,654	14.70
India	53,133,000	40,034,665	26.40
Brazil	40,784,014	39,194,830	4.05
United States	30,404,237	30,420,653	-0.05
Italy	18,620,221	18,266,318	1.94
Iran*			
Turkey	18,418,811	17,503,291	5.23
Spain	17,518,300	18,072,322	-3.07
Mexico	17,105,610	16,307,257	4.90
Indonesia	12,647,246	14,814,412	-14.43
Other	228,723,846	242,734,060	5.77
Total	621,915,870	613,869,842	1.31
Total EU-27	65,910,712	67,858,847	2.87
% of EU share Of Total	10.60		

No data available
Freshel, 2012.

Table 1. Fruit production by country.

Country	2008 (in T)	Average 03-07 (in T)	Var ø 03-07/08
China	228,208,174	363,905,398	-13.53
India	48,868,700	60,985,290	-19.87
United States	32,686,925	33,945,851	-3.71
Turkey	20,818,183	19,870,895	4.77
Russia	10,957,800	14,018,032	-21.83
Spain	10,386,800	10,848,069	-4.25
Italy	10,933,710	13,875,821	-21.20
Egypt	10,926,895	13,926,043	-21.54
Iran*			
Mexico	9.672,171	9,265,987	4.38
Other	142,390,924	177,893,627	-19.96
Total	525,850,281	627,300,111	-16.17
Total EU-27	52,747,418	60,543,485	-12.88
% of EU-share of Total	10.03		

No data available
Freshel, 2012.

Table 2. Vegetable production by country

a. Northumberland County Council, 2012.

b. TED, Wageningen, 2012.

Figure 2. Fruit and vegetable by-products: a. waste, possibly from a home or restaurant and b. apples discarded during processing.

The nutrient composition of selected by-products has been evaluated (Chedly and Lee, 2001). More recently, Mirzaei-Aghsaghali and Maheri-Sis (2008) reviewed nutrients in several fruit and vegetable by products. Results for research at the University of California, Davis indicated that tomato pomace, a by-product from the processing of various types of tomato products, contains 25.4 μg/g of α-tocopherol, 3.32 μg/g of lycopene and 3.89 μg/g of β-carotene (Assi and King, 2007). As shown in Table 3, other important nutrients in tomato pomace include crude protein, crude fat, vitamins and minerals.

Analysis	Tomato pomace[a]
AME_n	892.41 calories/lb (8.22MJ/kg)
Protein	26.9%
Crude fat	11.9%
Crude fiber	26.3%
Moisture	5.1%
Ash	3.5%

AME_n, gross energy of feed with corrections for gross energy of the excreta after correcting for retention of nitrogen by the body.

[a]King and Zeidler 2004, Assi and King 2008; Mansoouri et al, 2008.

Table 3. Selected nutrient content of tomato pomace.

Figures 3 shows *production* of fruits and vegetables by areas of the world while Figure 4 shows percentages of fruit and vegetable *loss* along the food supply chain from production to consumption in many locations.

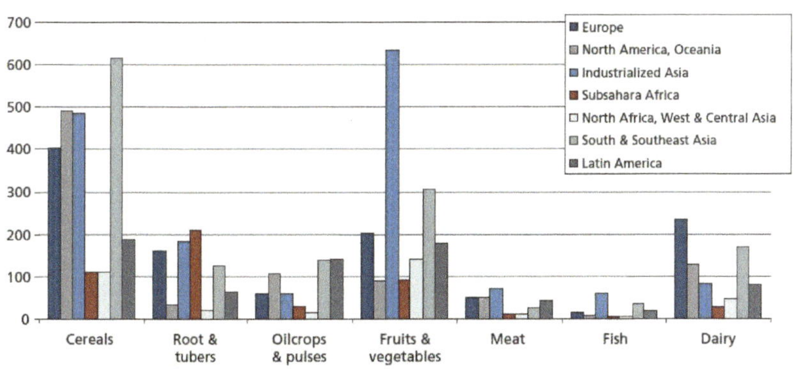

Bars for each commodity show countries as listed in the legion box (top to bottom) from left to right. Gustavasson et al. 2001.

Figure 3. Production volumes of each commodity group, per region (million tonnes).

Fehr and Romão (2001), using a 1998 report (discussing elevated food losses) by the Center for Packaging Technology (Centro de Tecnologia de Embalagem), estimated that for a typical Brazilian city of 440,000 people, the daily total loss of fruits and vegetables was 204 grams per person. This loss/person resulted in a total loss for the typical city of about 82 tonnes per day.

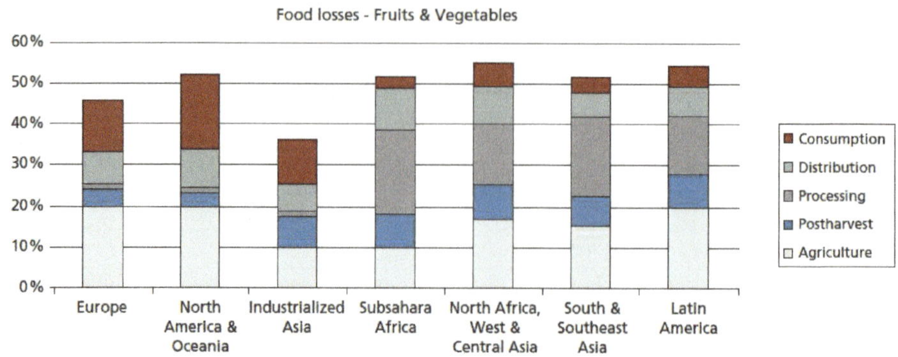

Percentage loss at various stages of the food supply chain in the bars is shown as listed in the legion box from top to bottom. Gustavasson et al. 2001.

Figure 4. Part of the initial production lost or wasted at different stages of the food supply chain for fruits and vegetables in different regions.

India, ranking second for worldwide production of fruits (10%) and vegetable (14%), can have loses as great as 30% to 100%, possibly due to climate, storage and handling conditions (Ashok, 1998; Gautam and Guleria, 2007). With Indian production of fruits and vegetables estimated at 150 million tonnes, the total waste could be 50 million tonnes per year (Gautam and Guleria, 2007).

According to Cuéllar and Webber (2010), in 1995, the United States Department of Agriculture reported waste for many agricultural commodities. Waste for products containing cellulose included grain (6.6 billion kg), raw and processed fruits (5.1 billion kg) and raw and processed vegetables (7.2 kg.). This loss amounted to 32.0%, 23.4% and 25.3%, respectively, of the food supply in each of the categories.

5. By-products in the environment

In many parts of the world, a majority of food waste clogs landfills and emits methane, a greenhouse gas in very rich to extremely poor countries. In addition to polluting the air, by-products can cause water pollution because it is difficult to burn materials with very high moisture content (Mirzaei-Aghsaghali and Maheri-Sis, 2008).

Although in the statement below, Apaak (2011) was referring to the transformation of by-products to energy, the statement indicates how waste, especially that from fruits and vegetables, is discarded in municipal areas.

> "A walk from Circle to the STC yard, a drive between the Achimota interchange and Lapaz, a visit to the vicinity of the landfill at Nungua, and a drive in sections of any of our regional capitals will prove that Ghana has a major waste management and disposal problem that can be a deterrent to economic development." Apaak, 2011.

6. Feed sustainability in the developing world – Use of fruit and vegetable by-products

In developed parts of the world, research is underway and technology is being used to convert fruit and vegetable waste to ethanol and remove coloring agents and antioxidants (Pap et al., 2004; Garcia-Salis et al.; 2010; Zhang, 2006; Yoshimurai 2012; Scalime Nutrition, 2012). Presently, the website for Scalime Nutrition promises one-stop shopping for determination of nutrients (especially polyphenols) from fruit and vegetable by-products and an analysis of the potential applications for nutrient use.

Certainly, in communities across the globe, some by-products will be composted or used for energy; however, in poverty stricken areas where protein consumption is limited, by-product use in the production of feed to produce more protein from small animals seems most urgent. For this discussion, these animals, easy to maintain in small numbers or large groups, include pigs, meat-type chicken, laying hens, turkeys and ducks. Presently, pig meat is the principal meat product consumed in the EU-27 (22.0 million tonnes in 2010) and in many parts of the world. Experts believe that consumption of chicken and turkey products will surpass that of pork in 2012 (Agricultural Products, 2011; Anon, 2011b).

Developing alternative feed sources for monogastric animals is especially important as traditional feed sources (especially corn and wheat) are converted into ethanol in the United States and other developed countries while greatly affecting the availability and price of these commodities in poor countries. Additionally, the cost of grain is expected to continue to rise in extremely food insecure countries due to consumption of traditional grain by humans, growing food needs of populations in emerging economies and harvesting problems associated with climate change (Yegani, 2007; FAO, 2011; Marcela and Ashitey, 2011).

> "One billion people suffer from hunger every day. Yet the earth's population is expected to increase from seven billion in 2011 to more than nine billion people by 2050. The Food and Agriculture Organization of the United Nations predicts that food production will need to increase by 70% over the next four decades to meet anticipated demand. Part of the increase will be driven by higher demand for **animal protein** (emphasis added), especially in developing countries with rising incomes." (Strumbos and Derieux, 2011).

7. A case for alternative feed sources for production of poultry and eggs in food insecure areas

In many poverty stricken rural areas of the world, women and girls comprise the bulk of the agricultural workforce. They are actively involved in backyard poultry rearing due to benefits of providing meat and eggs for meals and supplemental income for their families. Backyard and small-scale meat-type chicken and layer production (the combination hereafter referred to as poultry production) continues even though there are many constraints. Indeed, poultry production is promoted by humanitarian efforts to reverse poverty, increase food security and enhance nutrition/health of family members, especially children (ENAM, 2008). Some challenges affecting the success of nutrition programs are associated with education and communication structures. However, there is another challenge that is central to successful improvement of family nutrition by consumption of animal protein. The challenge is that many small scale farmers may not know how to produce alternative feed sources for poultry as the price of corn rises globally and becomes too costly for use. Indeed, lack of feed for small-scale production of poultry could have a significantly adverse effect on programs like ENAM.

Ghanaian poultry production, based on the use of corn and soy, makes a good case study. In their report, Ghana Poultry and Products Annual 2008, Flake and Ashitey highlighted the on-going condition for poultry production in the country. Due to government initiatives in the 1960's, the combined (small-, medium- and large-scale) poultry industry produced almost 95% of Ghana's chicken meat and eggs during the 1980's and 1990's. However, by 2001, a precipitous decline in production caused supply to slip to about 10 %. By 2008, low production was due to increased imports of cheaper poultry and the cost of grains, chiefly wheat bran or white and yellow corn. Whereas 50 kg of yellow corn cost $24.70 in 2007, it rose to $45.00 in 2008 due to the worldwide economic downturn. In 2008, feed cost represented up to 82% of variable production costs.

According to Marcela and Ashitey (2011), corn - a major staple food crop for the Ghanaian population – is presently produced in the country. Though data is sketchy, best estimates are that human consumption of all Ghanaian corn is at 85% while 15% is used to feed livestock with the greater quantity for poultry (~ 225,000 MT per year). From 2009 – 2011, the local wholesale price for 100 kg of white corn fluctuated between $30.00 to $34.00.

As Ghanaian corn production increases from an average annual supply of 1.5 million tonnes (Marcela and Ashitey, 2011) or remains steady, more of it may be substituted for malt in the growing brewing industry; ultimately less than 15% of the corn produced may be available at very high prices for the small poultry industry. As noted by the Food and Agricultural Organization (FAO), the International Fund for Agricultural Development and the World Food Program, "Food price volatility featuring high prices is likely to continue and possibly increase, making poor farmers, consumers and countries more vulnerable to poverty and food insecurity (FAO, 2011a)." If corn is not available for poultry feed during extremely stressful economic and environmental conditions, small-scale production certainly will not improve. It may not survive.

Flaherty et al. (2010) calculated that 1.5% of the Ghanaian higher education full time teaching equivalents were devoted to poultry research as compared to 93.1% for crops including cassava, cereals (maize, millet, and rice), cocoa, plantain, potatoes, oil palms, shea nuts and yams. If small family farmers in Ghana and other developing countries are to gain health and economic benefits from production of poultry meat and eggs, more research needs to focus on poultry production. In order to increase poultry production by small-scale farmers, research on alternative feed sources is needed.

Through international assistance in 2012/2013, Rwanda plans to develop Master's Degree programs in Animal and Plant Sciences at the National University of Rwanda (NUR) (USAID, 2012). A cooperative effort for experts in the Faculty of Agriculture at NUR to use fruit and vegetable by-products as well as that from other plant sources in developing feeding programs for mongastric animals would insure the availability of alternative feed sources for small-scale farmers in this country.

8. Challenges for use of alternative feed sources

As noted by Agudelo (2009), there are many challenges associated with use of alternative feedstuffs. These include anti-nutritional factors (disease or toxic factors), availability of raw material; composition and metabolizabilty of nutrients, consistency, cost (labor, processing, storage and transportation), health hazards, legality of use, palatability and overall effect of new feed on the resultant quality of the food. Each of these factors has to be considered and effectively minimized or totally eliminated when using alternative feed sources.

9. Endless possibilities for use of fruit and vegetable by-products in feeds

Two possibilities for inclusion of fruit and vegetable by-products in feedstuffs are discussed below.

9.1. Bovine feedstuffs

Chedly and Lee (2001) proposed a number of tropical by-products (containing cellulose) that could be used as ingredients in alternative feed sources for pigs (Table 4). Additional by-products from Sub Suharan African countries for consideration include discarded cereals (maize, millet, and rice), cocoa, plantain, oil palms, shea nuts and other crops in rural and urban agricultural areas (agriculture on open urban lands) that are deposited as waste (In an email from P. Akowua, October, 2011; akowua963@yahoo.com). Researchers have tested fruit and vegetable waste from shops in Salamanca, Spain and to a market place in Medellin, Columbia for use in bovine feedstuffs (Garcia et al, 2005; Angula et al., 2012).

9.2. Laying hen and poultry feedstuffs

High fiber agricultural products and by-products have been used as unconventional feedstuffs in non-feed-removal diets of laying hens. Diets containing 94% wheat middlings and various combinations such as 71% wheat middlings and 23% corn (Biggs et al., 2003, 2004) have been tested. Guar meal (10 and 15%) (Zimmerman et al., 1987), grape pomace with added thyroxin (Keshavarz and Quimby, 2002), various combinations of alfalfa (Landers et al., 2005; Donaldson 2005), and cottonseed meal (Davis et al., 2002) were also evaluated. Coconut meal and cumin seed meal were successfully used in layer and non-feed-removal molt diets in Asian and Middle Eastern countries (Ravindran and Blaire, 1992; Mansoori 2007).

Feed	DM (%)	Per kg DM			Per kg Fresh Matter			Inclusion rate fresh
		ME (MJ/kg)	CP (g/kg)	CF (g/kg)	ME (MJ/kg)	CP (g/kg)	CF (g/kg)	(kg/day)
Spent grain	22.0	8.2	260	130	1.8	57.2	28.6	5-20
Banana stems	9.5	5.5	20	210	0.52	1.9	20.0	5-10
Banana skin (ripe)	15.0	6.7	42	77	1.0	6.3	11.6	2-5
Rejected banana (ripe)	30.0	11.5	54	22	3.5	16.2	6.6	2-5
Cassava leaves	16.0	6.7	235	190	1.1	37.6	30.4	3-6
Cassava roots	28.5	12.5	16	52	3.6	4.6	14.8	5-15
Molasses	78.0	11.5	15	0.00	9.0	11.7	0	0.5-2
Breadfruit (ripe fruit)	29.8	10.8	57	49	3.2	17.0	14.6	4-8
Taro leaves	16.0	6.2	223	114	1.0	35.7	18.2	1-2
Taro Roots	25.0	13.2	45	20	3.3	11.25	5	2-5
Sweet potato (leaves)	12.0	5.8	200	145	0.7	24.0	17.4	10-20
Sweet potato (tuber)	30.0	13.5	70	25	4.1	21.0	7.5	5-10
Yam (leaves)	24.0	7.3	120	250	1.8	28.8	60	2-5
Yam (root)	34.0	13.5	80	25	4.6	27.2	8.5	2-5
Poultry litter	82	8.2	265	145	6.7	217.3	119.0	0.5-2
Olive cake	45.5	3.8	40	465	1.7	18.2	211.6	2-4
Olive leaves	56.8	5.7	105	300	3.2	59.6	170.4	3-6
Grape marc	37.1	4.9	138	410	1.8	51.2	152.1	1-3
Sugar-beet pulp	19.5	9.8	91	316	1.9	17.4	61.6	up to 20
Tomato pulp	22.5	8.0	215	350	1.8	48.4	78.8	up to 15
Wheat bran	89.1	8.1	160	137	7.3	142.6	122.1	1-3
Date palm fruit	87.6	12.0	32	50	10.5	28.0	43.8	0.5-1
Citrus pulp	23	10.3	75	200	2.4	17.3	46	up to 15

Key to columns: DM = dry matter. ME = metabolizable energy. CP = crude protein. CF = crude

Chedly and Lee, 2001.

Table 4. Tropical by-products proposed for ensiling.

In California, the producer of close to 50% of the fruits, nuts and vegetables grown in the United States, the number of by-products that could be used in feed for poultry and other monogastric animals is quite abundant. Tomato pomace is one of these by-products.

Patwardhan and King (2010) reviewed types of non-feed withdrawal diets and suggested that tomato pomace as well as safflower meal, an abundant by-product from processing of safflower seed (Davies, 2002), could be used in molt diets. When comparing weight gain, feed consumption, and egg production, results revealed the usefulness of tomato pomace in non-feed-removal molt diets (Patwardhan et al., 2011a). Due to a decrease in bone density of hens fed tomato pomace, it was recommended that post molt diets of hens previously fed this alternative feed source contain an additional source of calcium (Patwardhan and King, 2011b). Research on the palatability of tomato pomace indicated that commercial feed pellets containing pomace (10%) and invert sugar (10%) could support adequate egg production (King and Griffin, 2012a, unpublished data) for a period of 14-days.

10. Production of tomatoes

A very popular food, eaten fresh or in many commercial products, tomatoes are produced in several regions of the world with China leading production followed by the United States where over 95% of the tomatoes are grown in California (USDA, 2012). Two developing countries - Egypt and India - are among the ten countries that produce 65% of processed tomatoes and concomitant tomato pomace (Branthôme, 2010). Tunisia (engaged in recent tomato wars over the differential in prices between fresh tomatoes and those used for processing) produces processed tomatoes in Africa and the Mediterranean (Republic of Tunisia, 2011). Nigeria is ready for increased tomato production and processing also (IITA, 2011). Additionally, tomato production and processing is underway in EU countries, African countries, Southeast Asian countries (Vietnam, Cambodia, and Laos), Mexico, Central America and South America (Branthôme, 2010; AVRDC, 2011).

Tomato pomace - a mixture of cores, culls, trimmings, seeds, peels, liquor and unprocessed green tomatoes remaining after processing - is not in constant high demand for use in feed products in developed countries (Patwardhan et al., 2010) or developing ones (In an email from P. Akowua, October, 2011; akowua963@yahoo.com Akowua, 2011). It is usually treated as waste and discarded. Wet tomato pomace, combined with wheat, may be fed to ruminants to enhance their growth performance (Denek and Can, 2006; Abdollahzadeh, 2010). Holistic Natural Pet Food (2010) reported that tomato pomace improved the palatability of cat food.

11. A major deterrent to the increased use of tomato pomace and other by-products of plant origin

Tomato pomace is a high protein and high fiber by-product (Table 3). Furthermore, it contains α-tocopherol (Vitamin E, an antioxidant) and lycopene (an antioxidant and a coloring agent) that could be additional feed ingredients for poultry (King and Zeidler, 2004;

Al-Betawi, 2005; Assi and King, 2008). However, use of this by-product and many others that could be added to feed of monogastric animals is limited due to high fiber content (Assi and King, 2007, 2008; Patwardhan et al., 2011a, b). As with many by-products, the major fiber components - cellulose, hemicellulose and lignin - of tomato pomace are difficult to digest by non-ruminant animals like poultry because the greatly dispersed physical and chemical structure created by lignin polymers and their cross linking with other polysaccharides prevent free enzymatic access to cellulose and hemicelluloses (Figure 4).

Structurally, cellulose microfibrils are ordered polymer chains of β-1,4 linked D-glucose units with tightly packed crystalline regions (Figures 5). A diverse group of carbohydrates that are referred to as hemicelluloses, are soluble in strong alkali solutions, linear (no microfibril formation) and flat, with a β-1,4 backbone. Having relatively short side chains, hemicellulose, known as cross-linking glycans, form hydrogen bonds with cellulose (Saupe, 2011, Figure 5). Lignin, a cross-linked macromolecule composed of phenylpropanoid monomers, adds strength and rigidity to cell walls (Smook, 2002, Figure 5).

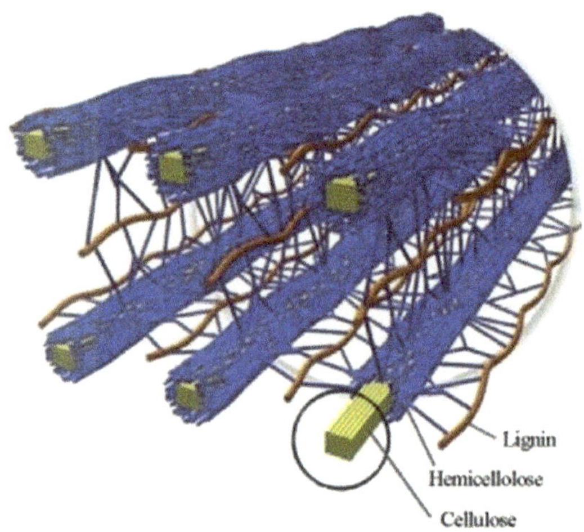

Doherty et al., 2010.

Figure 5. Cellulose strands surrounded by hemicelluloses and lignin.

12. Selected methods for treatment of lignocellulosics

Partial or total elimination of cellulose and accompanying polysaccharides in fruit and vegetable by-products is paramount for digestion by monogastric animals. Moreover, retention of nutrients for incorporation into feed is equally important. Polygastric animals can consume high fiber by-products which are often good sources of protein. However, consumption of these by-products by monograstric animals can result in diarrhea and ultimately weight loss. Ensiling, enzymatic conversion and use of calcium hydroxide under various condition are a few of the approaches discussed below.

12.1. Ensiling

Ensiling is generally used to store excess grain that might otherwise deteriorate over time. During ensiling, some bacteria secrete enzymes to degrade cellulose and hemicellulose to various simple sugars while others reduce simple sugars to acetic, lactic and butyric acids, causing a low pH that helps to preserve surplus grains. High lignin vegetable and fruit by-products, with or without ensiling, can be degraded to various nutrients by rumen herbivores (cattle, sheep, goat and deer) through digestion. These by-products cannot be included at high concentration in diets of monogastric animals like poultry. Therefore, probably, this treatment could only be used for low lignin (≤ 5%) by-products. Chedly and Lee (2001) proposed a list of by-products that could be ensiled or otherwise treated (Table 3). Another deterrent to ensiling is that when a specific ensiled by-product has been analyzed, the resultant material is usually lacking key vitamins, minerals or amino acids (Ly, 2006).

12.2. Enzymatic conversion by organisms

Bioconversion by enzymatic reactions during solid state fermentation by several organisms has been studied (Joshi and Sandhu, 1996; King and Assi, 2007, 2008, 2009; Demers et al, 2009). When comparing untreated and treated samples, Joshi and Sandhu (1996) found that after removal of ethanol, dried apple pomace had more protein, fat, vitamin C, zinc, manganese, copper and iron after solid state fermentation with *Saccharomyces cerivisae, Candida utilis, and Torula utilis*. Other components remained unchanged.

Pleurotus ostreatus (Assi and King, 2004, 2007, 2008) was used to convert tomato pomace amended without and with 487μM manganese/g substrate. In treated substrate without manganese, cellulose and hemicellulose were reduced. However, the mineral activator inhibited fungal growth and did not degrade lignin. An investigation was undertaken to determine if the O_2 consumption rate and CO_2 evolution rate of *P. ostreatus* were affected by manganese. Results indicated no delay for peak and cumulative CO_2 rates; thus, manganese may have reduced the metabolic activity of *P. ostreatus*. For future studies, researchers suggested use of (1) lower levels of manganese and (2) *P. ostreatus* in an O_2 atmosphere of > 20%.

12.3. Enzymatic hydrolysis by commercial enzymes

Enzymatic hydrolysis is not as quick as chemical catalytic hydrolysis; however, it (1) requires less time in the reactor, (2) can be performed at much lower temperatures which may be needed to conserve ingredients in feedstuffs, (3) has lower utility costs and (4) allows for a higher conversion rate than acid hydrolysis which is more equilibrium driven and can produce many end products (Purwadi, 2004; Purwadi et al., 2006; Demers et al, 2009). Although their main application was for production of biofuels rather than the release of valuable components for use in animal feeds, Demers et al. (2009) chose enzymatic hydrolysis to investigate the break down of available cellulose in apple pomace, wood shavings and swithchgrass feedstocks. Other investigators refined the use of enzymatic hydrolysis as well by utlilization of the cellulose enzyme complex of *Tricoderma viride* (Lindsey Wilke, 2011) and by the use of an immobilized β-glucosidase reactor (Issacs and Wilkes, 2011).

Work on enzymatic hydrolysis has been enhanced by use of catalysts needed to accelerate enzymatic activity. In their work on "cellulose enhancing factors" for oxidative degradation of biomass component, Quinlan et al. (2011) revealed that (1) GH61 enzymes need copper for maximal activity (2) that ascorbate and gallate (small molecule redox-active cofactors) enhance development of cellodextrin and oxidized products by GH61 and (3) methylated histidine was located in the coordination sphere of type II copper in the enzyme's active site. While most studies are focused on the final production of ethanol, hopefully, some research will focus on controlling enzymatic breakdown of cellulose in order to preserve protein and other nutrients for feed inclusion.

12.4. Heating and calcium hydroxide treatment

Various methods for solid state bioconversion and enzymatic conversion are proposed often for the ultimate breakdown of lignocellulosics to ethanol. However, nutrients needed for feed inclusion will be lost and even if modified when further developed, cost will prohibit their use by small-scale farmers in developing countries. Combinations of heat, treatment with calcium hydroxide and other methods to remove varying quantities of fiber may be more feasible for small- and large-scale use of by-products for feed ingredients.

Research has been conducted on the use of heat to reduce lignocellulosics in by-products (Squires, 1992; Carmen, 2003; Al-Betawi, 2005; Mondragon et al., 2010; King et al., 2012). According to Carmen (2003) and Mondragon et al. (2010), alkalinity and heat used during nixtamalization improved the availability of some nutrients in corn while reducing the amount of others. King and Griffin (2012b) are continuing studies that reduce cellulose, hemicelluloses and lignin after using combinations of heat (autoclave and microwave) and calcium hydroxide, which is less caustic than sodium hydroxide.

Calcium hydroxide (or inexpensive slaked lime) is available in many parts of the world (Goats, 2010). Where it is not readily available, other methods of producing it are possible. In communities near oceans, ash containing calcium hydroxide can be produced from burning pulverized oyster shell (without lead) or egg shells. In geographical zones away from the ocean, egg shells can be a source of calcium hydroxide.

In a recent study, Janker-Obermeier et al. (2012) found that at least 4–5 wt% of NaOH (~ 0.4 g/g biomass) was needed to dissolve hemicellulose (xylan) from straw at high amounts (73%) in the supernatant. Moreover, solubility of lignin was increased by microwave energy (between 120 and 735 J/g dry mass). In solid wheat straw, excessive amounts of hemicelluloses (>80%) and lignin (90%) could be removed without inordinate degradation of saccharide or solubilization of cellulose. These recent results suggest that, eventually, mild treatments for removal of lignocellulosics from fruit and vegetable by-products will be developed so that remaining nutrients can be incorporated into feed for monogastric animals.

As noted above, during processing of fruit and vegetable by-products for monogastric animals, some fiber and, ideally, all of the nutrients should be maintained. A schematic for processing of fruit and vegetable for this purpose is proposed (Figure 6). As mentioned above, procedures that combine various amounts of heat (no heat, autoclave and microwave) and calcium hydroxide are being investigated (King and Griffin, 2012). While cellulose in fruit and vegetable by-products could be used as fuel, potentially, it, along with associated nutrients, could be used as feed ingredients in the diets of animals. Procedures that maximize the amount of ligninocellulosics that can be removed while minimizing nutrient loss should be investigated thoroughly. Once fully developed these procedures, characterized by low technology input, can be made available to small-scale farmers in developed and developing countries where enzymatic reduction of high fiber content in by-products may not be a viable alternative.

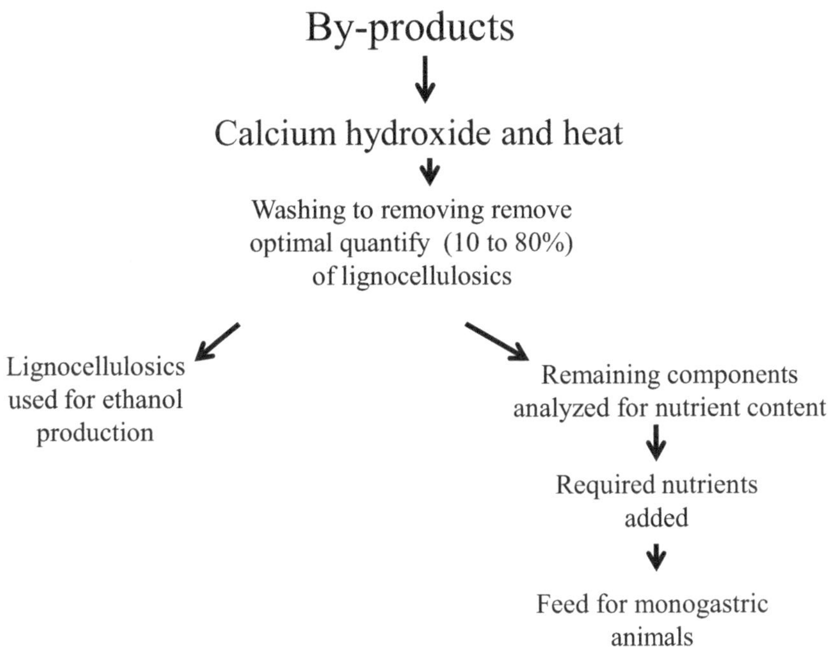

Figure 6. Removal of lignocellulosics from fruit and vegetable by-products.

Author details

Annie King
Department of Animal Science, University of California, Davis, USA

12. References

Abdollahzadeh F, Pirmohammadi R, Fatehi F, Bernousi T. Effect of feeding ensiled mixed tomato and apple pomace on performance of Holstein dairy cows. 2010. Slovak Journal of Animal Science. 43:31-35.

Agricultural products - statistics explained. 2011. Available from: http: www.epp.eurostat.ec.europa.eu/statistics_explained/.../Agricultural_product

Al-Betawi NA. Preliminary study of tomato pomace as unusual feedstuff in broiler diets. Pakistani Journal of Nutrition. 2005. 4:57-63.

Angulo J, Mahecha L, Yepes SA, Yepes AM, Bustamante G, Jaramillo H, Valencia E, Villamil T, Gallo J. Quantitative and nutritional characterization of fruit and vegetable waste from market tomato pomace: A potential use as bovine feedstuff. Journal of Environmental Management. 2012. 95:S203-S209.

Anon, University of York International Relations. University of York, Heslington, York, YO10 5DD, UK. 2011a. Available from: www.york.ac.uk › News and events › News › 2011 news › Research

Anon. Poultry challenges pork in worldwide meat consumption. 2011b. Available from: www.wattagnet.com/Poultry_challenges_pork_in_worldwide_meat.

Apaak CA. Keeping Ghana clean by profiting from our waste. 2011. Available from: www.ghanaweb.com/GhanaHomePage/NewsArchive/artikel.php?... - Cached

Ashok BS. New policy to woo corporate investment. The Financial Express (Newspaper). India 1998.

Assi JA, King, AJ. Assessment of selected antioxidants in tomato pomace subsequent to treatment with the edible oyster mushroom, Pleurotus ostreatus, under solid state fermentation. Journal of Agriculture and Food Chemistry. 2007. 55:9095 – 9098.

Assi JA, King AJ. Manganese amendment and *Pleurotus ostreatus* treatment to convert tomato pomace for inclusion in poultry diets. Poultry Science. 2008. 87:1889-1896.

Assi JA, King AJ, Vander Gheynst J. CO_2 evolution rate during solid-state fermentation of tomato pomace as a poultry feed ingredient. International Journal of Agricultural and Biological Engineering. 2009. 2:28 - 32.

AVRDC (Asian Vegetable Research and Development Center). Low cost processing technologies for vegetables in Southeast Asia. 2011. The World Vegetable Center. Available from: www.avrdc.org › ... › Enhancing consumption › Achievements

Biggs PE, Douglas MW, Koelkebeck KW, Parsons, CM. Evaluation ofnonfeed removal methods for molting programs. Poultry Science. 2003. 82:749–753.

Biggs PE, Persia ME, Koelkebeck KW, Parsons CM. Further evaluation of nonfeed removal methods for molting programs. Poultry Science. 2004. 83:745–752.

Biology blog. Effect of rising carbon dioxide on rangelands. 2011. Available from: httomato pomace://www.biology-blog.com/blogs/plant-science-blog.html

Branthôme F-X. Trends in tomato products consumption compared to total tomato consumption, World Processing Tomato Council, Avignon, France. 2010.

Carbohydrates - Cellulose. 2012. Available from: www.elmhurst.edu/~chm/vchembook/547cellulose.html

Carmen W. Nixtamalization, a Mesoamerican technology to process maize at small-scale with great potential for improving nutritional quality of maize based foods. 2nd International workshop – Food-based approaches for a healthy nutrition. 2003. Nov 23-28. Quagadougou, Burkina Faso.

Chedly K, Lee S. Paper 6: Silage from by-products for smallholders. 2001. FAO Corporate Document Repository. Available from: httomato pomace://www.fao.org/DOCREP/005/X8486E/x8486e01.htm

Cuéllar AD, Weber ME. Wasted Food, Wasted Energy: The Embedded Energy in Food Waste in the United States. Environmental Science & Technology. 2010. 44:6464–6469.

Das H, Singh SK. Useful byproducts from cellulosic wastes of agricultural and food industry – A critical appraisal. Critical Reviews in Food Science and Nutrition. 2004. 44:77-89.

Davis, K. UPC's comments to the FDA on the forced moulting of laying birds. 2004. Available from: http://www.upc-online.org/molting/commentstofda.htm

Demers A, Doane R, Guzman S, Pagono R. 2009. Enzymatic Hydrolysis of Cellulosic Biomass for the Production. Available from www.wpi.edu/Pubs/E-project/Available/E-project-04300

Denek N, Can A. Feeding value of wet tomato pomace ensiled with wheat straw and wheat grain for Awassi sheep. 2006. Small Ruminant Research. 2006. 65:260-265.

Doherty WOS, Mousavioun. P, Fellows CM. Value-adding to cellulosic ethanol: Lignin polymers. Industrial Crops and Products. 2011. 33:259–276.

Donaldson LM, Kim WK, Woodward CL, Herrera P, Kubena LF, Nisbet DJ, Ricke RC. Utilizing different ratios of alfalfa and layer ration for moult induction and performance in commercial laying hens. Poultry Science. 2005. 84:362–369.

ENAM (Enhancing Child Nutrition through Animal Resource Food Management), Hidden Hunger: The Story of the ENAM Project in Ghana and Child ...2011. Available from: www.partnership-africa.org › Blogs › hcrobert's blog (video)

Fadel J. Quantitative analyses of selected plant by-product feed stuffs, a global perspective. Animal Feed Science and Technology. 1999. 79:255-268.

FAO (Food and Agriculture Organization). 2011a. World hunger report 2011: High, volatile prices set to continue. 2011a. Available from: www.fao.org/news/story/en/item/92495/icode/

FAO (Food and Agriculture Organization). Cutting food waste to feed the world - Over a billion tonnes squandered each year. 2011b. Available from: www.fao.org/news/story/en/item/74192/icode/Cached

Fehr M, Romão DC. Measure of Fruit and Vegetable losses in Brazil - A Case Study. 2001. Environmental Development and Sustainability. 2001. 3:253-263.

Flaherty K, Essegbey GO. Asare R..Ghana's recent development in agricultural research, International Food Policy Research, Institute and Science and Technology Policy. 2010 Institute and the Food and Agricultural Organization, Rome Italy and Washington, DC.

Flake, L. and Ashitey, E. 2009. Ghana: Poultry and Products Annual. USDA Foreign Agricultural Service Global Alliance for Improved Nutrition Report. 2009. United States Department of Agriculture, Washington, DC.

Freshel, the European Fresh Produce Association. Production. 2011. Available from: www.enjoyfresh.eu/php/freshfacts/index.php?doc_id=21

Garcia AJ, Esteban MB, Márquez MC, Ramos P. Biodegradable municipal solid waste Characteristics and potential as animal feedstuffs. Waste Management. 2005. 25:780-787.

Garcia-Salas P, Morales-Soto A, Segura-Carretero A, Fernández-Gutiérrez A. Phenolic-compound-extraction systems for fruit and vegetable samples. Molecules. 2010. 15:8813-8826.

Gautam HR, Guleria SPS, Parmar YS, Guleria, YS. 2007. Fruit and vegetable utilization. Science Tech Entrepreneur. Available from: ebookfreetoday.com/utilization-of-fruit-waste~0.html

Goats B. 2010. Natural sources of calcium hydroxide/ Livestrong.com. 2010. Available from: www.livestrong.com/.../279343-natural-sources-of-calcium-hydroxid

Gustavasson, J., Cederberg, C., Sonneson, U., van Otterdijk, R., and Meybeck, A. Global Food Losses and Food Waste. 2001. Food and Agriculture Organization of the United Nations. Rome, Italy.

Holistic Natural Pet Food. 2011. Available from: www.holisticselect.com/.../ingredient-spotlight-the-nutritional-benefits-of-tomato-pomace/

IITA (International Institute of Tropical Agriculture). 2011. United States Agency for International Development, Washington, DC. 2011. Available from: www.usaid.gov

Isaacs. S. H. and Wilkes, C. R. 2011. Utilization of immobilized B-glucosidase in the enzymatic hydrolysis of cellulose.2012. Available from: http://escholarship.org/uc/item/3m42z0q5.pdf

Keshavarz K, Quimby FW. An investigation of different moulting techniques with an emphasis on animal welfare. Journal of Applied Poultry Research. 2002. 11:54-67.

King AJ, Griffin, JK. 2012. Unpublished data. 2012. University of California, Davis, California.

King AJ, Zeidler G. Tomato pomace may be a good source of vitamin E in broiler diets. California Agricultural Magazine. 2004. 58:59-62.

Landers KL, Woodward CL, Li X, Kubena LF, Nisbet DJ, Ricke SC. Alfalfa as a single dietary source for moult induction in laying hens. Bioresource Technology. 2005. 96:565–570.

Lindsey RR, Wilke CR. 2011. Process design and optimization of cellulose hydrolysis. 2011. Available from: http://escholarship.org/uc/item/2gz9q3dx.pdf

Ly NTH. The use of ensiled leaves for feeding pigs on-farm in central Vietnam. Workshop on forages for pigs and rabbits, August 21 – 24, 2006. http://www.mekarn.org/proprf/hoaly.htm

Marcela R.and Ashitey E. 2011. Ghana- 2011 Grain and Feed Annual. United States Department of Agriculture, Washington, DC.

Mansoori B, Modirsanei M, Farkhoy M, Kiaei M, Honarzad J. The influence of different single dietary sources on moult induction in laying hens. Journal of the Science of Food and Agriculture. 2007. 87:2555-2559.

Mansoori B, Modirsanei M, Radfar M, Kiaei M, Farkhoy M, Honarzad J. Digestible and metabolisable energy values of dried tomato pomace for laying and meat type cockerels. Animal Feed Science and Technology. 2008. 141:384 -390.

Mirzaei-Aghsaghali A, Maheri-Sis N. Nutritive Value of Some Agro-Industrial By-products for Ruminants - A Review. World Journal of Zoology 2008. 3:40-46.

Mondragón M, Bello-Pérez LA, Agama-Acevedo E, Betancur-Ancona D, Pena JLEffect of cooking time, steeping and lime concentration on starch gelatinization of corn during nixtamalization in starch Starch/Stärke. 2004. 56:248-253.

Northumberland County Council. 2012. How we manage your waste (Search: managing waste) . Available from: http://www.northumberland.gov.uk/default.aspx?page=13451.

Pap N, Pongrácz E, Myllykoski L, Keiski R. (2004) Waste minimization and utilization in the food industry: Processing of arctic berries, and extraction of valuable compounds from juice- processing by- products. In: Pongrácz E. (ed.): Proceedings of the Waste Minimization and Resources Use Optimization Conference. 2004. June 10. University of Oulu, Finland. Oulu University Press: Oulu. p. 159-168.

Patwardhan D, King AJ. Review: Feed withdrawal and non feed removal molt diets. World Poultry Science Journal. 2010. 67:253-268.

Patwardhan DS, King AJ, Mireles A. Tomato pomace and safflower meal as ingredients in non-feed-removal molt diets. Journal of Applied Poultry Research. 2011a. 20:291-302.

Patwardhan D, King AJ, Oberbauer AM, Holland TB. Bone measurements of molted layers fed low-salt corn and soybean diets containing safflower meal and tomato. Journal of Applied Poultry Research. 2011b. 20:190-196.

Purwadi R, Niklasson C, Taherzadeh MJ. Kinetic study of detoxification of dilute-acid hydrolyzates by $Ca(OH)_2$. Journal of Biotechnology. 2004.114:187-198.

Purwadi R. Continuous ethanol production from dilute acid hydrolyzates: detoxification and fermentation strategy. Ph.D. thesis. 2006. Department of Chemical and Biological Engineering. Chalmers University of Technology. Göteborg, Sweden.

Quinlan RJ, Sweeney MD, Leggio LL, Otten H, Poulsen J-CN, Johansen KS, Krogh KBRM, Jorgensen CI, Tovborg M, Anthosen A, Tryfona T, Walter CP, Dupree P, Xu F, Davies GJ, Walton PH. Insights into the oxidative degradation of cellulose by a copper meatalloenzyme that exploits biomass components. Proceeding of the National Academy of Sciences. DOI:2011: 10.1073/pnas.1105776108.

Ravindran V, Blair R. Feed resources for poultry production in Asia and the Pacific. II. Plant protein sources. World Poultry Science Journal. 1992. 48:205-231.

Ratledge C. Fermentation substrates. Annual Report of Fermentation Processes. 1977. 1: 4971.

Republic of Tunisia, Ministry of Industry and Technology. CEPI Brief N° 25. 2011. Tunis, Tunisia.

Rose Laboratory. 2012. Welcome to the Rose Laboratory: People, Cornell University, Ithaca, NY. 2012. Available from: http//labs.plantbio.cornell.edu/rose/

Saupe, S. G. 2011. Plant Physiology - Cell Walls: Structure & Function 2012. Available from: http://employees.csbsju.edu/ssaupe/biol327/lecture/cell-wall.htm

Scalime Nutrition. 2012. Available from: www.scalime.com/anglais/valorisationanglais.htm

Smook GA. Handbook of Pulp and Paper Technology. 3rd Edition. Vancouver. Angus Wilde Publications, Inc. 2002.

Squires MW, Naber EC, Toelle VD. The effects of heat, water, acid, and alkali treatment of tomato cannery wastes on growth, metabolizable energy value, and nitrogen utilization of broiler chicks. Poultry Science. 1992. 71:522-529.

Strumbos J., Derieux, R. 9 billion mouths to feed. CAES Outlook, College of Agricultural and Environmental Sciences. 2011. University of California Davis, Davis, CA.

TED Wageningen. (Search:food waste). 2012. Available from: http://tedxwageningen.org/2012/10/food-waste/

USAID (United States Agency for International Development) http://www.hedprogram.org/funding/rfa/WLP-RwandaAg2012.cfm.

USDA (United States Department of Agriculture) Economic Research Service. USDA ERS - Vegetables & Pulses: Tomatoes. 2012. Available from: www.ers.usda.gov/topics/crops/vegetables-pulses/tomatoes.aspx

Weiss WP, Frobose DL, Koch ME. Feeding value of wet tomato pomace for dairy cows. The Ohio State Extension Research Bulletin Special Circular 156 2011. Available from: http://ohioline.osu.edu/science156_25.html

Wright, S. Turning waste into power. 2006. Available from: http://dateline.ucdavis.edu/dl_detail.lasso?id=9083

Yegani M. Corn prices; where do we go from here? 2007. Available from: http://www.worldpoultry.net/weblog/poultry-research/corn-prices%3B-where-do-we-go- from-here%3F-id2978.html

Yoshimurai T, Hatakawa M, Takahashi F, Kawashima T. Study of bio-ethanol production from cellulosic waste (rice straw). Journal of Technology and Education. 2012. 19: 19-22.

Zimmerman PH, Hoof JA. Thwarting of behavior in different contextsand the gackel-call in the laying hen. Applied Animal Behavioral Science. 2000. 69:255-264.

Chapter 12

Structure - Properties Interrelationships in Multicomponent Solutions Based on Cellulose and Fibers Spun Therefrom

Ludmila Golova, Igor Makarov, Ludmila Kuznetsova,
Elena Plotnikova and Valery Kulichikhin

Additional information is available at the end of the chapter

1. Introduction

Cellulose is a natural polysaccharide with a practically inexhaustible raw source. The unique properties of cellulose, such as biodegradability, excellent mechanical properties, and relatively high hydrophilicity, make it possible to manufacture household and industrial materials on its base, which are useful for various applications.

Owing to its structural features, the transition of cellulose to the plastic state presents a challenge. Cellulose decomposes prior to the onset of melting; therefore, dissolution is the only way of preparing spinning systems for fiber and film formation. However, the regular structure of cellulose macromolecules, the presence of a developed system of H-bonds, and a relatively high rigidity of chains (the Kuhn segment is ~100 Å) sharply confine the range of possible solvents for cellulose. Intensive studies conducted by many research and industrial groups have led to the discovery of a new class of nonaqueous solvents for cellulose: oxides of tertiary amines, and among them *N*-methylmorpholine-*N*-oxide (NMMO) was found to be the most efficient since can dissolve up to 20% cellulose [1-5].

Cellulose fibers processed via "MMO process" and referred to as Lyocell fibers according to the resolution by the BISFA, are characterized by their highly ordered structural organization (high degree of crystallinity and orientation). On the one hand, the strength and modulus of the resultant fibers are high; on the other hand, their deformation characteristics are low, and the tendency toward fibrillation exists. As a result, processing becomes more difficult and the service characteristics of fibers deteriorate.

This poses an extremely important problem which, despite the efforts of researchers from many countries has not been solved yet. What are the ways to efficiently govern the fibrillation of Lyocell fibers?

Obviously, this situation requires fundamentally new solutions for controlling the structuring in cellulose and, hence, for the optimization of Lyocell fibers characteristics.

In our opinion, controlling the process of the structure formation of cellulose macromolecules should be solved on micro-and nanolevels by means of creation of multicomponent solutions via introducing to cellulose solutions of polar polymers of different nature or layered anisometric nanoparticles of aluminosilicate with high interfaces and surface energy. The idea of this way consisted in appearance of hindrance at crystallization of cellulose in gel-fibers to decrease their capability to fibrillation.

The basis for solving the problem associated with the structure management of cellulose on micro- and nanolevels in mixed solutions containing cellulose and various synthetic polymers or aluminosilicates is a new method of solid-phase dissolution of cellulose in NMMO, which allows to improve significantly the dissolving ability using high-melting hydrate forms of NMMO and obtaining highly concentrated dopes with concentration of cellulose up to 40% [6-7].

The aim of ongoing research is to develop approaches for controlling the structure of cellulose fibers obtained from solutions in NMMO. In this connection, it has been a challenge to study in detail the efficiency of NMMO as a solvent with respect to hydrophobic synthetic polymers of various nature, to investigate the mutual dissolution of cellulose and synthetic polymers in NMMO, and to characterize the rheological characteristics and structure-morphological features of mixed solutions in order to identify the evolution in the structure of cellulose in all stages from dissolution to spinning of oriented fibers. We investigate the mechanical properties of hybrid fibers, which were selected as a criterion for the choice of regimes for the formation of desired structure for cellulose fiber nanocomposites. All these tasks have been considered in this Chapter.

2. Objects

Baikal softwood cellulose with a degree of polymerization of 600 containing 8% of water (this level corresponds to its equilibrium sorption under normal conditions) was used. As polymeric additives, polar linear thermotropic LC alkylenearomatic polyesters with mesogenic triads based on fumaric and oxybenzoic acids with hexa- (HP-6) and decamethylene (HP-10) spacers, CPEs with different content of repeating units of the parent homopolymers, and CPEs with mesogenic triads based on terephthalic and oxybenzoic acids and decamethylene (PDTOB) spacers were chosen. Along with LC polyesters, the solubility of the amorphous aromatic polyamide poly (*m*-phenyleneisophthalamide) (PMPIA) in NMMO was examined as well.

Dissolving systems were monohydrate NMMO (13.3% of water) and high-melting hydrate form of NMMO with T_m = 120–160°C (water content was 8–10%).

As to aluminosilicate particles, the majority of experiments were carried out with two kinds of clays: Na-montmorillonite (Cloisite Na^{+}) and montmorillonite treated with dioxydecyldimethylammonium chloride (Cloisite 20A) produced by Southern Clay Product, USA.

3. Methods

3.1. Preparation of solutions

Polymer solutions in high melting NMMO crystal hydrate were prepared by mixing of powder components in definite way and subsequent heating in laboratory reactor equipped with stirring device. A completeness of the dissolution was monitored visually or with polarizing microscope by measuring the temperature of system transition to the fully isotropic liquid state.

All joint solutions of cellulose with synthetic polymers in NMMO were prepared according to the following procedure: preliminary solid-phase activation of the cellulose–NMMO system, subsequent addition of synthetic polymer, and homogenization at heating of three-component mixture by mechanical stirring. To inhibit thermo-oxidative degradation of NMMO and cellulose, 0.5% propylgallate was added. Under simultaneous action of temperature and deformation (e.g. at passing through an MV-3M capillary viscometer at 120°C), solid-phase composite melts and passes to flow state. A completeness of all components dissolution in NMMO was controlled by optical observation.

An original method for preparing solid-phase compositions of cellulose with layered aluminosilicates Cloisite Na^{+} and Cloisite 20A has been developed. The addition of clay to the previously obtained "solid" pre-solution of cellulose in NMMO provides a high dispersion and a uniform distribution of clay particles in a matrix cellulose phase. As soon as the system was additionally activated, solid-phase compositions containing 10, 14, and 18 wt% of cellulose, and Cloisite Na^{+} or Cloisite 20Å in amounts from 0,1 to 20 wt% were obtained.

Since dimensions of particles in raw clay Cloisite Na^{+} usually are equal to 3-7 mkm, a method of fractionation of them with extraction of nanodimensional particles by means of "aqueous activation/modification" by means of sedimentation and dynamic light scattering methods (centrifuge CPS 24000) the distribution curves were obtained. Choosing optimal regimes of aqueous activation and drying, particles of modified clay (M_2Cloisite Na^{+}) with dimensions 20-100 nm were prepared.

3.2. Research methods

The phase equilibrium and morphological features of polymer–NMMO systems were studied using Boetius polarization microscope (VEB Kombinat Nadema, former DDR).

DSC studies were performed on a Mettler 822e differential scanning calorimeter at a heating rate of 10 K/min.

X-ray measurements were carried out on DRON-3 and DRON-3M diffractometers (Ni-filtered CuK_α radiation) equipped with a high-temperature chamber (the temperature control allowed to keep accuracy within 1°C) and an IRIS-3.0 instrument (Ni-filtered CuK_α radiation, flat cassettes). Diffractograms and X-ray patterns of the samples were taken in the transmission mode.

The rheological characteristics of solutions were studied by MV-3M capillary viscometer of melt-indexer type and rotation rheometer PIRSP-2 steady-state and low-amplitude oscillation shear deformations using cone–plate operating unit. This equipment renders it possible to cover a wide interval of shear rates $\dot{\gamma} = 10^{-3}–10^{3}$ s^{-1} and shear stresses $\tau = 10–10^{6}$ Pa. The use of capillaries with different length, *L* to diameter, *D* ratio (5–40) showed that the correction for entrance effects is small (1.0–1.5) and for capillaries with high *L*/*D* ratio can be neglected. The viscosity of all systems was measured at 120°C. To visualize the flow morphology of solutions during deformation, the rotational rheometer was equipped with: (i) transparent plate–plate unit and glass optical windows, (ii) on a top plate the prism of full internal reflection was installed together with a polar (analyzer), and (iii) a lamp with a narrow focused light beam and polarizer was positioned in a hollow shaft of rheometer.

For estimation of the fibrillation capacity the obtained cellulose and hybrid fibers were treated in vibration mill M35L (oscillation amplitude is 4 mm, frequency is 48 Hz, power of electric engine is 1.7 kW) during 30 min in aqueous medium

Micrographs of composite materials on nanoscale were obtained using electron microscope LEO 912 AB OMEGA (Carl Zeiss, Germany) with a cathode based on LaB. The acceleration voltage is 100kV.

The tensile strength, the initial elastic modulus and elongation at break fibers and extrudates under consideration were measured with Instron 1122 testing machine at elongation rate of 10 mm / min.

4. Results and discussion

4.1. Crystal solvates of thermotropic alkylene-aromatic copolyesters and poly(*m*-phenyleneisophthalamide) with NMMO

4.1.1. CPE - NMMO Crystal Solvates

Solubility curves for CPEs and PMPIA in NMMO were constructed (Fig. 1) on the base of visual identification of the temperature corresponding to the transition of polymer blends with NMMO monohydrate (MH) to the fully single-phase state (T_{dis}).

As is seen (curves 1–3, 5), the solubility of the tested polyesters and copolyesters in NMMO MH varies in the following sequence HP-6 > HP-10 >PDTOB. Preparation of PDTOB solutions in NMMO MH with concentrations above 20% requires high dissolution temperatures (above 160°C, where CPEs undergo to degradation. Therefore, a higher melting NMMO hydrates should be used, which allow to dissolve up to 60% PDTOB (curve 4) [8].

Thus, the selection of various hydrate forms of NMMO as dissolving systems for copolymers under study leads us to conclusion, that in both cases: hydrophilic and hydrophobic polymers the dissolving ability of NMMO increases with a decrease in the water content and, accordingly, with increase of the melting point of NMMO.

The DSC data on dissolution of HPs and CPEs in NMMO have shown that the shape of DCS curves for all systems is the same. As an example, Fig. 2 shows the thermograms obtained for the system containing 40% HP-10 in NMMO MH.

As is seen(curve 1), during the first heating the low-temperature hydrate form of NMMO initially melts (the endo effect with a maximum at 36°C) and then the basic melting peak due to NMMO MH (at 76°C) appears. As follows from the DSC curve, processes accompanied by the exo effect proceed just after the melting of NMMO MH. During the second heating run, the profile of the curve changes drastically: only one new endothermic peak appears whose maximum (curve 2) is above the melting point of NMMO MH but below the melting temperature of the crystalline copolyester.

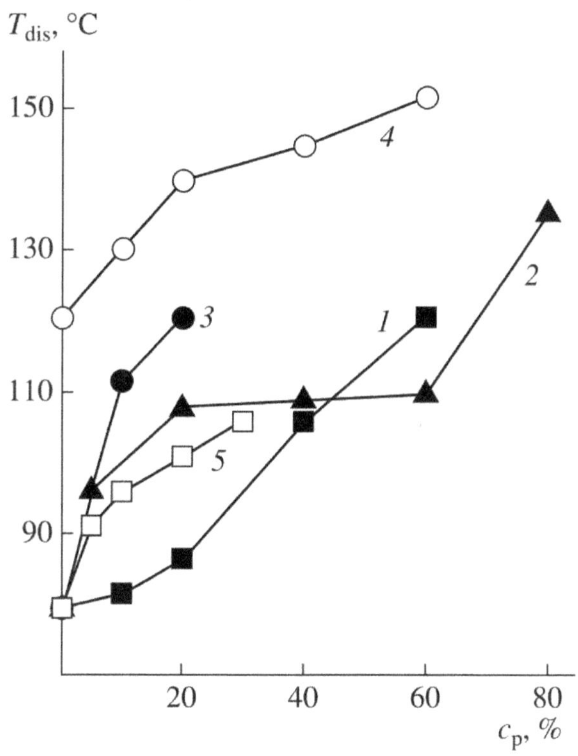

Figure 1. Solubility curves of (1) HP-6 in NMMO MH,(2) HP-10 in NMMO MH, (3) PDTOB in NMMO MH,(4) PDTOB in NMMO, and (5) PMPIA in NMMO MH.

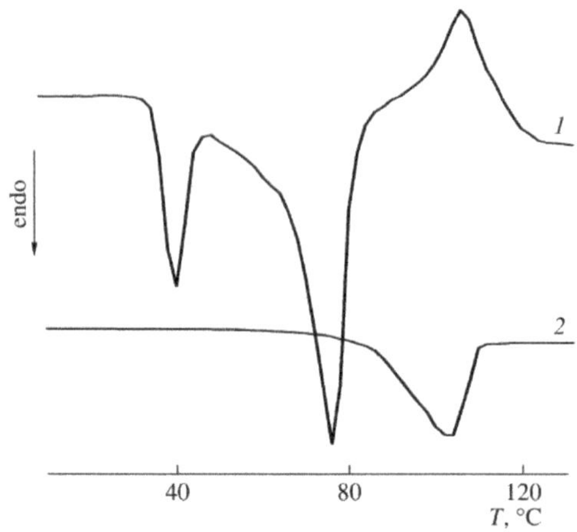

Figure 2. DSC curves of the blend containing 40% HP-10 and 60% NMMO MH under (1) first and (2) second heating runs.

This result allows us to assume that strong interactions between the CPE and solvent molecules exists giving rise to a new additive compound of polymer and solvent having crystalline structure.

Despite of evident information from DSC data concerning the new phase formation in the CPE–NMMO systems (the most probably, crystal solvate, CS), the structure of the novel adduct should be confirmed by X-ray diffraction measurements.

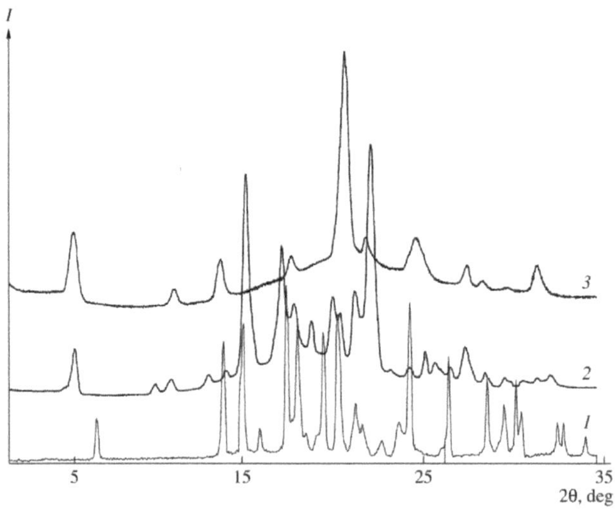

Figure 3. Diffractograms of (1) the NMMO MH solvent, (2) 40% HP-10 solution in NMMO MH, and (3) HP-10 homopolymer at 20°C.

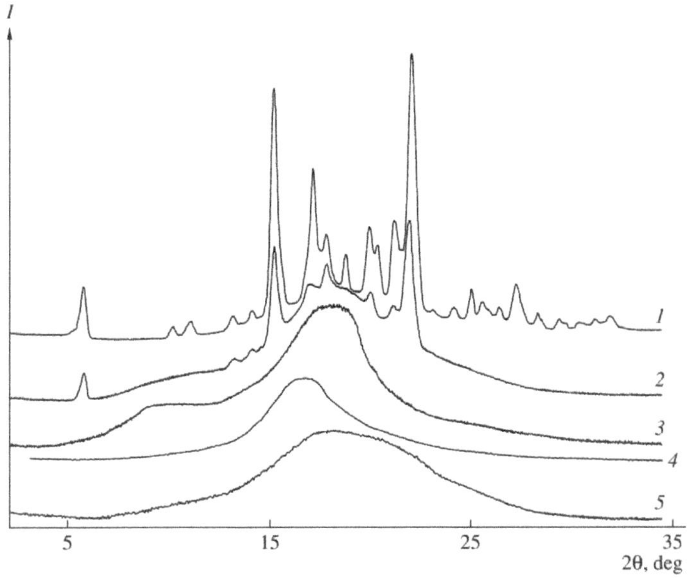

Figure 4. Diffractograms of 40% HP-10 solution in NMMO MH at (1) 20, (2) 110, and (3) 120°C; (4) NMMO MH at 90°C; and (5) HP-10 at 200°C.

Figure 3 displays the diffractograms of the parent compounds — HP-10, NMMO MH, and HP-10 solution in NMMO MH at 20°C. As is seen, the angular positions of $2\theta_i$ reflections measured for the solution differ appreciably from the corresponding $2\theta_i$ values for NMMO MH and HP-10. A specific feature of the diffractograms obtained for the solution and the parent HP-10 samples requires special attention. The first three reflections observed in the diffractogram of the neat HP-10 are manifested in the diffractogram of the solution. The same angular positions of these reflections in the diffractogram of HP-10 and its solution in NMMO MH suggest that NMMO MH molecules are embedded between macromolecules of the polymer causing the change its intermolecular order, at preserving intramolecular periodicity.

This assumption is supported by the X-ray data obtained upon solution heating (Fig. 4). As is seen, the diffractogram of the 40% HP-10 solution measured at 120°C shows three overlapping amorphous halos with maxima at 2θ = 11.6°, 18.1°, and 23.5° and the ratio of intensities I1 : I2: I3 = 61 : 100 : 38. Such a pattern of the diffractogram remains unchanged when the sample is heated. This circumstance indicates that the melting of the sample is not accompanied by its decomposition (congruent melting).

In the case of the solution, the center of gravity of the first two amorphous halos is markedly shifted to small angles (larger interplanar distances) compared with the HP-10 melt, for which the intensity curve of amorphous scattering contains a single amorphous halo with a maximum at 2θ = 19.1°. Such significant changes in the intensity distribution of amorphous scattering at passage from HP-10 to its solution in NMMO MH suggest that the statistics of distribution of intermolecular distances changes.

The X-ray pattern of the crystalline adduct of HP- 10 with NMMO MH exhibits a large number of Debye rings with a nonuniform intensity distribution (the presence of discrete spots along rings), thereby indicating a coarse-grained solvate structure with a crystallite size L > 1000 Å. These values of L are untypical for polymer crystals (for the L values, as a rule, do not exceed 400 Å). The X-ray patterns of HP-10, NMMO MH, and the proposed crystal solvate formed by HP-10 in the NMMO MH solution at a concentration of 40% are characterized by completely different crystal lattices. Hence, the structures of the crystalline phase are quite different.

The above experimental data allow us to conclude with a sufficiently high probability that, in the course of CPE dissolution in NMMO MH at elevated temperatures and subsequent cooling of these solutions, new structures—crystal solvates (CSs) of CPEs with NMMO are formed. The phase composition of crystal solvates depends on the polymer-to-solvent molar ratio. The temperature position of the endo peaks in DSC curves due to the melting of CSs changes and achieves a maximum and constant value at the molar ratio of the components that corresponds to the equilibrium composition of the crystal solvate. Thus, the DSC profiles of the systems containing 20–60% HP-10 in NMMO MH are characterized by the presence of constant heat effects at 100–105°C, thus suggesting the equilibrium nature of the crystal solvate phase formed in this concentration range. The formation of CSs of a constant molar composition is also proved by the presence of a plateau in the solubility curve of HP-10 in NMMO MH (Fig. 1, curve 2) at 100–105°C, that is, at a temperature corresponding to T_m of HP-10/ NMMO MH crystal solvates.

These results are in good agreement with X-ray measurements. The diffractograms of systems containing 20–50% HP-10 in NMMO MH are practically identical. Consequently, in the range of HP-10 concentrations in NMMO MH from 20 to 40%, a polymer-to-solvent molar ratio equal to 1 : 5 that corresponds to the equilibrium composition of.

The formed CSs are distinguished by improved stability to water since they preserve their structure when more than 80% water is added to the system. At this content of water, the NMMO multi hydrates occur in the liquid state. Most probably, macromolecules of the hydrophobic polymer protect water-sensitive NMMO molecules; as result, hydrophobic crystal solvates of HP-10 with NMMO MH are formed.

Optical microscopy was applied to analysis of morphological features of CSs. The known morphological kinds of CSs belong to faceted structures (rhombs and parallelograms), spherulites, and shish kebab.

Figure 5 demonstrates micrographs of CSs in crossed polars for the co-crystallization of HP-6, CPE, and HP-10 copolymers with NMMO MH. All crystal solvate structures belong to various types of branched crystals—dendrites. The development of the CSs morphology is a multifactor process influenced by the components nature and crystallization conditions. Thus, the concentration of solution, its prehistory, the time of solution aging at a temperature above the melting point of the CS, and the rate of cooling affect not only the size of the structures being formed by also the details of their morphology.

Figure 5. Dendrite structures in systems: (a) 40% HP-6 solution in NMMO MH, (b) 30% CPE solution in NMMO MH, (c) 40% HP-10 solution in NMMO.

4.1.2. PMPIA–NMMO Crystal Solvates

The aromatic polyamide (PMPIA), like the thermotropic alkylene-aromatic CPEs, dissolves rather easy in NMMO MH. As follows from Fig. 1 (curve 6), the resulting solutions contain more than 30% PMPIA. The DSC data for solutions of PMPIA showed that though general tendencies observed for solutions of CPEs in NMMO, namely, formation of additive compounds (CSs) are preserved, the character of structure formation occurring during dissolution of PMPIA in NMMO is more complicated. Thus, the DSC curves measured for solutions containing 5–10% PMPIA in NMMO MH show two exothermic peaks with maxima at 86–94°C along with endothermic peaks corresponding to the melting of NMMO bihydrate (T_m = 36°C) and NMMO MH (T_m =76°C). The appearance of the exothermic peaks indicates on formation of new additive compounds of PMPIA with NMMO MH. (Fig. 6).

The profile of the DSC curves obtained for PMPIA solutions prepared in the high-melting NMMO (T_m = 110–120°C) is identical to that of the DSC curves of PMPIA solutions in NMMO MH and only positions of exothermic peak maxima corresponding to CSs formation are different. For CSs, the values of T_m vary in a wide range from 103 to 135°C.

X-ray analysis of 5–20% PMPIA solutions in NMMO MH and in high-melting NMMO has shown that various types of ordered solvate systems may be formed depending on the hydrate form of the starting NMMO. In Fig. 7, the diffractograms of 5 and 15% PMPIA solutions in NMMO MH as well as individual components at various temperatures are compared. The diffractogram of PMPIA (Fig. 10, curve 6) exhibits two overlapping amorphous halos with maxima at 2θ* ~ 13.9° and 23.3° with the ratio of integral intensities equal to 1 : 5 ($I_1 : I_2$ = 1 : 5). The diffractogram of the individual NMMO MH at 20°C (Fig. 7, curve 1) shows a set of reflections whose angular positions are in agreement with the data from. The scattering picture of the NMMO MH melt (Fig. 7, curve 7) is characterized by the presence of a single amorphous halo at 2θ* ~ 16.9°. In the case of PMPIA solutions in NMMO MH, the scattering pictures turn out to be qualitatively different (Fig. 10, curves 2,

3). These pictures show an amorphous peak at $2\theta_1 \sim 6.8°$ along with a certain set of rather well-resolved reflections in $2\theta_2$ region 17°– 30°. When the solutions were heated to 95°C, the amorphous halo observed at $2\theta_1$ was conserved despite of a small shift toward small angles (to 6.5°), while all reflections at $2\theta_2$ are disappeared and the amorphous halo peaking at $2\theta_2 \sim 16.8°$ is appeared again.

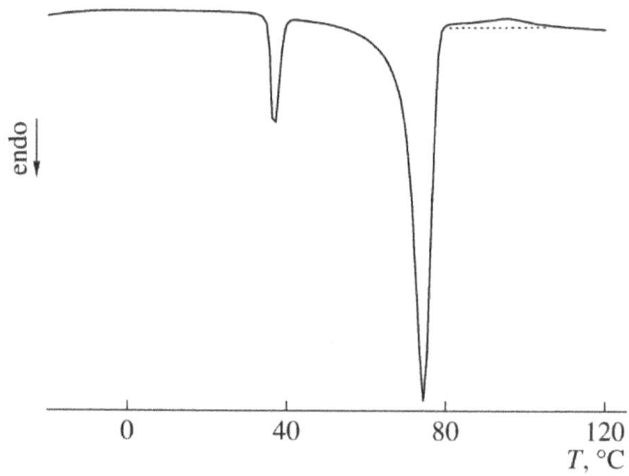

Figure 6. DSC curves of the blend containing 5% PMPIA in NMMO MH.

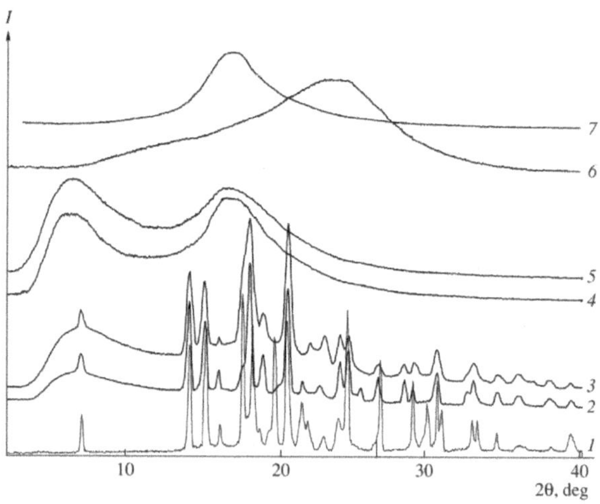

Figure 7. Diffractograms of (1) NMMO MH; (2) 5 and (3) 15% PMPIA solutions in NMMO MH at 20°C; (4) 5 and (5) 15%PMPIA solutions in NMMO MH at 90°C; (6) PMPIA; and (7) NMMO MH melt at 95°C.

A comparison of the intensity distribution of amorphous scattering for the individual PMPIA and NMMO MH and their solution leads us to suggestion that the resulting solutions are not single-phase. The amorphous phase of solutions is predominantly formed by PMPIA macromolecules solvated via specific interactions with solvent molecules. Another phase of solutions is enriched with solvent molecules and is characterized by scattering at $2\theta_2 \sim 16.8°$. The formation of such phase structures with intermolecular distances d = 13.0–13.6 Å, as calculated from the angular position of the first amorphous halo at $2\theta_1$ was proved by an increase in the ratio of integral intensities of the first and second amorphous halos ($I_1 : I_2$) with increasing content of PMPIA in solution.

The diffractograms of PMPIA solutions in the high melting NMMO display a single amorphous halo in the large-angle region with a maximum at $2\theta^* \sim 19.1°$ and a set of discrete Bragg reflections localized against its background. Under heating to 125°C, the melting of the ordered phase takes place, the intensity of the amorphous halo grows, and the angular position of its maximum shifts toward small angles - to $2\theta^* \sim 17.7°$. Note that the $2\theta^*$ value for the PMPIA solution at 20 and 125°C differs substantially from the corresponding value for the NMMO melt ($2\theta^* \sim 16.9°$). Consideration of this fact coupled with features of the diffractogram measured for the individual PMPIA, such as presence of two overlapping amorphous halos with maxima at $2\theta^* \sim 13.9°$ and 23.3° ($I_1 : I_2 = 1 : 5$), results in conclusion that the solution may be considered as single-phase. The melting point of the CS prepared in the high-melting NMMO is much higher (120–123°C). Thus, at a smaller content of water in NMMO, a high-melting crystal solvate phase of PMPIA with NMMO is formed.

The morphology of the crystal solvates formed in the PMPIA–NMMO system are different. Thus, CSs based on NMMO MH are typical spherulitic with superposition of ring textures of amorphous solvate layers (Fig. 8a), while CSs prepared from the high-melting NMMO are dendrite spherulites (Fig. 8b).

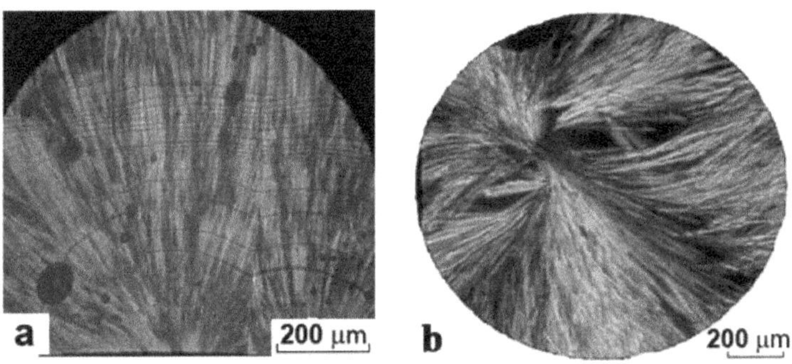

Figure 8. Micrograph of the PMPIA–NMMO crystal solvate prepared in NMMO MH (a) and in the high-melting NMMO (b).

Thus, the experimental evidence showed that the unique properties of NMMO as a high polar donor solvent ensure its high dissolving ability not only with respect to hydrophilic polymers but also with respect to hydrophobic liquid-crystalline CPEs and aromatic polyamides. As a consequence, dissolution processes are accompanied by the formation of CSs of various natures. A high solubility of cellulose and of synthetic polymers under study in NMMO gives us grounds to expect their compatibility in NMMO solutions on molecular level in a certain temperature–concentration range.

4.2. Solutions of Cellulose and its blends with Synthetic Polymers or Layered Aluminosilicates in NMMO

4.2.1. Solutions of Cellulose in NMMO

A highly efficient interaction of NMMO with OH-groups of cellulose is ensured due to presence of a semipolar N —›O bond with two unshared electron pairs at the oxygen atom in the NMMO molecule which can interact with two proton-containing groups. However, despite the high dissolving capacity of NMMO, the preliminary aqueous activation of cellulose is needed to facilitate the access of solvent molecules to the functional groups of cellulose to accelerate dissolution.

The schematic representation of cellulose dissolution in NMMO (diagram in Fig. 9) renders it possible to follow the evolution in the phase composition of the three component cellulose–NMMO–H_2O system during dissolution [5,9-10].

The first stage of the process (swelling of cellulose in aqueous solution of NMMO) proceeds in two stages. The first stage includes the treatment of cellulose with aqueous NMMO solution to produce a homogeneous pulp that contains, as an example, 35% H_2O, 9% cellulose, and 56% NMMO (point C in the diagram). At the second stage, an excess of water is removed to form a homogeneous suspension of the following composition: 20% H_2O, 13% cellulose, and 67% NMMO (point B). After further removal of water excess, dissolution occurs at the following approximate contents of the components: 14% cellulose, 10% H_2O, and 76% NMMO (point A).

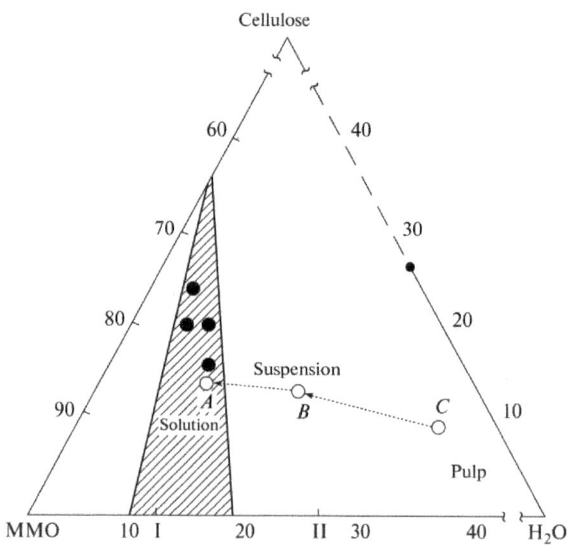

Figure 9. Schematic phase diagram of the cellulose–NMMO–H_2O system: (line CBA) variation in the composition of the system during cellulose dissolution via the traditional NMMO process; (closed circles) composition of the system during cellulose dissolution via the solid phase NMMO process.

A new method of solid-phase dissolution of cellulose in NMMO [6] allows to improve significantly the dissolving ability of NMMO MH using higher activity of high-melting hydrate forms of NMMO. This process involves the stage of solid-phase activation of cellulose by crystalline NMMO, which proceeds under triaxial compression, shearing, and forced plastic flow. Under the conditions of all-hydrostatic compression and shear, the mechanochemical activation of cellulose by crystalline NMMO occurs via formation of H-complexes. At further simultaneous action of temperature and shear rate, solid H-complexes (or solid precursors of solutions of cellulose in NMMO) melt and transform into highly concentrated flow solutions with a high level of homogeneity.

Thus, the solid phase process is distinguished by the invariable phase composition of the system during cellulose dissolution, beginning from the stage of preparing solid presolutions to their melting and transition to the liquid state, as illustrated by the position of corresponding points in the schematic diagram (Fig. 9). Four points shown in the diagram correspond to solutions with different concentrations of cellulose ranging from 16 to 25% prepared via solid-phase activation.

Preparation of cellulose solutions in NMMO in a sufficiently wide concentration range gave us the supplementing information on the phase state of the NMMO–H_2O–cellulose system and completing the phase diagram in the range of cellulose concentrations from 0 to 50% (Fig. 10) [11].

The lines of phase equilibrium are constructed from melting points from DSC curves, X-ray data, and polarizing microscopy data. Highly concentrated solutions of cellulose in NMMO are inclined toward supercooling, and they do not solidify at room temperature after heating for several months. These circumstances make attainment of the equilibrium state extremely difficult.

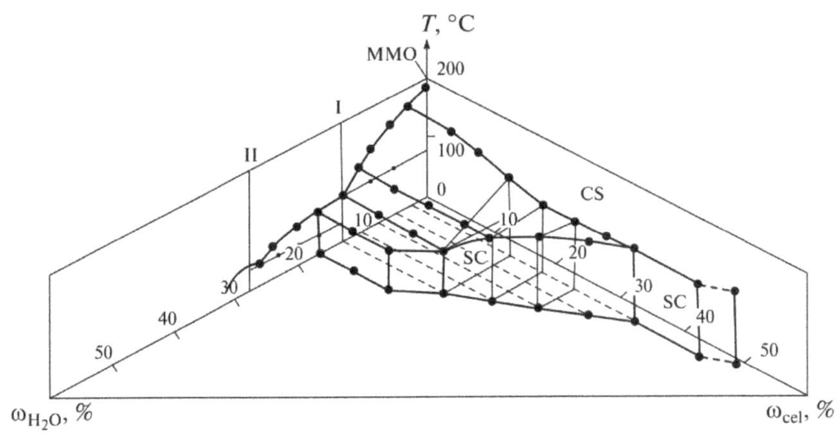

Figure 10. 3D state diagram of the cellulose–NMMO–H_2O system: SC is the solid H-complex, CS is the solution of cellulose in NMMO. (I) NMMO 2.5-hydrate and (II) NMMO monohydrate.

Although the above phase diagram is demonstrates clear conditions for the formation of highly concentrated solid complexes of cellulose with NMMO and the temperature–concentration regions of their transition to the flow state. As is seen from the diagram, the position of liquidus lines reflecting the melting temperatures of solid complexes helps us to choose appropriate variables to obtain dopes of different composition. This tendency becomes more pronounced with an increasing concentration of cellulose in solution.

4.2.2. Phase equilibrium in mixed solutions of cellulose with synthetic polymers in NMMO

As was shown above, the unique properties of NMMO as a highly polar donor solvent are responsible for its high dissolving ability not only for hydrophilic but also for hydrophobic polymers. Numerous studies have been devoted to the problem of solubility of polymers in ternary systems composed of two polymers and one solvent. The number of polymer pairs that can be dissolved in one common solvent is very limited, and all polymer solutions tend to undergo phase separation even at low concentrations [12]. Therefore, complete compatibility of polymers, which is primarily controlled by the nature of a solvent, is the exception rather than the rule.

The phase state of ternary HP(CPE)–cellulose–NMMO systems was studied by polarizing microscopy, turbidimetry, visual observations, and rheological studies over a wide range of concentrations of polyesters and cellulose when the overall content of polymers was 35%. All solutions are optically isotropic and do not show any birefringence.

To estimate the limiting compatibility of cellulose and polyesters in common solvent, phase state diagrams are constructed for two systems: cellulose–CPE–NMMO and cellulose–PDTOB–NMMO [13]. Taking into account all experimental implications in the construction of 3D phase state diagrams, we studied the phase equilibrium on the cross section of the 3D diagram at a constant temperature of 120°C.

Figure 11 presents the fragment of this cross section. The upper point of the triangle corresponds to 100% NMMO, whereas the left-hand and right-hand corners correspond to 100% of cellulose and 100% of CPE, respectively. The binodal is the boundary line describing amorphous equilibrium. Under binodal the transition of the cellulose–CPE–NMMO system from the single-phase to the two-phase state takes place. Above the binodal, the system is thermodynamically compatible and has below the binodal, the system undergoes phase separation.

As was shown in [24], among all CPEs under study, PDTOB has the lowest solubility in NMMO; however, according to the phase diagram, the cellulose–PDTOB–NMMO system (binodal 2) is characterized by a higher level of compatibility between components than the cellulose–CPE–NMMO system (binodal 1). Since CPE has higher solubility in NMMO, one could expect that the corresponding phase equilibrium curve in ternary system should be located under the bimodal 2; however, contrary to expectations, the binodal 1 appears to shift towards lower concentrations of polymers.

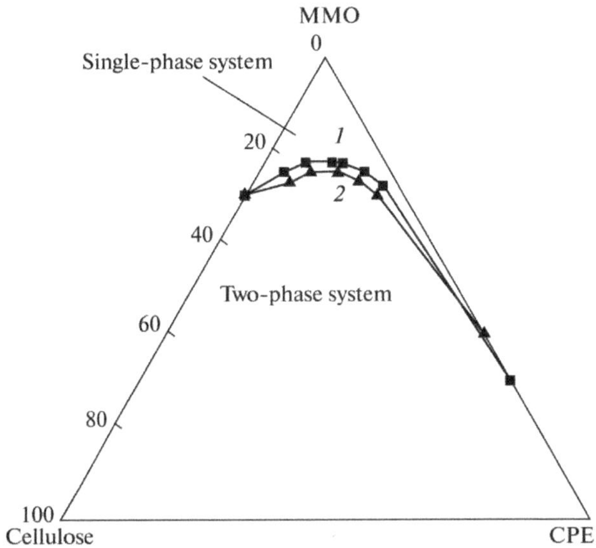

Figure 11. Phase state diagram for ternary systems at 120°C: (1) cellulose–CPE–NMMO and (2) cellulose–PDTOB– NMMO

The heterogeneous character of the single-phase mixed solutions of cellulose and CPE was studied by spectroturbidimetry (Fig. 12).

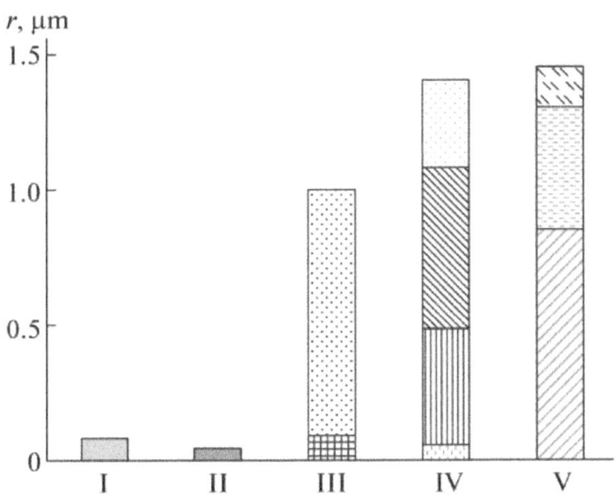

Figure 12. Histogram of heterogeneity in the cellulose–CPE–NMMO system: (I) 10% cellulose solution in NMMO; (II) 15% solution of CPE in NMMO, and (III–V) mixed cellulose–CPE solutions in NMMO: (III) 10 : 5, (IV) 10 : 10, and (V) 10 : 15%.

As compared with solutions of neat components, the measurements have demonstrated a significant increase in heterogeneity of the mixed solutions. Upon the addition of CPE, for example, to cellulose solution one can reasonably suggest that due to higher rate of dissolution in NMMO this component quickly transforms into solution and cellulose now became dissolved not in pure NMMO but in its polymer-containing solution. Since the thermodynamic quality of the "dissolving system" with respect to cellulose is reduced, the interaction between cellulose macromolecules increases, leading to formations agglomerates and aggregates that causes increasing the heterogeneity of the mixed solution. As the concentration of CPE is further increased (the region below the binodal), the system undergoes amorphous separation and the two-phase emulsion is formed. At higher concentrations of CPE, the dissolving potency of NMMO with respect to cellulose dramatically decreases and fraction of undissolved cellulose appears in a system.

In contrast to the mixed solutions of cellulose and CPE in NMMO, which are single-phase systems over a wide concentration interval up to an overall content of polymers in solution of ~20%, all mixed solutions of cellulose and PMPIA in NMMO are biphasic. Thermodynamic incompatibility of solutions results in development of fascinating morphological effects in two-phase solutions of PMPIA–cellulose mixtures in NMMO. Optical observations showed that cellulose solutions in NMMO containing 1–5% of PMPIA are emulsions. The droplets of the dispersed phase are nearly spherically shaped, and their size distribution is rather wide (Fig. 13a). All droplets are exceptionally labile and already at weak shear deformation, break down into smaller-sized droplets, which can be extended and form threadlike fibrous structures (Fig. 13b). As a result, highly regular fibrillar morphology with high periodicity in the arrangement of fibrils in the bulk phase is formed.

This phenomenon is most pronounced upon the extrusion of mixed solutions through a capillary rheometer. The entrant region, where convergent flow and stretching takes place, provides more favorable conditions for the development of fibrillar morphology. The use of detachable capillaries allowed us to study flow morphology by visualizing the evolution of optical patterns of the samples collected from a capillary and from various regions of the inner chamber of micro viscometer. The collected samples were studied using a polarizing microscope. As was found, in the mixed cellulose–PMPIA–NMMO systems, the formation of fibrils was observed even in the chamber of the viscometer at a distance of ~10 mm from the entrance to the capillary channel (Fig. 13c). Fibrils achieve their maximum level of regularity in the capillary (Fig. 13d) and, especially, upon drawing of extrudate in the exit zone (Fig. 13e). Under the action of extension and shear stresses, the dispersed phase formed by flexible-chain PMPIA structures divides (disintegrates) the cellulose solution matrix. Shear stresses induced on interfaces between two incompatible solutions lead to the orientation of the forming macrofibrils. As a result, of the above phase morphological transformations in the mixed cellulose–PMPIA–NMMO systems, especially at a low PMPIA content, a highly ordered fibrillar morphology forms, which breaks down under the action of temperature. For example, when mixed solutions containing 5% of PMPIA are heated at temperatures above 160°C, liquid threads begin to break down and form a necklace of droplets. When in the mixed cellulose–PMPIA–NMMO system, polyamide is the

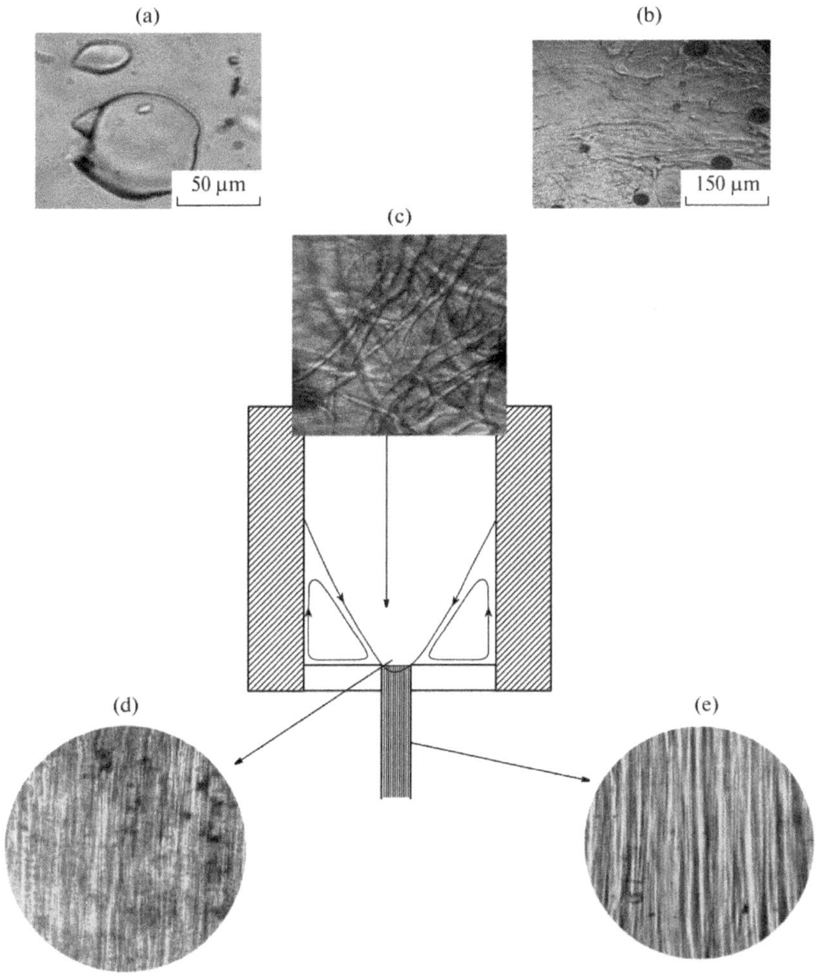

Figure 13. Micrographs of the mixed cellulose–PMPIA solution in NMMO with a concentration of 18% and with a ratio between components of 17 : 1: (a) the test sample is sandwiched between glasses on a hot table of polarizing microscope, (b) the same sample under deformation, and (c–e) specimens collected from different regions of capillary viscometer.

predominant component, the fibrillar structure becomes less ordered and more movable. Upon heating, the threads completely degrade already at temperatures below 120°C. As further increase of temperature, the above droplets undergo coalesce.

The homogeneity of cellulose mixed solutions with polyesters in NMMO and the heterogeneity of solutions with PMFIA results in a significant difference in their rheological properties.

4.2.3. Rheological characteristics of mixed solutions of cellulose and synthetic polymers in NMMO

Comparative analysis of viscous characteristics of CPE–cellulose solutions in NMMO let us to conclusion, that the introduction of CPE into cellulose solutions does not change drastically profiles of flow curves or the concentration dependences of the viscosity inherent for binary solutions cellulose-NMMO. At low and intermediate shear rates, mixed solutions demonstrate Newtonian character of flow; at high shear rates, a marked viscosity anomaly is observed. The introduction of CPE into cellulose solutions increases their viscosity in whole range of shear rates. The concentration dependences of the viscosity of cellulose solutions show positive deviations from values corresponding to the logarithmic additivity rule. This indicates on strong intermolecular interactions between molecules of solvent and cellulose as well as between cellulose macromolecules themselves. The introduction of CPE macromolecules into the cellulose–NMMO system gives rise to even higher positive deviations of the concentration dependence of viscosity.

Due to the specific structural features of cellulose and even despite its strong electron donor–acceptor interaction with NMMO, cellulose solutions in NMMO are known to be highly structured. Due to presence of CPE in NMMO, solvent quality decreases and interaction between cellulose macromolecules appears to be enhanced; as a result, the viscosity of the mixed solutions is increased. Therefore, the difference between binary and ternary solutions is concerned with the level of structural organization of cellulose macromolecules.

Hypothetically, the high level of structuring of the mixed solutions increases with increasing overall concentration of polymers and variation in the ratio between components. As a result, mixed solutions are metastable. This assumption was proven by rheological studies performed for mixed solutions cellulose and CPE in NMMO over a wide concentration interval.

For example, at passing to 25% mixed solution concentration, which according to ternary phase diagram is located in vicinity of the binodal, the character of rheological behavior changes, especially in the regions of limiting concentrations (20% of cellulose/5% of PDTOB and 5% of cellulose/20% of PDTOB). For example, in the case of the 20/5 mixture viscosity anomaly at high shear stresses increases dramatically. Flow curves of the 5/20 mixture can change their position depending on prehistory of the sample and deformation conditions. Therefore, deformation of the solutions in their pre-transition region increases phase instability and can lead even to the phase separation. One can expect that for the mixed systems located directly on binodal, the effect of deformation will be the most pronounced.

Figure 14 presents two flow curves for the cellulose-PDTOB (13/15) mixture. Curve 1 corresponds to short deformation times (strain is 10–100 rel. units), and in these conditions the system does not undergo phase separation. With increasing time of deformation action (strain is above 100 rel. units) the shear stress does not grow monotonically at each shear rate, but after reaching the definite strain decreases (stepwise curve). This kind of behavior

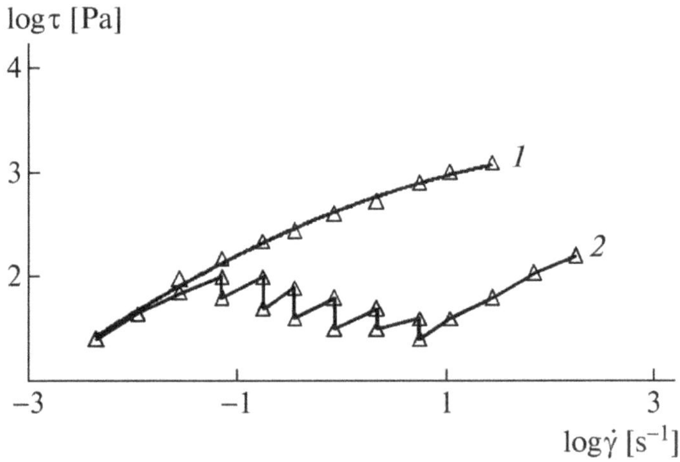

Figure 14. Flow curves of the mixed cellulose–PDTOB solutions in NMMO; overall concentration of polymers is 28%, and cellulose : PDTOB ratio = 13 : 15: (1) small strains, (2) high strains.

reflects transition through binodal induced by strain. By other words, after relaxation at stopping the previous shear rate, the system is homophase but deformation at flow acts as thermodynamic factor changing the bimodal location, and the system becomes heterophasic. Cessation of flow at each shear rate leads to stress relaxation and reverse transition into one-phase region, but new loading causes the same transformation of the phase state. These results indicate on two circumstances: significant changes in the rheological behavior of mixed solutions on passing from the single-phase to the two-phase state, and influence of flow on location of the binodal curve.

Mixed cellulose–CPE solutions in NMMO with phase compositions corresponding to single phase region have been studied over a wide range of shear stresses with attempt to see morphological changes, they have not been revealed.

In contrast to the above systems, two-phase cellulose–PMPIA systems in NMMO are characterized by a well-pronounced heterogeneous morphology, which is controlled by the composition of the sample and deformation conditions. First, let us consider the specific features of the rheological behavior of mixed cellulose–PMPIA–NMMO systems under capillary flow. All solutions that contain 14, 18, and 20% of polymers show a non-Newtonian behavior or, in other words, their viscosity almost linearly decreases with increasing shear stress and follows the power-law flow regime.

The concentration dependences of the viscosity (log τ = 3.6 Pa) were constructed for equi-concentrated mixed solutions; the overall concentration of polymers in solutions was 14 or 18%. As follows from Fig.15, the viscosity of all the solutions lies below the line of logarithmic additivity.

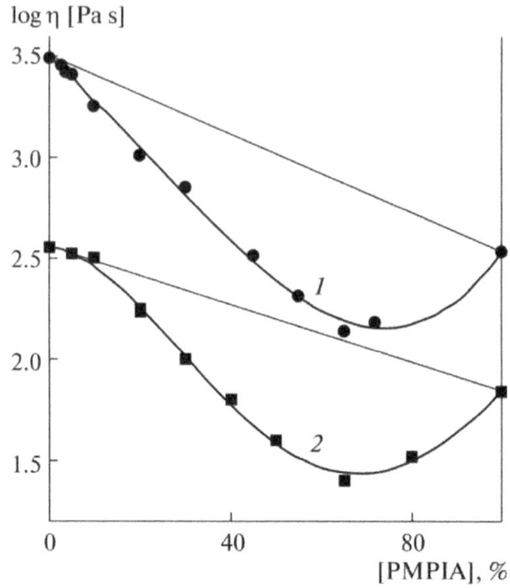

Figure 15. Concentration dependences of viscosity of the mixed cellulose–PMPIA solutions in NMMO. The overall concentration of polymers is (1) 14 and (2) 18%.

On passing through the entrance zone, droplets of PMPIA solution (the dispersed phase) are elongated and transformed into long liquid threads oriented along the flow direction, forming a fibrillar system. As a result, inside the capillary a complex profile of linear velocities is realized (a multi-parabolic one). Due to the superposition of multiple profiles, the overall flow rate decreases.

In addition to capillary viscometry, the rheological behavior of the mixed cellulose–PMPIA solutions in NMMO was studied on a rotation rheometer with its cone–plate working unit. Measurements were performed under steady-state shear by measuring dependences of tangential (τ) and normal (N_1) stresses on the shear rate and in the low-amplitude oscillatory deformation regime by measuring the frequency dependences of storage modulus G' and loss modulus G''. In the case of the first difference of normal stresses, their coefficient connecting N_1 with shear rate was calculated: $N_1 = \zeta \dot{\gamma}^2$.

Figure 16 presents the concentration dependences of the effective viscosity η and the coefficient of normal stresses. As follows from this figure, in contrast to the concentration dependences of viscosity constructed from data of capillary flow, the above mentioned dependences have S-shaped profiles with positive deviation at low concentrations and negative deviation at high concentrations. A slight positive deviation from additivity can be explained by different flow kinematics in rotation and capillary regimes leading to different orientation of the disperse phase droplets.

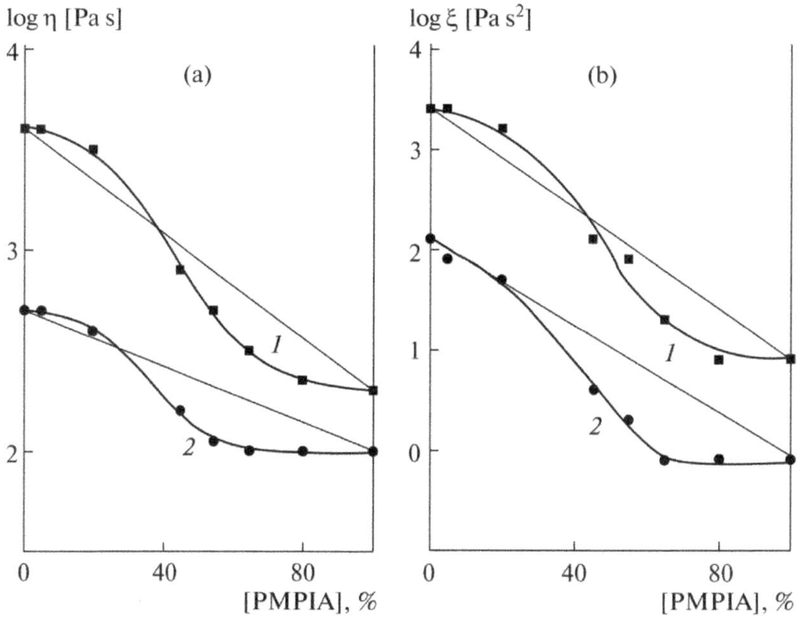

Figure 16. (a) Viscosity η and (b) coefficient of normal stresses ζ vs. composition for the mixed cellulose–PMPIA solutions in NMMO at log τ = (1) 3 and (2) 4 Pa. The overall concentration of polymers is 18%.

The use of the transparent cell as the working unit in the rotation rheometer allowed us to perform the direct visual observation of the behavior of mixed systems under flow. Upon deformation in an uniform shear field, mixed solutions containing from 5 to 40% of PMPIA are transformed into a microheterogeneous system with a highly developed interfacial surface. In the case of 5% content of PMPIA solution, the shear deformation is insufficient to provide any deformation of droplets and they play a role of filler. When the content of the PMPIA phase is increased, the dimensions of droplets increase and the deformation induce the transformation of the droplet-matrix morphology into the fibrillar one. The formation of liquid continuous threads of the dispersed low-viscous phase is accompanied by negative deviations of the concentration dependences of the viscosity from the additive straight line, similar to the case of the capillary flow.

The concentration dependences of G' and G'' as well as steady-state viscosity are S-shaped. According to the traditional concept on the character of flow of incompatible polymer systems, the transition the positive to the negative deviations of viscosity from the additivity line is treated as phase inversion. In our case, the revealed dependences are controlled by changes in the morphology of the flowing mixed system and by its ability for orientation at different by intensity deformation actions.

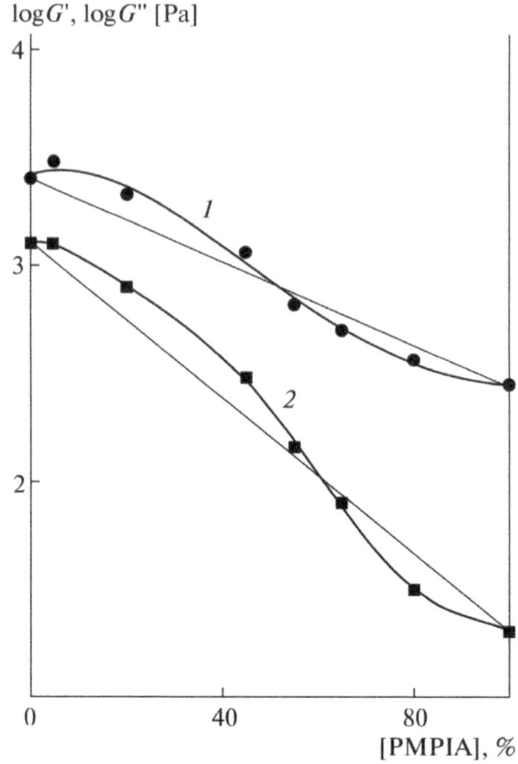

Figure 17. (1) Storage G' and (2) loss moduli G" vs. composition at $\omega = 1\ s^{-1}$ for the mixed cellulose–PMPIA solutions in NMMO. The overall concentration of polymers is 18%.

Scales of positive and negative deviations can be selected as criteria of the deformability of PMPIA droplets. Depending on droplets deformability at different flow regimes, the interval of inflection shifts along the concentration axis. Maximum positive deviation is for the oscillatory regime. In this case, the point corresponding to the transition to negative deviations is observed at 50–60% PMPIA content. In a homogeneous shear field (under steady-state flow), positive deviations of viscosity are much lower and take place in the concentration interval below 30–40%. Finally, under capillary flow, no positive deviations are observed and maximal viscosity is achieved at ~65% of PMPIA.

Hence, solutions of cellulose and LC copolyesters in NMMO far from binodal change their rheological characteristics according to the traditional mechanism of flow of mixed systems with high specific interaction between components. However, in vicinity of bimodal deformation induces phase transition from the homogeneous to the heterogeneous system leading to unstable flow with oscillation of shear stress. In the case of the mixed cellulose–PMPIA–NMMO systems, the rheological behavior is controlled by transformations of morphology dependent on solution composition and flow regime.

4.2.4. Rheological Characteristics of Mixed Solutions of Cellulose with Natural and Modified Clays

In order to estimate the effect of aluminosilicate additives on the rheological behavior of cellulose solutions in NMMO, rheological properties of filled solutions containing 18 and 10% cellulose were chosen as matrices [14-15]. It was shown that flow of sufficiently viscous 18% cellulose solutions filled up to 9% with different clays is determined mainly by rheological behavior of the polymer matrix. Using less concentrated 10% cellulose solutions allowed to increase the clay content in the solutions up to 20%. The flow curves for 10% cellulose solutions filled to various degree with particles of Cloisite Na^+ are presented in Fig. 18. As is seen from the figure, already at a low content of solid phase the flow of the filled composition becomes a non-Newtonian.

With increasing clay concentration the degree of the viscosity anomaly increases, and at 20% content of clay in the system the shape of flow curve indicates on existence of the yield stress (viscoplastic behavior) due to the formation by filler particles the percolation network. At high shear rates the structure of clay is destroyed, and as a result the rheological behavior becomes similar (pseudoplastic behavior) to cellulose solution in NMMO.

Introducing to the 10% cellulose solution hydrophobic clay - Cloisite 20A is caused some intrinsic features of rheological behavior. The identified features of the rheological behavior of these compositions are presented in Fig. 19 as the generalized concentration dependence of the relative viscosity for both 18% and 10% of matrices containing hydrophilic Cloisite Na^+ and organoclay Cloisite 20A.

For easier comparison, the function is the relative viscosity ($\eta_r = \eta/\eta_{cel}$) , and the argument is a fraction of clay in composition. At the solid content up to 9%, changes of relative viscosity with concentration for all compositions can be described by a single curve, irrespectively of the clay nature and the matrix viscosity. With further increasing of the solid phase concentration the dependence becomes more complex. The single the viscosity-composition curve for solutions filled with Cloisite Na^+ particles does not change, but introduction to solution particles of Cloisite 20A clay leads to a splitting of the curve into two branches. A high viscous branch (curve 1) practically coincident with the concentration curve of systems with natural clay, a low viscous branch lies much lower (curve 2), that is why the generalized dependence for these compositions has a minimum localized at 10% of clay in solution. Starting with concentration corresponding to the splitting, the flow curves of filled solutions containing 15% - 20% of Cloisite 20A demonstrate the presence of the yield stress.

Unexpected at first sight the experimental fact of reduction of the viscosity of filled solutions with Cloisite 20A (curve 2) seems to be explained by different prehistory of their preparation. Within variety of many factors that determine the specific features of systems under investigation, as the main parameter the moisture content in "initially hydrophobic" clay was chosen, since all other components are hydrophilic containing almost equilibrium moisture, meanwhile the sorption of water by hydrophobic clay was unknown.

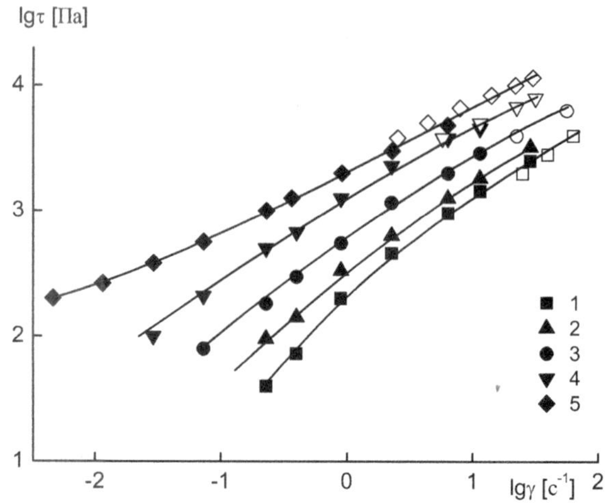

Figure 18. Flow curves for 10% solutions of cellulose in NMMO with Cloisite Na^+. Content of clay: 0 (1), 5 (2), 10 (3), 15 (4), 20 wt.% (5). Filled points - rotational viscometry data, open points - data of capillary viscometry.

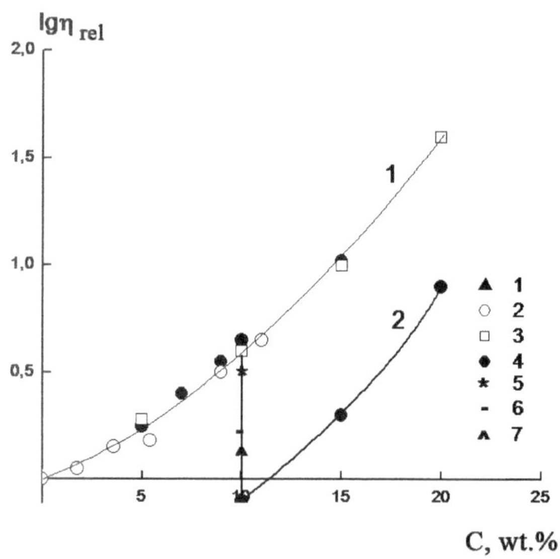

Figure 19. Concentration dependence of the relative viscosity filled solutions of cellulose in NMMO with Cloisite Na^+ and Cloisite 20A at log $\tau = 3.0$ [Pa]. 1-18% solution of cellulose + Cloisite Na^+; 2 - 18% solution of cellulose + Cloisite 20A, 3 - 10% solution of cellulose + Cloisite Na^+; 4 -10% solution of cellulose + Cloisite 20A. Figures 5, 6 and 7 correspond to 6, 13 and 16% of moisture content in Cloisite 20A.

Model experiments on influence of moisture content in Cloisite 20A on the viscosity values using pre-moistened clay with a fixed water content were carried out. Obtained results have confirmed the assumption that a decrease in the viscosity of filled solution is caused by increase of water content in the clay. By means of X-ray diffraction method, a comparative structure analysis of the neat clay (~ 2% of water) and extremely Cloisite 20A moistened after prolonged sorption up to 16% of water it was shown that the basal reflection of neat clay at 2θ, equal to 3,2° transforms to diffuse asymmetric reflection with maxima at 2θ 2,6° and 3.9° for moistened clay (Fig. 20). Such change of the diffraction pattern indicates on specific ability of clay hydrophobized with quarterly ammonium bases to absorb water. We can expect that water molecules should come to the polar Na^+ center of ionic surfactant localized on negative clay platelets. On the one side, this can lead to increase of the interplanar spaces, i.e., to decrease of the intrinsic scattering angle, but on the other side, presence of polar water molecules in vicinity of hydrophilic clay elementary platelets could cause hydrophobic interaction (repulsion) that change conformations of hydrophobic tails of modifier leading to their shrinking. As a result of dual action of water molecules, the interspace width change in non-homogeneous manner (extension-shrinkage).

Rheological data show that 10% content of clay in the system is critical, at which heterophase structure of the interplanar spaces forms and can initiate distortion of clay crystalline structure with formation of irregular layers. Shear action can cause fragmentation of clay platelets ensembles accompanied by a drop in the viscosity of the system as whole.

Introduction to the cellulose solutions of nanoparticles of M_2Cloisite Na^+ in amount not exceeding 0.1% leads to decrease of the solutions viscosity more than in two times (Fig. 21).

Filled solutions of cellulose in NMMO with addition of nanoparticles of M_2Cloisite Na^+ are characterized by a pronounced fibrillation. Apparently, the reason of reducing the viscosity is directly related to the fibrillar structure of cellulose solutions in presence of clay nanoparticles and extremely high surface energy of highly developed interfaces.

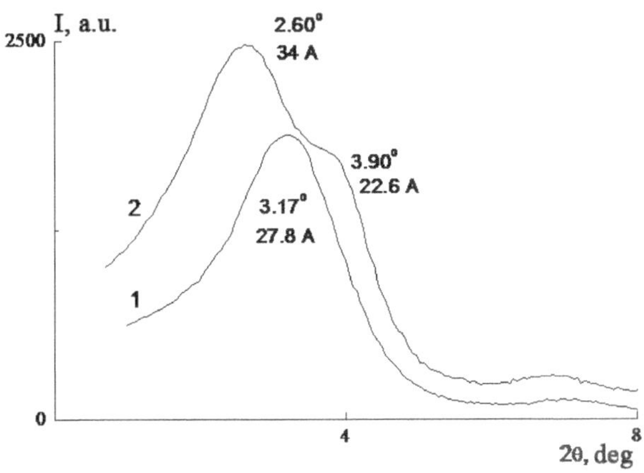

Figure 20. Diffraction patterns of the neat Cloisite 20A (1) and moistened Cloisite 20A (2).

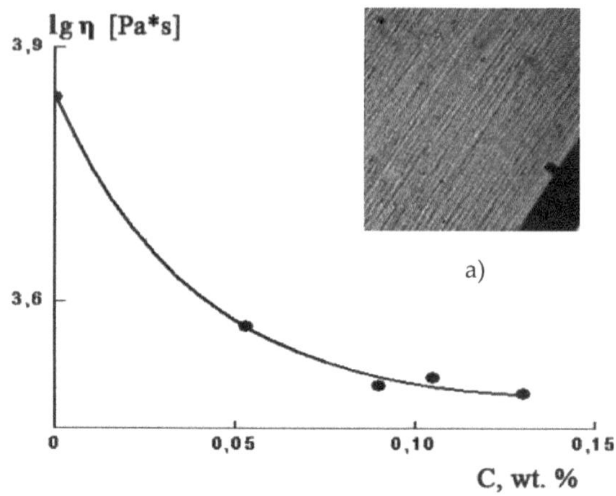

Figure 21. Dependence of the viscosity on concentration of M_2Cloisite Na^+; a) Microphotography of 18% cellulose solution with 0,03% M_2Cloisite Na^+.

The rheological behavior of filled solutions of cellulose with microparticles of Cloisite Na^+ and Cloisite 20A is similar to the behavior of traditional polymer filled systems, while in presence of nanoparticles the flow mechanism changes from traditional segmental to stratified one, i.e., stream morphology becomes regular.

5. Evolution of cellulose structure at transition from dopes to cellulose and blend oriented fibers

5.1. Structure of cellulose and its blends with synthetic polymers

X-ray data demonstrate that solutions containing 18% cellulose in NMMO are amorphous and isotropic, as evidenced by the short-range order in the arrangement of macromolecules. The bulk extrudate obtained without drawing after full removal of the solvent is likewise isotropic, as is seen from X-ray patterns shown in Fig. 22a.

Spinning of fibers from 18% cellulose solution leads to substantial changes in the cellulose structure. The as-spun fiber containing the residual solvent is characterized by nonequilibrium metastable modification of cellulose. The X-ray pattern of this sample (Fig.21b) exhibits well-defined equatorial reflections. Presence of only one intense reflection on the equator of the diffractogram implies that the intensity abruptly declines with the diffraction angle 2θ. This effect is usually attributed to an increase in the disorder owing to the formation of disruptions along chains. A similar character of diffractograms, as a rule, is typical for *2D* columnar mesophases. During complete phase separation, a crystalline phase of the cellulose is formed (Fig. 22c).

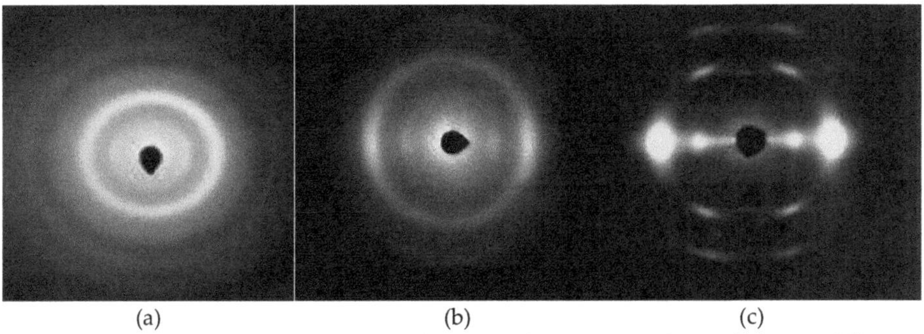

Figure 22. Flat camera X-ray patterns of 18% solution of cellulose in NMMO (a), cellulose partially washed-out from NMMO (b) and completely coagulated oriented cellulose fibers (c).

During spinning of the cellulose fiber from NMMO solutions, there appears a crystalline phase in which long crystallites are extended in the longitudinal direction [16-18]. On the one hand, this situation indicates on high order and increased density of the structure being formed. On the other hand, this leads to decrease of the probability of transverse hydrogen bonds with neighboring crystalline clusters formation enhancing the tendency of the system toward fibrillation. The transformation of structure in the cellulose–NMMO system during the spinning of model fibers is schematically shown in Fig. 23.

A comparison of X-ray pattern taken for the crystalline fiber (Fig. 22c) and the gel fiber (Fig.22b) of cellulose leads us to suggestion that the package of macromolecules in the plane of the 2D mesophase and the ac plane of the crystal are similar. This conclusion is supported by the fact that the angular positions of the equatorial reflections of the crystal and the mesophase are similar. The transition of the cellulose–NMMO system to the mesophase is probably related to the structural features of cellulose as a high-regular polymer with a developed system of H-bonds.

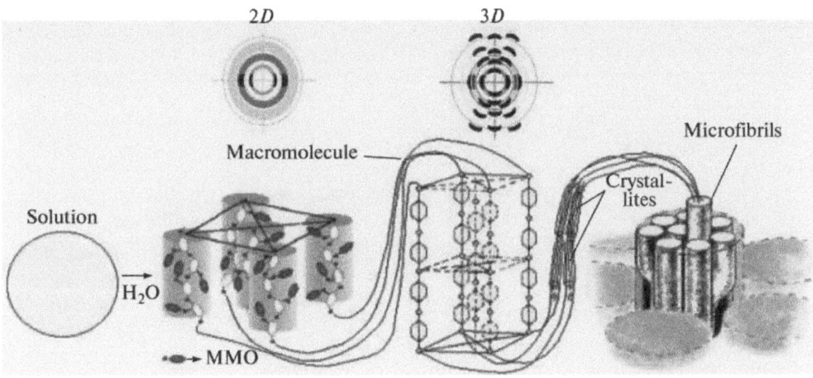

Figure 23. Schematic representation of the evolution in the cellulose structure during spinning of fiber samples from cellulose solutions in NMMO

Let us consider the evolution of the cellulose structure in blends. According to the X-ray data, all NMMO solutions of cellulose blended with copolyesters, PMPIA or clays, are amorphous; however, the character of structural transformations in them during isolation of the polymer phase is appreciably different from that in cellulose solutions [19]. Our experiments showed that although the chemical nature of the polymer introduced in cellulose solutions and the phase states and morphologies of blend solutions are different, the resultant structure of the formed blend polymer phase is of the same type. Since the evolution of the structure is the most pronounced for the cellulose–PMPIA system, let us examine step-by-step formation of the structure of cellulose composites.

After removal of the solvent, cellulose composites containing 5% PMPIA, as is seen from the diffractograms shown in Fig. 24, are not amorphous even in the absence of drawing, and demonstrate a marked increase in the intensity of scattering in the range of angles $2\theta = 12°$ in the equatorial direction and the absence of redistribution between equatorial and meridional scatterings at $2\theta > 15°$ (Figs. 24, 25a).

Presence of two peaks on the diffractogram indicates that the packing with two mean intermolecular spacings occurs. These spacings are appreciably different. It is quite probable that the molecules located in the closest proximity more easily form the system of inter- and intramolecular hydrogen bonds that is necessary for formation of ordered domains capable to be oriented. Therefore, the layered 1D packaging is formed at a small draw ratio ($\lambda = 6$) (Fig. 25a).

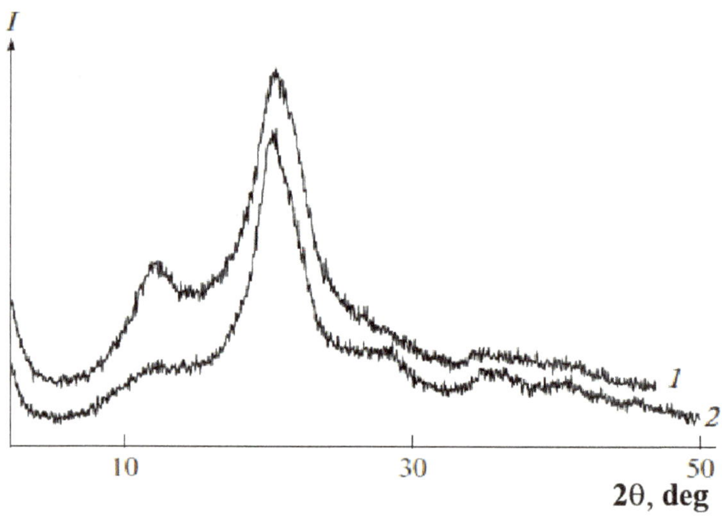

Figure 24. Equatorial (1) and meridional (2) diffractograms of blend extrudates composed of 95% cellulose and 5% PMPIA

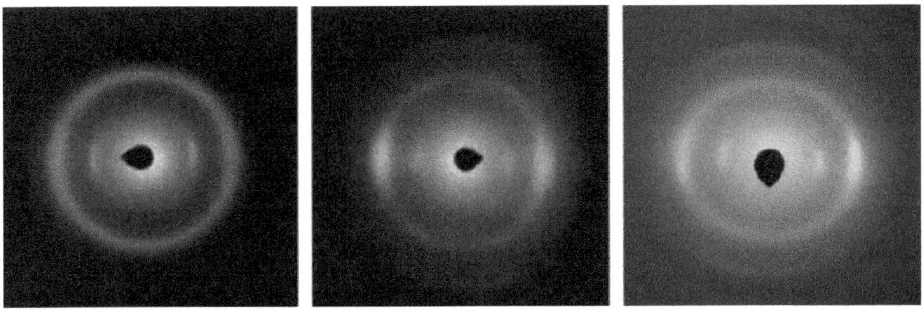

(a) (b) (c)

Figure 25. X-ray patterns of blend fiber samples composed of 95% cellulose and 5% PMPIA at various draw ratios: λ = (a) 6, (b) 13, and (c) 17.

Formation of composite fibers under conditions of orientation drawing accompanies by further evolution in the scattering pattern. As is seen from Fig. 25b, the redistribution of intensity covers now the entire region of scattering. As in the case of the purely cellulose mesomophic gel fiber, two reflections are observed on the equator (22 b). As the order of packing of cellulose macromolecules in the basal plane (at the intermolecular level) is improved, the transition of the layered 1D packing to the 2D columnar mesophase occurs. In this case, the position of equatorial reflections on the X-ray patterns of the oriented blend sample coincides with the position of equatorial reflections of the mesophase cellulose gel fiber and with the position of main basal reflection 101 of the cellulose crystal.

An increase in the draw ratio of the composite sample to λ = 17 facilitates further perfection of its structure. On the X-ray pattern of the fiber (Fig. 25c), the meridional reflection appears at 2θ =35°C. The features of scattering in the vicinity of the equator, which have been previous assigned to ordering in the basal plane, are preserved. At the same time, there are no reflections in the quadrants of the X-ray pattern.

A comparison of the oriented fibers of cellulose and the composite having the same draw ratios convincingly proves this finding. In fact, scanning of the composite fiber samples reveals no near-meridional reflections or other quadrant reflections (Fig. 26). The absence of quadrant reflections is the distinguishing feature of the composite fiber. At the same time, the angular positions and the half-widths of equatorial and meridional reflections for the cellulose and composite fibers are almost the same at a small content of the polymer additive (4%).

The above described picture of scattering obtained for the composite fiber with λ = 17 shows that the system contains two independent levels of order: 2D order in the basal plane (the intermolecular level) and 1D order along the fiber axis (the intramolecular level), that is, on the whole, non-crystalline 3D order. Fig. 27 schematically represents the most pronounced structural transformations occurring during isolation of the polymer phase from blend solutions in NMMO.

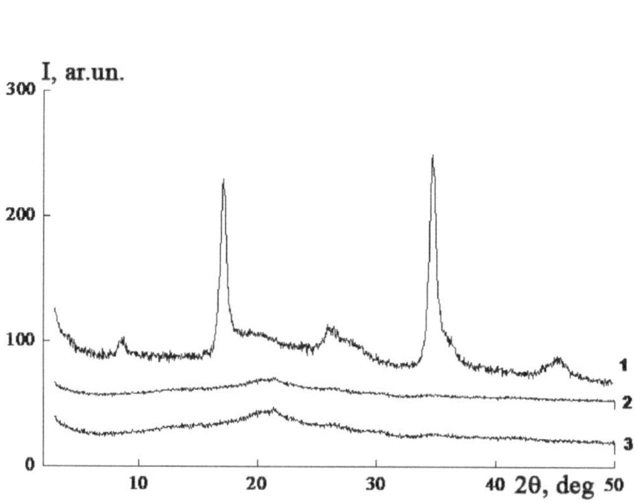

Figure 26. Diffractograms of the composite cellulose fiber containing 4% PMPIA during scanning (1) along the meridian and at angles to the meridian of (2) 10° and (3) 20°.

Thus, it has been shown for the first time that the introduction of copolyester and PMPIA into cellulose solutions in NMMO makes it possible to govern the processes of cellulose structuring and to stop them at the stage of formation of the 2D mesophase and, thus, to avoid further perfection of the structure and formation of the crystalline phase of cellulose.

On the basis of the above evidence, evolution in the structure of cellulose during isolation of the polymer phase from its own solutions and solutions of cellulose blends in NMMO may be depicted as follows. If the cellulose fiber is spun from NMMO solutions under the conditions of uniaxial drawing, the structure of cellulose forms in two stages. At the first stage, when the solvent is incompletely removed from solution (~95%) and, accordingly, the phase separation is incomplete, regular interchain periodicity develops in the main bulk of the cellulose phase. NMMO molecules remaining in the polymer phase are linked via strong electron donor–acceptor bonds with protons of hydroxyl groups of cellulose and, thus, favor restoration of intrachain hydrogen bonds that are necessary for the formation of classical three-dimensional ordering. As a consequence, conformational distortions appear in chains and only the two-dimensional ordering of the 2D mesophase type is implemented in the system. At the second stage, when the solvent is fully removed and, accordingly, when the phase decomposition in the cellulose phase is completed, the traditional ensemble of inter and intramolecular hydrogen bonds forms and cellulose passes to the crystalline phase.

Copolyester and PMPIA are thermodynamically incompatible with cellulose but possess high affinity for NMMO. These polymers dissolve in NMMO to form solutions that in the case of copolyester are compatible with the solution of cellulose. However, in the case of PMPIA, the inverse situation is observed: solutions of cellulose and PMPIA in the same solvent are incompatible. During isolation of the polymer phase from blend solutions, the redistribution of specific interactions and hydrogen bonds, but conformational transformations of cellulose macromolecules associated with incompatibility of polymers most probably create steric hindrances to the development of intermolecular ordering that is traditional for cellulose.

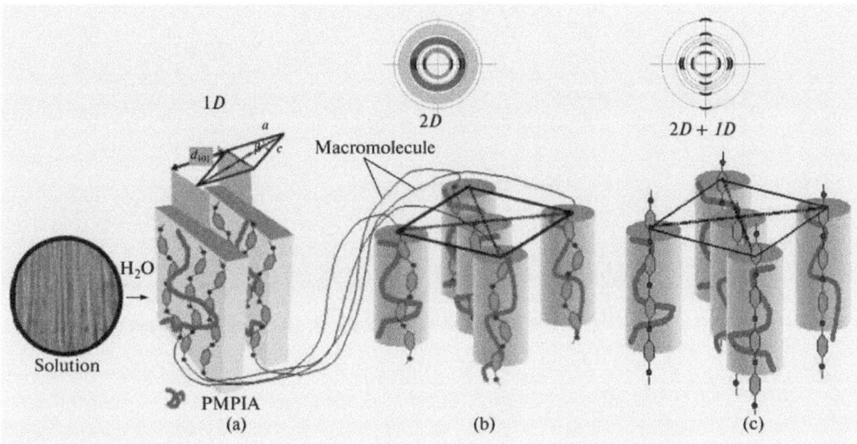

Figure 27. Schematic representation of the structure evolution during spinning of composite fibers from the cellulose–NMMO–

PMPIA blend system: (a) 1D order in the basal plane (layered, intermolecular); (b) 2D order in the basal plane (intermolecular); and (c) 2D order in the basal plane (intermolecular) + 1D order along the fiber axis (intramolecular), that is, →3D noncrystalline order.

5.2. Structure of cellulose and its blends with aluminosilicates

The evolution of structure transformations in composite systems based on cellulose and layered aluminosilicates Cloisite Na^+ and Cloisite 20Å, obtained through a stage of solid-phase dissolution of cellulose in NMMO, has been investigated by X-ray diffraction (XRD) technique [20-22]. A solid-phase process provides high dispersion and uniform distribution of particles of clay in a matrix cellulose phase.

Diffractograms in Fig.28 reflect interaction between components of the system "cellulose-Cloisite Na^+-NMMO" realized on the first stage of the process—the solid-state activation.

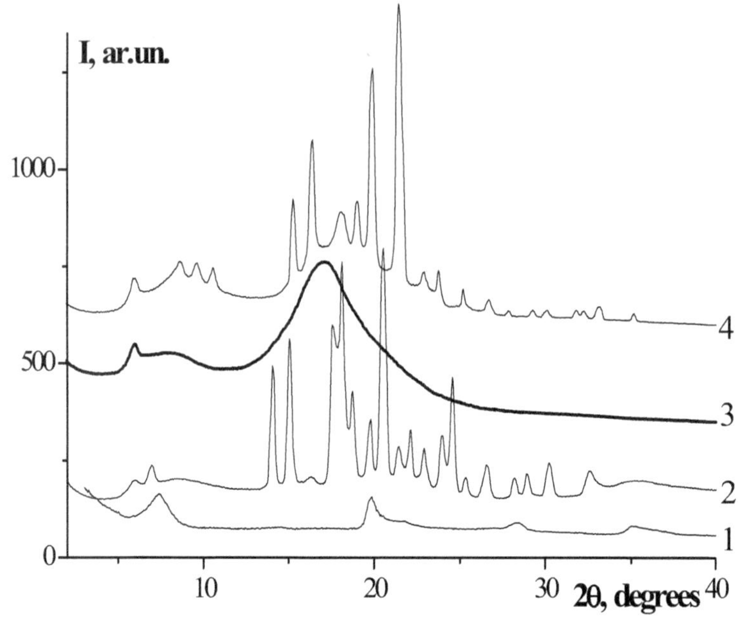

Figure 28. Diffractograms of Cloisite Na^+ (1), solid-activated mix system: 16,2% cellulose - 1,8% Cloisite Na^+ - 82% MMO at 20 (2), 120 (3) and 20°C after cooling from 120°C (4)

Comparison of the X-ray characteristics indicates that after introducing the clay into the composition no changes in the angular position of the basic layer peak of the clay at $2\theta_L$ take place. However, an additional peak appears shifted a little to a small-angle side in comparison with the reflection at $2\theta_L$. The shift of the additional new peak corresponds to the increase of interlayer spacing for 3 Å. This additional peak becomes the basic one on heating up to 120°C (curve 3) and subsequent cooling (curve 4). It seems that the weak changes in clay interlayer spacing are caused by water migration from the system to the clay due to their high affinity to each other. So, the XRD pattern of the solid-phase treated by composition corresponds to the superposition of the two patterns: Cloisite Na^+ and the solid pre-solution of cellulose in NMMO.

The scanning at a low speed with high acquisition reveals no reflexes in the angular region of $2\theta < 3°$, that did not enables us treat this system as an intercalated one.

Nanosize particles (M_2Cloisite Na^+), received as a result of activating dispersion of Cloisite Na^+, are discrete tacktoids with the same interlayer distance as in the initial clay. Possibly, such type of a clay structure is caused by steric restriction for the penetration of bulk H-complexes into interlayer regions of the polar montmorillonite Cloisite Na^+.

Essentially different structure of activated solid-phase mixed system is realized in Cloisite 20 A. As seen at the equatorial XRD pattern of the composition (Fig. 29), the basic layer peak of Cloisite 20A shifts to the small-angle region from the initial angular position $2\theta_L$=3.3º to $2\theta_L$=1.76°, (curve 1).

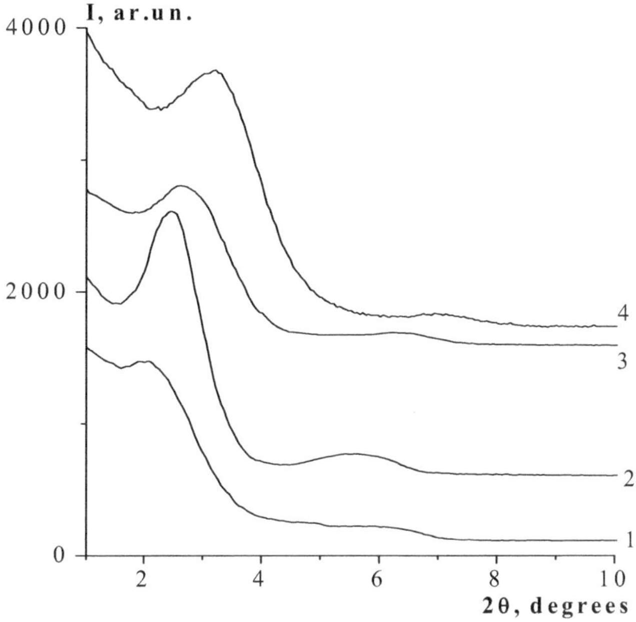

Figure 29. Diffractograms of the solid-phase activated system 16,2% cellulose-1,8% Cloisite 20A - 82% MMO at 20 (1), 120 (2), 20°C after cooling from 120°C (3) and Cloisite 20A (4).

That is caused by the influence of shear stress. Molecules of cellulose and NMMO penetrate into interlayer regions of the clay crystal lattice causing to the shift one-dimensional expansion by more than 20Å. It happens in spite of the hydrophobic nature of Cloisite 20A conditioned due to the presence of non-polar side modifying substituents. A heating and, accordingly, melting of mixed "solid" pre-solutions, they transfer into a viscous state that leads to the changes in the XRD pattern (curve 2): the intensity of the basic layer peak of the modified clay sharply increases and the angular position of this peak slightly shifts towards wide- angle side up to $2\theta_L$=2.04º. The similar changes in the character of the XRD pattern occur also on heating the individual Cloisite 20A. This is apparently connected with conformational changes of its side chains. Conformational reorganization happened on heating promotes the formation of more perfect and homogeneous structure also after cooling the sample. This manifests itself by the intensive growth of the peak at $2\theta_L$ (curve 3).

The above cited set of experimental data allows us to come to the following conclusions. Under the solid-phase activation of a cellulose-NMMO-Cloisite Na^+ system the solid-phase complex formation between molecules of cellulose and NMMO occurs only. As for the system cellulose-NMMO- Cloisite 20A, herein a reaction of solid-phase intercalation of molecules of cellulose and NMMO into the interlayer regions of the modified clay happens alongside with processes of H-complex formation between cellulose and NMMO.

Under simultaneous influence of temperature and velocity gradient solid-phase pre-solutions melt and passes into a viscous state. As seen in Fig. 30, the XRD patterns of mixed solutions of cellulose in NMMO, containing Cloisite Na^+, are characterized by the slight shift of the basic layer peak of the clay towards the small-angle side, i.e. identically to what happens in solid-phase state.

Moreover, the basic layer peak of the clay at $2\theta_L$ is characterized by higher intensity on the equator in comparison with the meridian (Fig. 30). Such character of azimuthal redistribution of peak intensity is obliged to high-degree orientation of discrete clay particles along the flow axis direction. It is achieved under the influence of longitudinal and shear stresses arising in the pre-entered zone of a capillary. Schematic representation of the realized structure is presented in Fig. 30a.

The structures of the mixed solutions of cellulose in NMMO, containing Cloisite 20A, and the activated solid-phase compositions are completely identical and are characterized by pronounced intercalation of the cellulose macromolecules solvated by NMMO into interlayer regions of the clay.

The revealed structural peculiarities of the mixed systems under study demonstrate themselves the most essentially at the last stage of the process – allocation of a cellulose-silicate phase from NMMO solutions.

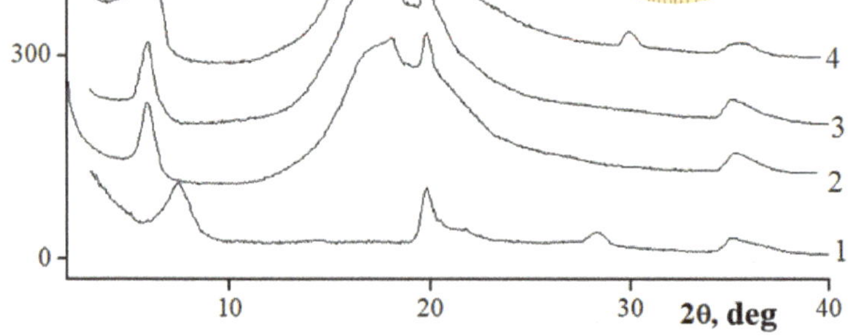

Figure 30. Diffractograms of Cloisite Na^+ (1), 18% cellulose solution in NMMO, containing 10% Cloisite Na^+, in film form (2), in extrudate form (3 – meridional, 4 – equatorial) and schematic representation of the solution extrudate structure (a).

The XRD patterns of oriented samples of the cellulose-Cloisite Na^+ compositions (Fig. 31) are characterized by the practically similar slight increase of the clay interlayer spacing d_L up to 15.0Å, like in the case of the activated solid-phase mixtures and solutions described ealier. Such a character of the XRD patterns is indicative of complete absence of intercalation.

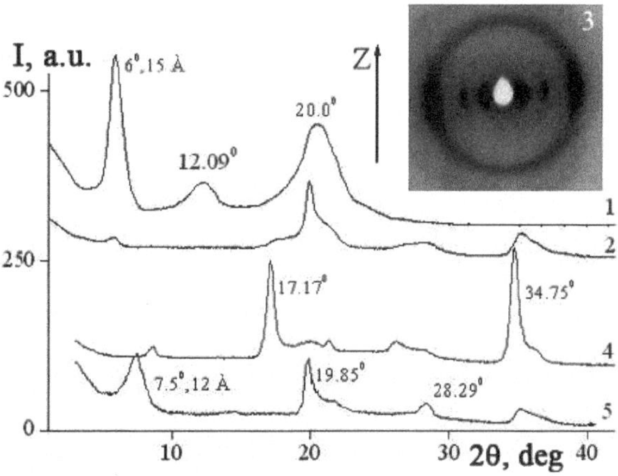

Figure 31. Equatorial (1) and meridional (2) diffractograms and X-ray pattents (3) of composite sample of 90% cellulose + 10% Cloisite Na^+; meridional diffractogram of cellulose fibers (4) powder diffractogram of Cloisite Na^+ (5).

Analysis of the X-ray patterns of oriented samples allows us to obtain more exact information about realized structure. So, two equatorial peaks rather legibly manifest themselves on the equator of pattern of a mixed sample (Fig. 31a). One of the peaks located in at wide angles corresponds to the oriented cellulose. The angular positions of the equatorial peaks of cellulose on the patterns for mixed system (Fig. 31a) and oriented cellulose partially washed-out from NMMO (22b) coincide. As has been mentioned above, the presence of only these equatorial peaks indicates that cellulose macromolecules form the columnar type 2D-mesophase. The peak of oriented clay is also present on the equator of the mixture pattern. Is has been also established that the degree of orientation for both components depends on a content of nanocomposites. The increase of content and, accordingly, the degree of structuredness of cellulose phase in solution promotes the rise of an orientation order for both, cellulose itself and clay. The X-ray patterns of composite with Cloisite 20A and composite of cellulose with Cloisite Na^+ as additive are practically identical.

Thus, a cellulose phase under the influence of shear stress is oriented and this initiates the implementation of certain orientational order of anizodiametric clay particles. The oriented clay particles give an additional impact on orienting the cellulose phase. In other words, in mixed solutions and respectively in the composites develop a unique synergy of orientation of the matrix phase and the filler. At the same time, particles of the clay immobilized in the matrix phase retain their size and do not aggregated.

Apparently, under the influence of shear of mixed systems at capillary flow anisodiametric solid particles of the filler embedding in a less structured region of the matrix phase, orient in the direction of extrusion. This leads to a perturbation of the stress and velocity fields of a viscoelastic matrix phase of cellulose and, consequently, to its local deformation. The stressed state of the matrix at the micro level does not allow the cellulose macromolecules rearrange with the formation a three-dimensional crystal structure. As a result, two-dimensional but not three-dimensional structure is formed, and it is characterized by a regular packing of conformationally disordered macromolecules of cellulose in the basal plane. Schematic representation of the evolution in the cellulose structure during spinning of fiber samples from mixed solutions of cellulose in NMMO with the addition of nanoparticles of clay is presented in Fig. 32.

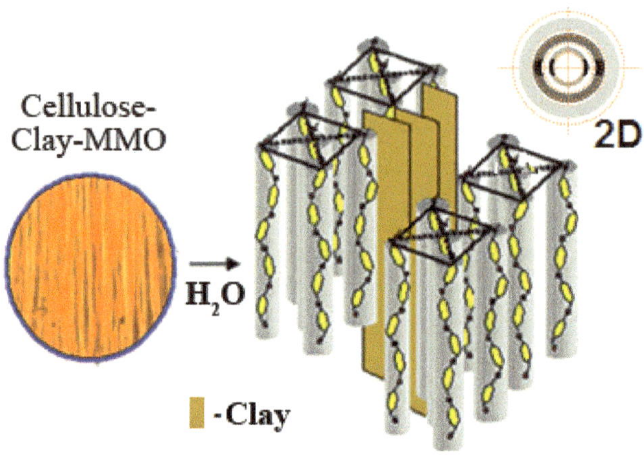

Figure 32. Schematic representation of the structure evolution during spinning of composite fibers from the cellulose–clay–NMMO blend system.

So, it has been shown that the introduction of copolyester, PMPIA and clays into cellulose solutions in NMMO makes it possible to govern the processes of the cellulose structuring and to finish them on the stage of the *2D* mesophase formation and, thus, to avoid further perfection of the structure and the formation of the crystalline phase of cellulose. In other words, the structural situation is the same as in the case of the as-spun cellulose fibers. The only difference is that, in purely cellulose fibers, the resulting mesophase is not at equilibrium, while in blend systems, the *2D* mesophase forms after the complete phase decomposition of the system and, correspondingly, is at equilibrium.

6. Fibrillization and mechanical properties of cellulose and synthetic polymers and layered aluminosilicates micro- and nanocomposite fibers

The following structure peculiarity of the process under discussion was found. This is the formation of mesamorphic matrix cellulose phase due to introducing synthetic polymers or micro- and nanoadditives of layers alumonosilicates into cellulose solutions. This effect allows us to solve two important concerns of the processing of cellulose from solution in NMMO: it prevents fibers fibrillation and provides high mechanical properties of novel composite materials. Photomicrographs of cellulose and mixed fibers obtained before and after vibrating mill treatment are shown in Fig. 33.

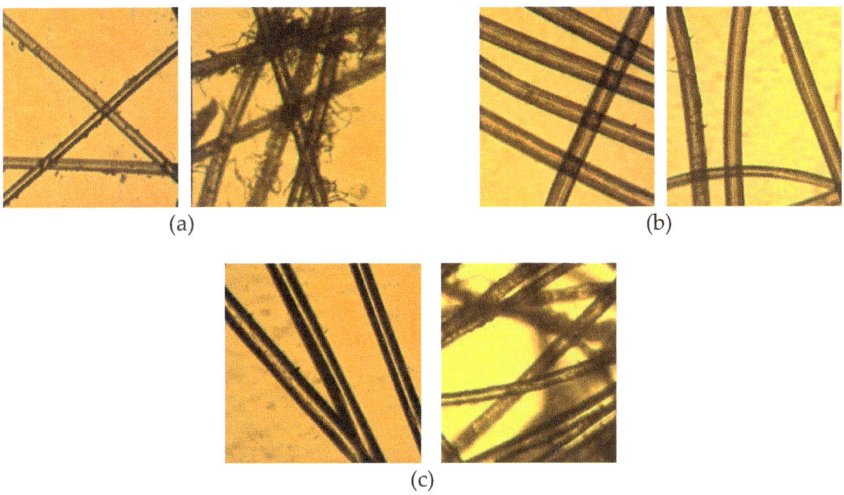

Figure 33. Micrographs of cellulose (a) and mixed fibers: cellulose/PMPIA (b) and cellulose/M_2CloisiteNa$^+$(c) before and after mechanical treatment in a vibrating mill in the aquatic environment (diameter of the fibers is 14-16 μm).

The obtained micro- and nanocomposite fiber cellulose/CPE (PMPIA) and cellulose/clay have high mechanical properties, which by 1.5–3 times exceed the strength properties of cellulose fibers (Table1). At the same time, the deformation parameters of composite fibers are not reduced, and, in some cases, are substantially increased in comparison with the standard cellulose fibers.

System	Ø, μm	s,MPa	ε, %	*E, GPa*
Cellulose 100 %	34	420	10	11
Cellulose 95% - PMPIA 5%	28	710	12	22,6
Cellulose 90% – Cloisite Na$^+$ 10%	30	530	11	14
Cellulose 99,97% – Cloisite Na$^+$ 0,03%	18	1520	10	30

Table 1. Mechanical properties of micro- and nanocomposites

7. Conclusion

A method for the preparation of solid-phase compositions of cellulose with synthetic polymers and layered aluminosilicates Cloisite Na^+ and Cloisite 20A has been developed. It provides high dispersion and uniform distribution of a polymer or micro- and clay nanoparticles in the matrix cellulose phase.

The addition of polymers and clay to cellulose solutions in NMMO, regardless of the nature of a polymer as well as phase and/or structure peculiarities of mixed solutions, allows us to control the process of structurization of the matrix cellulose phase, leading it to the formation of 2D mesophase, thereby excluding a possibility of crystallization.

Novel nanostructures composed of cellulose/clay and cellulose/polymer fibers possess high mechanical properties: their strength and modulus are 1,5–2 times higher than those of cellulose.

Our studies have made it possible to approach the solution of an extremely important issue that consists in the regulation of the cellulose structuring. Until now this did not have any direct solution. Thus, it becomes possible to manufacture cellulose-based fibers that have the desired combination of characteristics.

Author details

Ludmila Golova, Igor Makarov, Ludmila Kuznetsova,
Elena Plotnikova and Valery Kulichikhin
A.V. Topchiev Institute of Petrochemical Synthesis, Russian Academy of Sciences, Moscow, Russia

8. References

[1] Franks N. F. and Varga J. K. 1979, US Patent, No. 4290815
[2] Chanzy H., Noe P., Paillet M., Smith P. Swelling and dissolution of cellulose in amine oxide/water systems. Journal of Applied Polymer Science: Applied Polymer Symposium 1983; 37 239-259.
[3] Kruger R. Cellulosic filament yarn from the NMMO process. Lenzinger Berichte 1994; 4 49-52.
[4] Golova L. K., Kulichikhin V. G., and Papkov S. P. Mexanism of dissolution of cellulose in nonaqueous solvents. Vysokomol. Soedin 1986; A(28) 1795-2016.
[5] Cousey H. A., Smith S.B. The Formation and Structure of a new Cellulose Fibre. Lenzinger Bericht 1996; 7 51-63.
[6] Golova L. K. 1992, Russia Patent, No. 1645308
[7] Golova L. K. A new method of processing cellulose via highly concentrated "solid solutions". Khim. Volokna 1996; 1 13-21.

[8] Golova L. K., Makarov I. S., Matukhina E. V., Kuptsov S. A., Shambilova G. K. and Kulichikhin V. G. Crystal Solvates of Thermotropic Alkylenearomatic Copolyesters and Poly(*m*-phenyleneisophthalamide) with *N*-Methylmorpholine-*N*-oxide . Vysokomol.Soedin 2008; A (50) 665 – 678.

[9] Amstrong R.N., Varga J.K., McCorsley C.C. Spinnable solutions of cellulose dissolved in amine oxides. Proceeding of the Fifth International Tappi Dissolving Pulp Conference, Vienna; 1980.

[10] Chanzy H., Dube V., Marchessault R.H. Crystallization of cellulose with NMMO: a new method a texturing cellulose. Journal of Polymer Science: Polymer Letters Edition 1979; 17 (4) 219-226.

[11] Golova L. K., Borodina O. E., Kuznetsova L. K. Hard Phase MMO-Process. Khim. Volokna 2000; 4 14-21.

[12] Papkov S. P. Physicochemical Principles of Polymer Solution Processing. Khimiya, Moscow. Russian; 1971.

[13] Weigel P., Gensrich J., Fink H.P. The structural format of cellulose fibres from amine oxide solutions. Lenzinger Berichte 1994; 9 35-41.

[14] Makarov I.S., Golova L.K., Plotnikova E. P., Matukhina E.V., Rebrov A.V., Kulichikhin V.G. RHEOLOGICAL PROPERTIES AND STRUCTURE OF NANOCOMPOSITES BASED ON CELLULOSE WITH NA-MONTMORILLONITE. 2nd International Polysaccharide Conference. Wageningen – Holland; 2011.

[15] Plotnikova E. P., Golova L.K., Makarov I.S., Kulichikhin V.G.Rheological properties of solutions of cellulose containing particles of aluminum silicates in MMO. Vysokomol.Soedin; 2012. (in press).

[16] Schurz J., Lenz J. Investigations on the structure of regenerated cellulose fibrer. Macromol. Symp. 1994; 83 273-289.

[17] Andersen E.M., Mitchell G.R. In Situ X-ray scattering investigation of solution of cellulose in NMMO during shear flow. Polymer 1998; 39 (26) 7127-7129.

[18] Golova L. K., Makarov I. S., Plotnikova E. P., Tereshin A. K, and Kulichikhin V. G. Solutions of Mixtures of Cellulose and Synthetic Polymers in *N*-methylmorpholine-*N*-oxide,Polymer Science . Vysokomol.Soedin 2009; A (51) 414-420.

[19] Golova L. K., Makarov I. S., E Matukhina E.V., Kulichikhin V.G. Solutions of Cellulose and Its Blends with Synthetic Polymers in N- Methylmorpholine-N-oxide: Preparation, Phase State, Structure, and Properties . Vysokomol.Soedin. 2010; A 52(11) 1209-1219.

[20] Makarov I.S., Golova L.K., Matukhina E.V., Kulichikhin V.G. Cellulose nano- and microcomposites with natural and modified clays. 10th Annual Conference, YUCOMAT Herceg Novi, Montenegro; 2008.

[21] Makarov I.S., Golova L.K., Matukhina E.V., Kulichikhin V.G. Biodegradable composite materials on the basis of cellulose with synthetic polymers and laminated aluminosilicates. The 1st EPNOE Conference "Polysaccharides as a Source of Advanced Materials".Turku, Finland; 2009.

[22] Makarov I.S., Golova L.K., Rebrov A.V., Matukhina E.V., Kulichikhin V.G. Biodegrable nanocomposites based on cellulose with Na-montmorillonite. AERC 2011. 7th Annual European Rheology Conference .Suzdal - Russia; 2011.

Chapter 13

Analysis of Relaxation Behavior of Free Radicals in Irradiated Cellulose Using Pulse and Continuous-Wave Electron Spin Resonance

Hiromi Kameya and Mitsuko Ukai

Additional information is available at the end of the chapter

1. Introduction

Food is irradiated to destroy harmful bacteria and parasites that might be inadvertently present. Irradiation at lower dose levels also extends shelf-life and can be used to control insects. Irradiation extends the shelf-life of food by retarding maturation in vegetables and reducing spoilage organisms that can grow even under refrigeration. Irradiation can also be used in place of fumigants and other quarantine procedures to import or export fruits and vegetables without the risk of transporting harmful insects.

Electron spin resonance (ESR) studies have reported on induced radicals in irradiated plant foods (Ukai et al., 2006; Ukai & Shimoyama, 2003a, 2003b, 2005; Nakamura et al., 2006; Shimoyama et al., 2006). There are two types of ESR signals from irradiated plant food: one is a singlet with a g-value of nearly 2.0, and the other appears symmetrically as two side signals on both sides of the singlet. The singlet signal has been identified as an organic free radical (Ukai & Shimoyama, 2005).

Clear side signals have been reported in pepper following gamma-ray irradiation (Ukai & Shimoyama, 2003a, 2003b). Similar side signals in wheat flour that contains cellulose and starch are more complicated (Shimoyama et al., 2006; Ukai & Shimoyama, 2005). Side signals have been observed in ginseng that contains a significant amount of fiber (Nakamura et al., 2006) and in irradiated strawberry seeds whose main component is cellulose (Raffi & Stocker, 1996; Raffi & Agnel, 1989). However, only one side signal at the lower magnetic fields has been detected (Raffi & Agnel, 1989). The side signal in the higher magnetic field was considered covered by other stronger ESR signals. These side signals have been named after the "cellulotic radical," because they are observable in ESR spectra of samples containing a significant amount of cellulose (Raffi et al., 2000). Ehrenberg et al. have reported that the side signals appearing from irradiation were derived from peroxide radicals (Loftroth et al., 1964; A. Ehrenberg et al., 1962). There have been many recent reports on the side signals (Lee et al., 2009; Lee et al., 2008; Yordanov et al., 2009; Raffi et al., 2009; Yamaoki et al., 2009; Sanyal et al., 2008; Cutrubinis et al., 2007; Polovka et al., 2007; Jo & Kwon, 2006; Butz & Hildebrand, 2006).

Lee et al. have reported on irradiated sesame seeds using ESR (Lee et al., 2009; Lee et al., 2008). The report concluded that a pair of ESR lines appears on both sides of the central signal in ESR spectra due to the cellulose radicals formed by ionizing radiation. Yordanov et al. have reported a pair of signals in the irradiated dry medical herbs (Yordanov et al., 2009). A central line and two weak satellite lines were detected. The authors named these after the "cellulose-like" and "carbohydrate-like" triplets (Yordanov et al., 2009). However, their spectra were undefined. Raffi and Yordanov have reported on aromatic herbs, spices, and fruits after irradiation (Raffi et al., 2009). They have also reported a relatively weak triplet.

Polovka et al. have reported on irradiated spices (Polovka et al., 2007). Simulation of ESR spectra for the black pepper samples revealed the formation of three paramagnetic species, i.e., the triplet, doublet assigned to "carbohydrate" radical structures, and the typical three-line "cellulotic" signal. The paramagnetic signals identified in individual irradiated spice samples, along with the spin Hamiltonian parameters, have been employed in quasi-empirical simulations.

A triplet line has been reported from the irradiated flesh of kiwifruits, with the extra signals resulting from cellulose radicals on both sides of the endogenous signal (Jo & Kwon, 2006). Studies on the irradiated shells of walnuts and pistachios have shown that a sample can be unambiguously identified, if beside the central signal satellite peaks, a separation of 6.0 ± 0.05 mT is detected (Butz & Hildebrand, 2006). However, there is little explanation on the origin of the side signals.

With the CW-ESR method, irradiation can be performed using a weak (continuous) microwave field without disturbing the steady state spin system. The pulsed-ESR method employs instantaneous microwave radiation and reorients the spin systems. This can be observed in the absence of a microwave field. Pulsed ESR determines the relaxation behavior of radicals. The relaxation times, T_1 and T_2, were evaluated using the decay of echo signals as a function of time intervals between microwave pulses. However, strict measurement conditions must be met for an accurate analysis of pulsed ESR.

In this study, gamma ray radiation was applied on filter paper that was made of pure cellulose and on an oblate of pure starch which has structure similar to cellulose. We analyzed the ESR side signals originating from these polymers after irradiation. Using theoretical calculations, we simulated the side signals by assuming a triplet signal from the radical at the C(5) position of the glucose unit (Fig. 1 A). This revealed the molecular mechanism of the radical formation. Furthermore, we examined irradiated black pepper using pulsed ESR. Black pepper contains many cellulose and many researchers reported the radicals in irradiated black pepper using ESR. But there is no report of the measurement of irradiated black pepper containing much cellulose using pulsed ESR. We employed an electron spin echo (ESE) using the pulsed ESR method, i.e., the $\pi/2$- and π-pulse sequence. We measured the relaxation of the spin system and various magnetic parameters such as the g-value and hyperfine couplings.

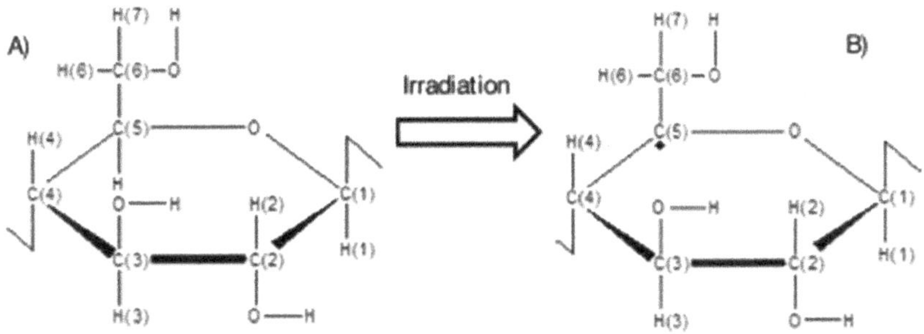

Figure 1. The chemical structure of the cellulose radical. A) before irradiation and B) after irradiation. The triplet ESR spectrum of cellulose is due to the interactions between the two hydrogen atoms at the C6 position of the glucose unit with the unpaired electron formed by the removal of the hydrogen atom at the C5 position of the glucose unit by irradiation

2. Method

2.1. Materials

Filter paper (ADVANTEC MFS., INC) and oblate (Kokko-Oblate Co., Ltd.) were used as samples. The quality of filter paper and oblate used was more than 90% (Japanese Industrial Standards Committee, 1995) cellulose and 99.2% (Kokko-Oblate Co. Ltd.) starch. Black pepper powder which is one of the foods containing much cellulose, commercially available in Japan, was used in the present study. It was bottled in glass and stored at +4.0 °C. After the samples were obtained, they were carefully sealed in a quartz sample tube (99.9% purity; Eiko, Tokyo, Japan). We prepared sample tubes in an argon (Ar) atmosphere and sealed the tubes to remove any oxygen effects (Nakamura et al., 2006). First, the ESR sample tube was degassed for 5 min, and Ar gas was substituted. Then, Ar gas was purged into the ESR sample five times for 2 min each time. The inside pressure of the ESR tube filled with Ar gas was ca. 1 mmHg.

2.2. Irradiation treatment

Irradiation was carried out at the Japan Atomic Energy Research Institute (Takasaki Research Institute) at room temperature (ca. 300 K). We used the gamma-ray originated from ^{60}Co as the irradiation source. The dose rate was 2 kGy/h, and the dose level was controlled by the irradiation time. The irradiation dose levels used were 1, 10, 25, 50, and 100 kGy.

2.3. CW-ESR measurements

CW-ESR measurements were performed with JES-FA100 (JEOL) at 25 °C using the microwave X-band frequency (9.45 GHz). The field modulation frequency was 100 kHz. For the determination of g-values (Swartz et al., 1972), an Mn marker (JEOL) was employed for the correct microwave frequency. In all ESR measurements pertaining to the present study, the magnetic field was swept in the range of 250.0 ± 250.0 mT with a field modulation of 1.0 mT, time constant of 0.03 s, and sweep time of 4 min. When focusing on the single peak at g = 2.00, the magnetic field was swept over the range of 336.5 ± 7.5 mT, with a time constant of 0.03 s and sweep time of 4 min. The side peaks were detected using smaller noise conditions, comprising a field modulation of 1.0 mT. ESR signal parameters were analyzed using the WIN-RAD (Radical Research) software program.

2.4. Pulse-ESR measurements

All pulse-ESR measurements (ESP-380E, Bruker Biospin, Yokohama, Japan) were performed at 25 °C. To detect the pulsed electron spin-echo signals, we used a $\pi/2$–π pulse sequence. We used 16 ns for the $\pi/2$ pulse and 24 ns for the π pulse. The pulse interval was 200 ns; recycle delay, 1 ms; and microwave frequency, 9.480 GHz. The electron spin-echo envelope modulation (ESEEM) was measured at the center of the magnetic field (341.5 mT) of the main peak. Theoretically, when the $\pi/2$ pulse is assigned as 16 ns, the corresponding π pulse is 32 ns. We used a shorter pulse, 24 ns, to compensate for the effect of the pin diode on pulse quality. To obtain the echo-detected ESR spectra, the intensity of the electron spin echo was measured as a function of the magnetic field.

2.5. Magnetic interaction tensors

The hyperfine data for two H_α in H_2C_α-C_β fragments were obtained from ENDOR data of irradiated malonic acid (Sagstuen et al., 2000). In the first step, we transformed the tensors into the XYZ axes (Z⊥XY-plane) in terms of the two tensor elements A_{XX} and A_{YY}. These were obtained by standard procedures, assuming an angle (30° slightly idealized).

$$A_{XX} = A_1\cos^2 30^0 + A_2\sin^2 30^0 = -17.95\ G$$

$$A_{YY} = A_1\sin^2 30^0 + A_2\cos^2 30^0 = -29.05\ G$$

In the second step, rapid rotation about either X or Y was taken into account.

X: The two H_ζ-tensors both became axially symmetric about X, with $A_\parallel$ = -17.95 G. One had Trace (A) = $A_\parallel + 2A_\perp$= (trace invariant under similarity transformation) = $A_1 + A_2 + A_3$ = -69.8 G. Therefore, $A_\perp$= -25.9 G.

Y: The two H_ζ-tensors became axially symmetric about Y, with $A_\parallel$ = -29.05 G. By the similar calculation, $A_\perp$= -20.4 G.

Powder spectra for the two models were simulated by an updated version of the KVAPOL program (Thuomas & Lund, 1976).

2.6. Observation of relaxation times of CW-ESR

To determine T_1 and T_2, we used Lund's program (Lund, 2009). For a Gaussian envelope of Lorentzian spin packets, previous treatments have shown the EPR line shape can be expressed as

$$g(r) \propto \frac{B_0\beta}{\Delta B_G}\int_{-\infty}^{\infty}\frac{e^{-(a^2r'^2)}dr'}{t^2+(r-r')^2} \tag{1}$$

which is a convolution of a Gaussian line of width t/a and a Lorentzian line of width t, resulting in a Voigt line profile. Here, β is the transition probability of the line g(r) centered at B_0. The variables r and r′ are defined by the corresponding magnetic fields B and B′ as

$$r = \frac{B-B_0}{\Delta B_L} \qquad r' = \frac{B'-B_0}{\Delta B_L} \tag{2}$$

The parameters a and t^2 affecting the shape of the saturation curve are given by

$$a = \frac{\Delta B_L}{\Delta B_G} \qquad t^2 = 1+\gamma^2 B_1^2\beta T_1 T_2 = 1+s^2 \tag{3}$$

ΔB_L and ΔB_G are the full widths at half maximum value (FWHM) of the unsaturated Lorentzian and Gaussian line shapes, respectively, and can be expressed in terms of the peak-to-peak widths λ_L and λ_G of the corresponding 1st derivatives as

$$\Delta B_L = (\sqrt{3}/2)\lambda_L \qquad \Delta B_G = \lambda_G/\sqrt{2} \tag{4}$$

The saturation factor s contains the gyromagnetic ratio γ, amplitude of the (left-handed) rotating microwave magnetic field component in the cavity, B_1, and spin-lattice and spin-spin relaxation times T_1 and T_2, respectively. Note that the amplitude B_1 is one-half of the amplitude of the linearly polarized field B_1 in the resonance resonator, as employed experimentally.

The amplitude B1 of the rotating microwave magnetic field component at the sample position of a microwave resonator is related to the input microwave power by an expression of the type

$$B_1 = k\sqrt{Q_L P} = K\sqrt{P} \tag{5}$$

where the constant K depends on the type of resonator and its quality factor Q_L (loaded Q) with the sample in place. It may often be difficult to estimate its value precisely.

Substituting the experimentally measured microwave power P and introducing the spin-relaxation-dependent parameter

$$P_0 = \frac{1}{K^2\gamma^2\beta T_1 T_2} \tag{6}$$

the absorption line shape of equation (1) can be recast to

$$g(r) \propto C\frac{\beta\sqrt{P}}{t}u(ar, at) \tag{7}$$

where the Voigt profile u is the real part of the complex error function w

$$w(z) = exp(-z^2)\, arfc(-iz)\ (w = u + iv, z = at + iar)$$

and

$$t = \sqrt{1 + \beta P/P_0}$$

The procedure to evaluate the line shape function g(r) (equation (7)) numerically, by expanding the function u(ar,at) as the real part of the complex error function, was outlined previously, and was also used in the present work. The code for the Gautschi algorithm, to calculate a Voigt profile as the real part of the complex error function, is a FORTRAN translation of the original ALGOL procedure. To measure the saturation behavior of a single, inhomogenously broadened line, the transition probability β is set to 1 as in a simple two-level system.

Experimentally, the first derivatives of the absorption spectra are recorded, and therefore, the function g(r) has to be differentiated with respect to the variable r (or magnetic field). This was more conveniently done numerically in this work.

The spin-spin relaxation time T_2 is given by definition as $1/\gamma\Delta B_L = 2/\gamma\lambda_L\sqrt{3}$. The estimate of the spin-lattice relaxation time T_1 follows from equation (6) and depends on the factor K to calculate the B_1 field from the corresponding microwave power according to equation (5).

2.7. Observation of relaxation times of pulse-ESR

We used two π/2–τ–π pulse sequences at 16 ns and 24 ns for taking echo measurements. The pulse interval was 200 ns, and the recycle delay was 1.0 ms. For each relaxation measurement, we employed a specific pulse sequence. To determine T_2, we used a two-pulse sequence:

π/2 pulse–τ–π pulse–τ–(echo)

To determine T_1, we used a three-pulse sequence (i.e., the inversion recovery method);

π pulse–τ–π/2 pulse–τ–π pulse–τ–(echo)

We used the same parameter sets as for the spin echo measurements, and fixed τ at 200 ns, to obtain relaxation measurements.

2.8. g-value determination

The g-value was determined independently using the position of the magnetic field and the microwave frequency. For accuracy, the values of the magnetic field and frequency were considered out to the fourth and fifth decimal places, respectively.

3. Results

3.1. Appearance and disappearance of the ESR signals

Figure 2 shows ESR spectra (0 to 500mT) of the filter paper and oblate before and after gamma-ray irradiation. No ESR signals were detected from the two types of samples before irradiation. This suggests the high purity of those samples without an irradiation history.

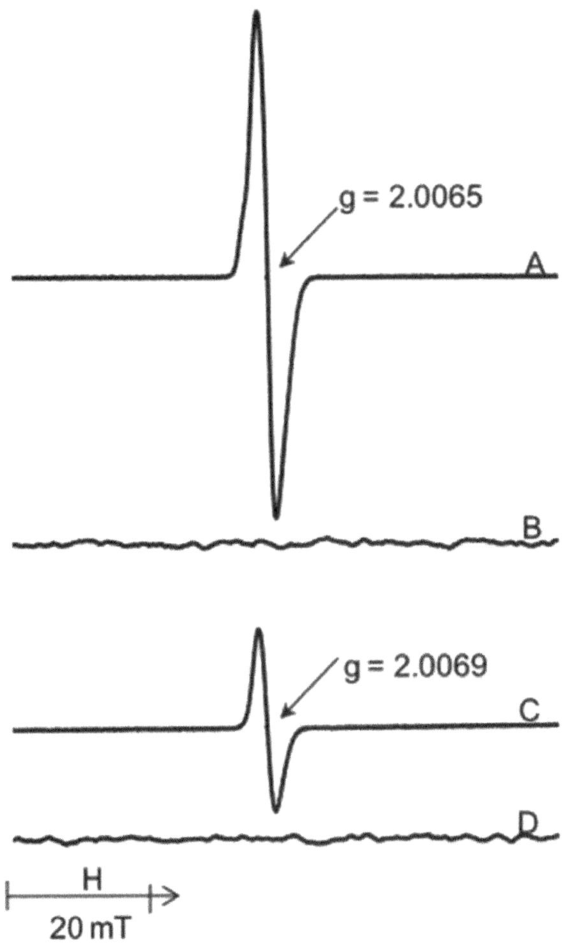

Figure 2. ESR spectra of filter paper and oblate before and after gamma-irradiation. A: irradiated filter paper, B: filter paper, C: irradiated oblate, and D: oblate.

A strong and sharp singlet signal (P_1) was observed near g = 2.0 from the irradiated samples. The g-values of the P_1 signal were identified as g = 2.0065 for the filter paper and g = 2.0069 for the oblate. We have already reported on the ESR singlet signal at g = 2.0 using irradiated ginseng (Nakamura et al., 2006) and black pepper (Ukai & Shimoyama, 2003a, 2003b). These irradiated samples showed two kinds of radicals by ESR analysis: Fe^{3+} at g = 4.0 and Mn^{2+} hyperfine splitting (Swartz et al., 1971). However, irradiated filter paper and irradiated oblate showed neither the Fe^{3+} signal nor Mn^{2+} hyperfine splitting. The signal intensity of P_1 measured from the irradiated filter paper was stronger than that from the irradiated oblate.

Figure 3 shows ESR spectra of the filter paper and oblate followed by irradiation at a magnetic field strength of 329 to 344 mT. In the case of the filter paper, S_1 and S_2 signals were observed at both sides of the P_1 signal. The side signals (S_1 and S_2) were detected at symmetrical positions around the P_1 signal. The g-value of the side signals was found to be 2.0241 for S_1 and 1.9799 for S_2. However, corresponding side signals were not observed in the oblate case.

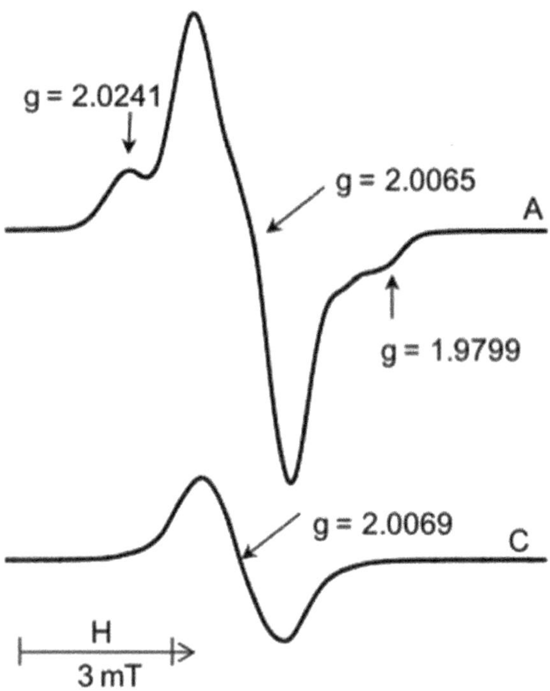

Figure 3. ESR spectra of filter paper and oblate after irradiation. A: irradiated filter paper and C: irradiated oblate.

Experimental evidence (Figs. 2, 3) shows that side signals only appear in cellulose molecules after irradiation. The two side signals occur simultaneously in the ESR spectrum of cellulose. We observed the side signals disappear simultaneously when increasing temperature (Ukai & Shimoyama, 2003b).

Furthermore, according to a monograph on the ESR studies of irradiated polymers (Rånby, 1977), the majority of references have concluded that, among other candidate radicals, a radical at the C(5) position of the glucose unit is the most plausible one in irradiated cellulose. In general, hydrogen abstraction occurs at various chemical bonds by the gamma-irradiation. All the proton sites undergo various chemical reactions simultaneously. However the radical produced at the C(5) position of the glucose unit was comparably stable, so its ESR signal was observable. The stability of the radical was enhanced by the nearby oxygen and alkyl group at the C(6) position of the glucose unit. The ESR signal of the radical at the C(5) position of the glucose unit yields a triplet line shape due to hyperfine interactions between the unpaired electron of a carbon atom and two protons.

Two side signals (S_1 and S_2) were detected in irradiated filter paper (containing cellulose), but not in oblate (containing starch). Therefore, we considered that the side signals were from the cellulose and not from starch. We observed side signals induced by irradiation in botanical foods (Ukai & Shimoyama, 2003a, 2003b, 2005; Shimoyama et al., 2006) and crude drugs (Nakamura et al., 2006) containing cellulose.

Filter paper and oblate consist of cellulose and starch, respectively, and both of them are polysaccharides, represented by the chemical formula $(C_6H_{10}O_5)_n$. Cellulose is a glucose polymer, a β-glucose molecule polymerized by a glycoside bond. Conversely, starch is a polymer produced by polymerizing α-glucose molecules. These structures are different from each other. Cellulose has a sheet-like structure, whereas starch has a helical structure because of the difference in 1, 4 bonds. Thus starch molecules do not form regular sheets and H-bonding is very different. Even though cellulose and starch have similar chemical formulas, their radical formation process may differ due to their structures (Rånby, 1977). It has been reported that the triplet ESR spectrum of cellulose (pure cotton) is due to the equal interaction of the two hydrogen atoms at the C6 position with the unpaired electron formed by the removal of the hydrogen atom at the C5 position of the glucose unit (Rånby, 1977; Arthur, 1971; Arthur & Hinojosa, 1971; Arthur et al., 1966; Baugh et al., 1967) (see Fig. 1 B). The bond breakage at the C5 position may occur only in cellulose. Thus, induction of the side signals should be caused by the strong bonding by β-1, 4 bonds in cellulose.

Upon irradiation, the C5 bond was broken and a radical formed in the cellulose ring (Rånby, 1977). The equivalent two protons located near the unpaired electron at the C5 position of the glucose unit. Thereby, hyperfine interactions occur between the electron spin and the two protons, and the triplet ESR line resulted. The side signals are a part of the triplet. Some interaction between the two protons and the unpaired electron sites at the C2, C3, and C4 positions of the glucose unit are possible, and can be predicted in the cellulose structure (Rånby, 1977; Arthur et al., 1966; Baugh et al., 1967). One needs to consider the physical three-dimensional structure and energy levels of cellulose bond sites. The irradiated starch samples showed a doublet ESR signal (Rånby, 1977; Adamic, 1968). The unpaired electron in the C5 position of the glucose unit produces the doublet-line spectrum by the hyperfine interaction with H5 (Rånby, 1977; Adamic, 1968).

We revealed the origin of the ESR side signals detected from irradiated filter paper containing pure cellulose. Using theoretical calculations, we further revealed the molecular mechanism of the radical formation of irradiated glucose polymers. The side signals are found to be a precise indicator for irradiation effects in cellulose. They originated from neither the peroxide radical of glucose polymers nor the so-called "cellulotic" radicals. By a simulation method, we proved that the side signals originate from a triplet consisting of a hyperfine interaction with two protons, although the main peak is invisible by an overlapping organic free radical at g = 2.0. Note that we reconfirmed that the simulated spin concentration coincides with the experimental ESR value of 1.7×10^{15} spins/g.

3.2. Molecular motions of radical by simulation

Figure 4 shows experimental ESR as well as the simulated ESR spectra of the radical at the C(5) position of the glucose unit. We simulated the ESR spectra under three different conditions of the radical molecule based upon its molecular structure at the C(5) position of the glucose unit. The first is a spectrum at the rigid limit. In the second case, both C_α-C_β and C_β-H bonds undergo rapid rotations. For the simulation, we employed a set of magnetic parameters such as hfc and g-value tensors as shown Table 1. We used g = 2.0065 and A = 3.0 mT for the radical at the C(5) position of the glucose unit. Because the experimental ESR spectrum indicated a signal with modified hyperfine values, rather than a powder pattern at the rigid limit, we postulated that the radical undergoes rotational motions. In fact, we found through a simulation that the C_α-C_β and C_β-H bonds do undergo simultaneous rotations.

	Rigid limit	C-C rotation	C-C rotation and C-H rotation
g-values	$g_x = 2.0071$	$g_\perp = 2.0064$	$g_\perp = 2.0064$
	$g_y = 2.0065$	$g_\perp = 2.0064$	$g_\perp = 2.0064$
	$g_z = 2.0059$	$g_\parallel = 2.0067$	$g_\parallel = 2.0067$
Hyperfine splitting (mT)	$A_x = 3.46$	$A_\perp = 2.00$	$A_\perp = 2.00$
	$A_y = 1.24$	$A_\perp = 2.00$	$A_\perp = 2.00$
	$A_z = 2.28$	$A_\parallel = 2.99$	$A_\parallel = 2.99$
Line shape ratio (L:G)	0:100	0:100	0:100

Table 1. Spectral parameters for simulations

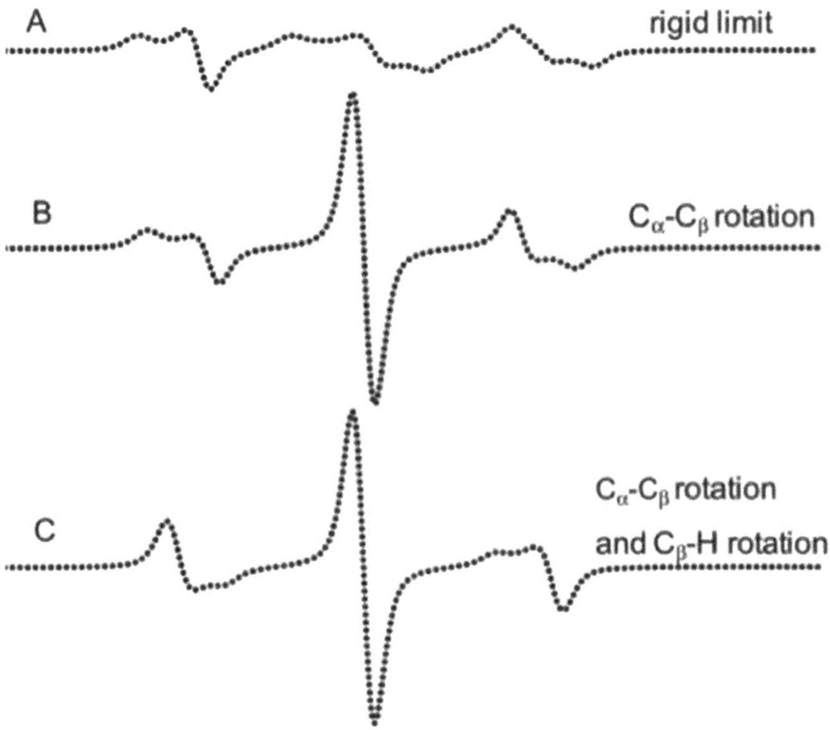

Figure 4. The simulated ESR spectra originated from the radical at the C(5) position of glucose unit under three different conditions of the radical molecule. Experimental ESR spectra of filter paper and oblate after irradiation. A: simulated spectra of rigid limit, B: simulated spectra of C_α-C_β rotation, and C: simulated spectra of C_α-C_β rotation and C_β-H rotation.

3.3. Field-swept signal of ESE and hyperfine interaction detected by ESEEM

Figure 5 shows a field-swept echo spectrum (shown from 330.0 to 350.0 mT) of the radicals from gamma-ray-irradiated black pepper. We found three peaks: a main peak and two side peaks. An integrator was not used in the present pulse ESR experiment; therefore, the line width of the main peak detected by pulse-ESR was broader than that of the main peak detected by CW-ESR. The signal intensity of the main peak, at g = 2.005, increased with the radiation dosage. Two side peaks were observed when the irradiation doses were higher than 25 kGy. Black pepper contained much cellulose. So, these peaks are due to irradiated cellulose.

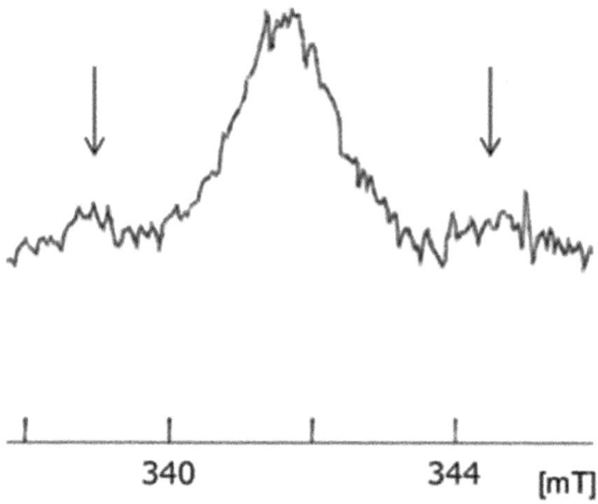

Figure 5. Field-swept (330.0 to 350.0 mT) echo spectrum of gamma-ray-irradiated black pepper at 100 kGy analyzed by pulsed EPR with a microwave frequency of 9.480 GHz. Black pepper powder commercially available in Japan was used. The initial tau value of time between pulses was 200 ns.

The CW-ESR spectra showed 6.0 mT of hyperfine separation between the two side peaks. The field-swept echo yielded a splitting of ca. 6 mT, as shown in Fig. 5. We concluded that both the ESR and the field-swept echo showed the same radical species. We have already reported the corresponding ESR peaks with other specimens using the CW-ESR measurements (Ukai et al., 2006; Ukai & Shimoyama, 2003a, 2003b, 2005; Nakamura et al., 2006; Shimoyama et al., 2006). We observed the same information in the spectra of both pulse- and CW-ESR.

Figure 6 shows a decay signal from the electron spin echo of the gamma-ray-irradiated black pepper sample at the magnetic field position of the main peak (341.5 mT). This echo decay shows ESEEM during the initial time range. ESEEM is also caused by the weak hyperfine interaction between the radiation-induced radicals and remote matrix protons. The two-pulse ESEEM spectra in the present study showed a decay of the observed main peak. The main peak is attributed to oxidized hydroquinone generating paramagnetic semiquinone that exists as an anion or neutral radical. The central peak of the cellulose radical exists in the same position (Kameya et al., 2011).

The peaks in the Fourier-transformed ESEEM spectrum shown in Fig. 6 appeared at 14 MHz and 28 MHz. These peaks are considered to originate from the matrix protons (Astashkin et al., 2000; Gramza et al., 1997). The protons are situated around the radical without being chemically connected. Because this pulse ESR system can detect the interaction between radicals located within a distance of 5 Å, the matrix protons should be situated within 5 Å.

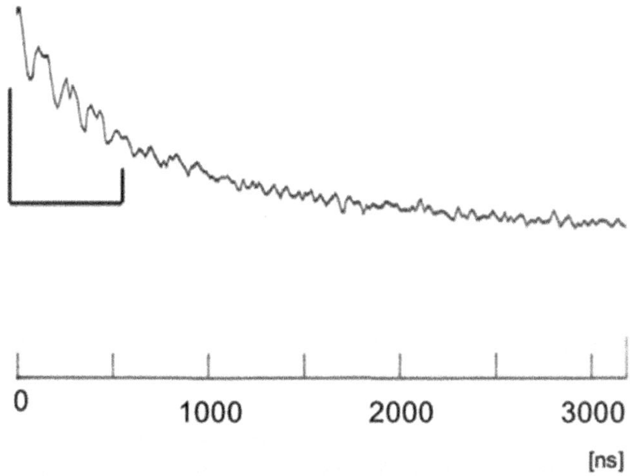

Figure 6. Two-pulse ESEEM time domain spectrum of gamma-ray-irradiated black pepper at 100 kGy with a microwave frequency of 9.480 GHz. Black pepper powder commercially available in Japan was used. The initial tau value of time between pulses was 200 ns.

3.4. Relaxation times, T_1 and T_2

We measured the relaxation times (T_1 and T_2) of the radical at 341.5 mT in the black pepper using the pulsed echo sequence. Table 2 summarizes the relaxation times (T_1 and T_2) of the irradiation-induced radicals in irradiated black pepper. T_1 was calculated using exp(-t/T) as the reverse of the spectrum that corresponds to 1-exp (-t/T). T_1 of the radical produced by 1 kGy of irradiation was 29.7 μs, and T_1 of the radical produced by 100 kGy was 36.0 μs. Relaxation time T_2 was calculated using exp (-t/T), and found to be 276 ns for 1 kGy of irradiation, and 437 ns for 100 kGy of gamma-ray-irradiation. Both the T_1 and T_2 values were enhanced by higher irradiation dosage levels., The relaxation time of the electron spin lattice T_1 depends on several parameters, and this relaxation can occur in various ways, e.g., spin diffusion. We expected the relaxation times to decrease because the dipole interaction (the distance between the radicals) weakened with increasing concentration of irradiation-induced radicals. However, the value of T_1 increased slightly, possibly because of an energy flow through the chemical bonds resulting from radical formation through irradiation. Relaxation time T_2 reflected the interaction among spins. We expected T_2 to decrease with increasing concentration of radicals, i.e., interaction between the radicals by gamma-ray irradiation. However, the values of T_2 were found to increase with the irradiation dose levels. We believe that this interaction between radicals decreased with conformational changes occurring because of bond breakage, caused in turn by the gamma-ray irradiation.

	Pulse-ESR value		CW-ESR value	
Irradiated dose (kGy)	T_1 (×10 ∫s)	T_2 (×10^2 ns)	T_1 (∫s)	T_2 (×10^2 ns)
0	3.0	2.8	3.2	1.4
1	3.0	2.9	3.3	1.5
10	3.0	3.5	3.3	1.5
25	3.2	3.5	3.4	1.5
50	3.3	4.2	3.4	1.6

Table 2. Relaxation times (T_1 and T_2) of the radiation induced radicals in irradiated black pepper

In this research, we observed relaxation behaviors of the main singlet peak at 341.5 mT. However, from the theoretical analysis of side peaks, we found that the main singlet peak overlaps with the center peak of the triplet (Kameya et al., 2011). These observations suggest that the carbon-centered radical is responsible for the overlapping of peaks and affects the relaxation times, T_1 and T_2. The relaxation times were also calculated theoretically using the CW-ESR parameter (Lund, 2009). We revealed that T_1 and T_2 from pulsed-ESR and CW-ESR were changed similarly before and after irradiation.

Relaxation times (T_1, T_2) of radicals in black pepper were also measured using pulsed-ESR. T_1 and T_2 were theoretically calculated using the CW-ESR parameter. We used pulsed-ESR and CW-ESR to calculate T_1 and T_2. T_1 and T_2 values increased according to irradiation. We revealed that T_1 and T_2 from pulsed-ESR and CW-ESR were changed similarly before and after irradiation. Before gamma-ray irradiation, no signals were observed in the cellulose. However, after gamma-ray irradiation, a singlet at $g = 2.0$ was observed, and a pair of side signals appeared simultaneously. Analysis of our theoretical spectra simulation revealed that the hyperfine interactions between the electron spin and the two protons resulted in a triplet ESR line. We compared the experimental spectra with the simulation spectra, and the results corresponded quite closely. We concluded that the twin peaks from the ESR spectra of irradiated cellulose were due to radicals produced at the C(5) site of molecular cellulose.

4. Conclusion

1. We detected ESR side signals from irradiated filter paper containing pure cellulose, and revealed the molecular mechanism of the radical formation of irradiated glucose polymers using theoretical calculations. The side signals are a precise indicator for irradiation effects in cellulose.

2. We postulated that the radical undergoes rotational motions. In simulation, the C_α-C_β and C_β-H bonds undergo simultaneous rotations.
3. We detected the same information in the spectra of both pulse- and CW-ESR. We observed a decay signal from the electron spin echo of the gamma-ray-irradiated black pepper sample at the magnetic field position of the main peak (341.5 mT). This echo decay immediately shows ESEEM during the initial time range.
4. Relaxation times T_1 and T_2 were calculated theoretically using the CW-ESR parameter. We used pulsed-ESR and CW-ESR to calculate T_1 and T_2. We revealed that T_1 and T_2 from pulsed-ESR and CW-ESR were changed similarly before and after irradiation.

Author details

Hiromi Kameya*
National Food Research Institute, Tsukuba-shi, Ibaraki, Japan

Mitsuko Ukai
National Food Research Institute, Tsukuba-shi, Ibaraki, Japan
Hokkaido University of Education, Hakodate-shi, Hokkaid, Japan

5. Acknowledgement

We would like to extend our deepest gratitude to Prof. Anders Lund (Linko¨ping University) and Prof. Sergei A. Dzuba (Institute of Chemical Kinetics and Combustion) for their valuable advice and guidance.

6. References

Adamic, K. (1968). EPR Study of Radicals Generated in Starch by Microwaves or by Conventional Heating. *Starke*, Vol.20, p. 3

Arthur, J. C. Jr. (1971). Cellulose and Cellulose Derarzves. *High Polym.*, Vol.5, pp. 977-990

Arthur, J. C. Jr. & Hinojosa, O. (1971). Oxidative Reactions of Cellulose Initiated by Free Radicals. *J. Polym. Sci.*, Vol.36, pp. 53-71

Arthur J. C.; Mares, J. T.; & Hinojosa, O. (1966). ESR Spectra of Gamma-irradiated Cotton Cellulose I and II. *Text Res. J.*, Vol.36, p. 630

Astashkin, A. V.; Mader, M. L.; Pacheco, A.; Enemark, J. H.; & Raitsimring, A. M. (2000). Characterization of Chloride-depleted Human Sulfite Oxidase by EPR Spectroscopy: Experimental Evidence for the Role of Anions in Product Release. *J. Am. Chem. Soc.*, Vol.122, pp. 5294-5302

* Corresponding Author

Baugh, P. J.; Hinojosa, O.; Arthur, J. C. Jr. (1967). ESR Study of Post-irradiation Reactions of Cellulose and Acrylonitrile. *J. Appl. Polym. Sco.*, Vol.11, p. 1139

Butz, B. & Hildebrand, A. (2006). Bleichen von Walnüssen und Pistazien: Einfluss auf den Bestrahlungsnachweis mittels ESR und GC/MS (Bleaching of Walnuts and Pistachios: Influence on Detection of Irradiation by ESR and GC/MS). *Deutsche Lebensmittel-Rundschau*, Vol.102, No.4, pp. 154-157

Cutrubinis, M.; Chirita, D.; Savu, D.; Secu, C. E.; Mihai, R.; Secu, M.; & Ponta, C. (2007). Preliminary Study on Detection of Irradiated Foodstuffs from the Romanian Market. *Radiat. Phys. Chem.*, Vol.76, No.8-9, pp. 1450-1454

Ehrenberg, A.; Ehrenberg, L.; Loftroth, G. (1962). Radiation-induced Paramagnetic Centers in Plant Seeds at Different Oxygen Concentrations. *Abh. Dtsch. Akad. Der Wiss. Berlin K1. Med.*, Vol.1, pp. 229-231

Gramza, M.; Hilczer, W.; Goslar, J.; & Hoffmann, S. K. (1997). Electron Spin Relaxation and ESEEM Spectroscopy of the Glycine Radical in Diglycine Nitrate Single Crystal. *Acta Chem. Scand.*, Vol.51, pp. 556-561

Japanese Industrial Standards Committee (1995). *Filter Paper for Chemical Analysis*, JIS P3801

Jo, D. & Kwon, H. J. (2006). Detection of Radiation-induced Markers from Parts of Irradiated Kiwifruits. *Food Control.*, Vol.17, No.8, pp. 617-621

Kameya, H.; Nakamura, J.; Ukai, M.; & Shimoyama, Y. (2011). Electron Spin Resonance Spectroscopy of Gamma-Irradiated Glucose Polymers. *Appl. Magn. Reson.*, Vol.40, p. 395

Kokko-Oblate Co., Ltd.: http://www.kokkooblate.co.jp/index.html (2012.3.30)

Lee, J.; Kausar, T.; Chung, H. W.; Jeong, I. Y.; Bhatti, I. A.; & Kwon, J. H. (2009). Properties of PSL, TL and ESR to Identify the Irradiated Sesame Seeds After Steaming. *Food Sci. Biotechnol.*, Vol.18, No.2, pp. 374-378

Lee, J.; Kausar, T.; Kim, B. K.; & Kwon, J. H. (2008). Detection of γ-Irradiated Sesame Seeds Before and After Roasting by Analyzing Photostimulated Luminescence, Thermoluminescence and Electron Spin Resonance. *Jour. Agric. Food Chem.*, Vol.56, No.16, pp. 7184-7188

Loftroth, D.; Ehrenberg, A.; & Ehrenberg, L. (1964). Analysis of Radiation Induced Electron Spin Resonance Spectra in Plant Seeds. *Radiar. Bot.*, Vol.4, pp. 455-467

Lund, A. (2009). Relaxation Time Determination from Continuous-wave Microwave Saturation of EPR Spectra. *Rad. Res.*, Vol.172, pp. 753–760

Nakamura, H.; Ukai, M.; & Shimoyama, Y. (2006). ESR Detection of Wheat Flour Before and After Irradiation. *Spectrochimica Acta*, Vol.63, No.4, pp. 883-887

Polovka, M.; Brezova, V.; & Simko, P. (2007). EPR Spectroscopy: A Tool to Characterise the Gamma Irradiated Foods. *J. Food Nutr. Res.*, Vol.46, No.2, pp. 75-83

Raffi, J. & Agnel, J. P. (1989). Electron Spin Resonance Identification of Irradiated Fruits. *Radiat. Phys. Chem.*, Vol.34, pp. 891-894

Raffi, J. & Stocker, P. (1996). Electron Paramagnetic Resonance Detection of Irradiated Foodstuffs. *Appl. Magn. Reson.*, Vol.10, pp. 357-373

Raffi, J.; Yordanov, N. D.; Chabane, S.; Douifi, L.; & Gancheva, V. (2000). Identification of Irradiation Treatment of Aromatic Herbs, Spices and Fruits by Electron Paramagnetic Resonance and Thermoluminescence. *Spectrochimica Acta*, Vol.63, No.4, pp. 409-416

Raffi, J.; Yordanov, N. D.; Chabane, S.; Douifi, L.; Gancheva, V.; & Ivanova, S. (2009). Identification of Irradiation Treatment of Aromatic Herbs, Spices and Fruits by Electron Paramagnetic Resonance and Thermoluminescence. *Spectrochimica Acta*, Vol.56, pp. 409-416

Rånby, B. (1977). *ESR Spectroscopy in Polymer Research*, Springer-Verlag, Berlin and Heidelberg, pp. 235-240

Sagstuen, E.; Lund, A.; & Maruani, J. (2000). Weakly Coupled Proton Interactions in the Malonic Acid Radical: Single Crystal ENDOR Analysis and EPR Simulation at Microwave Saturation. *J. Phys. Chem.*, Vol.104, pp. 6362-6371

Sanyal, B.; Sajilata, M. G.; Chatterjee, S.; Singhal, R. S.; Variyar, P. S.; Kamat, M. Y.; & Sharma, A. (2008). Identification of Irradiated Cashew Nut by Electron Paramagnetic Resonance Spectroscopy. *J. Agric. Food Chem.*, Vol.56, No.19, pp. 8987-8991

Shimoyama, Y.; Nakamura, H.; & Ukai, M. (2006). Free Radical Irradiated Wheat Flour Detected by Electron Spin Resonance Spectroscopy. *Spectrochimica Acta*, Vol.63, No.4, pp. 888-890

Swartz, M. H.; Bolton, J. R.; & Borg, C. D. (1972). *Biological Applications of Electron Spin Resonance*, John Wiley & Sons, Inc., New York, pp. 23-29

Thuomas, K. Å. & Lund, A. (1976). Analysis of EPR with Large Quadrupole Interaction. *J. Magn. Reson.*, Vol.22, pp. 315-325

Ukai, M.; Nakamura, H.; & Shimoyama, Y. (2006). An ESR Protocol based on Relaxation Phenomena in Irradiated Foods. *Spectrochimica Acta*, Vol.A63, pp. 879-882

Ukai, M. & Shimoyama, Y. (2003). An Electron Spin Resonance Study of Evolution of Organic Free Radicals in Irradiated Pepper. *J. Food Sci.*, Vol.68, No.7, pp. 2225-2229

Ukai, M. & Shimoyama, Y. (2003). Free Radicals in Irradiated Pepper: An Electron Spin Resonance Study. *Appl. Magn. Reson.*, Vol.24, pp. 1-11

Ukai, M. & Shimoyama, Y. (2005). Free Radicals in Irradiated Wheat Flour Detected by Electron Spin Resonance. *Appl. Magn. Reson.*, Vol.29, pp. 315-324

Yamaoki, R.; Tsujino, T.; Kimura, S.; Mino, Y.; & Ohta, M. (2009). Detection of Organic Free Radicals in Irradiated Foeniculi fructus by Electron Spin Resonance Spectroscopy. *J. Nat. Med.*, Vol.63, No.1, pp. 28-31

Yordanov, N. D.; Lagunov, O.; & Dimov, K. (2009). EPR Spectra Induced by Gamma-irradiation of Some Dry Medical Herbs. *Radiat. Phys. Chem.*, Vol.78, No.4, pp. 277-280

Chapter 14

Cellulose Nanofibers and Its Applications for Resin Reinforcements

Mariko Yoshioka, Yoshiyuki Nishio, Satoru Nakamura, Yoshiyuki Kushizaki, Ryo Ishiguro, Toshiki Kabutomori, Takeo Imanishi and Nobuo Shiraishi

Additional information is available at the end of the chapter

1. Introduction

It is widely recognized that technologies that can convert biomass resources into commercially viable materials are needed. Cellulose is a candidate among biomass due to its abundance in nature. The characteristics of cellulose, which include no thermoplasticity and being insoluble in ordinary solvents, have limited its applications. With the aim of widening its application possibilities, several works have been documented on mechanochemical treatments of cellulose in the dry state and in the wet states.[1-6] Endo et al. [1-4] developed novel cellulose composites by ball milling mixtures of cellulose and poly(ethylene glycol) (PEG). The composites are reported to have formed by insertion of PEG molecules among the cellulose molecular chains. [3,4]

Works of Kondo et al. [5] and ours [7] appeared as patent publications. In the former case, fine cellulose powder (average powder length and width: 28 and 11 μm, respectively) was pulverized in aqueous suspension by counter collision at a pressure of 200 MPa, being done once or repeatedly up to 60 times or more, using an ultra high-pressure homogenizer, Star Burst System HJP-25005(Sugino Machine Ltd.). In our case [7], cellulose micronized powder (KC flock W-400G, average particle size 24 μm) was used in the same way at a pressure of 245 MPa, being done once or repeatedly up to ten times, using a Star Burst System HJP-25080. These patents appeared independently within the period of 13 months. Kondo et al. [5,6] claimed that pulverization by an ultra high counter collision successfully decomposed the interaction within the assembly structures of cellulose molecules without any damage of the molecular structure, and finally liberated the components into various sizes in water to provide a transparent and homogeneous component−water system after repeated treatments (60 times). While having some claims in common with those of Kondo et al., [5] our work [7] concentrated more on utilization of the ultra high-pressure counter collision-treated cellulose.

In fact, our research interests were stimulated by the works of Endo et al., [1-4] and we started to research suitable methods for efficient wet mechanochemical treatment of biomass. Works were performed with the intention of determining pretreatment methods for successive efficient chemical reactions for cellulose in water. The scope of our preliminary works included using a planetary ball mill (Fritch Japan Co.,Ltd.), a bead mill (Ashizawa Finetech Ltd.), a vibration mill (Chuo Kakoki Co.,Ltd.) as well as ultra high-pressure homogenizers. The last of these, the Star Burst System HJP-25080 (Sugino Machine Ltd.) [7], was found to be the most effective and acceptable for successive chemical reaction in water.

It can be said that this counter-collision treatment at ultra high pressure is an extension of the microfibrillation of cellulose done by ITT Rayonier research group, in which dilute slurries of cellulose fibers were subjected to the repeated mechanical actions of "homogenizing" machine (Gaulin) [8,9]. The slurry was pumped at high pressure and fed through a spring-loaded valve assembly [8]. The aqueous cellulose slurry was pumped at 0.5−5 l/min. The operation pressure was 55 MPa, which is 20%−25% of those for the Star Burst System models.

Because cellulose is composed of aggregated micron-sized microfibrils within the cellulose fiber, the above finding suggests that the cellulose fibers were rapidly expanded in the surface area and grew into their substructural microfibrils by mechanical action and heat. Cellulose nanofiber (CNF) formation or its primordials is considered to be attained.

Starting from these fabricating cellulose nanofibers, authors have been conducting trials for developing related biomass composites. This review chapter addresses some of our findings, which are reported in the three following sections, each having their own experimental section and results and discussion section.

2. Fabricating cellulose nanofibers through ultrahigh pressure (245 MPa) counter collision treatments

As described in the above introduction section, Kondo et al. [5] and authors [7] found a method for fabricating cellulose nanofiber through ultrahigh pressure (200 or 245 MPa) counter collision treatments of cellulose powders in an aqueous slurry state. In our case, the morphologic features of the obtained cellulose nanofiber (CNF) were explored by use of freeze–drying technique and SEM observations. Obtained nanofiber are tend to cohere by themselves during the melt-blending with thermoplastic polymers, oligomers or prepolymers. Thus, the blending or fabricating methods have been explored on the one hand, and chemical modification of starting cellulose in order to prevent the agglomeration of the CNF on the other hand. The chemical modification can place restraint on the inducement of self-assembly of CNF by steric hinderance, electrostatic repulsion and so forth. A recent finding was that the species of the cellulose affects a lot for this fabrication of the nanofiber. These obtained features of CNF are discussed first.

2.1. Experimental

2.1.1. Materials

Micronized cellulose powder (KC flock W-400G, average particle size: 24 μm) was supplied by Nippon Paper Chemicals. "Avicel", Crystalline Cellulose was from Asahi Kasei Co. Deionized water was used throughout, including in suspension of sample powders.

2.1.2. Methods

Micropulverization of cellulose powders

Approximately 500 g of cellulose micronized powder was suspended in 20 times weight of deionized water assisted by a powerful stirring device and placed in the feed tank of the Star Burst HJP-25080 (Sugino Machine Ltd.), which was used for slurry jet counter collision. The aqueous cellulose slurry from the feed tank was divided into two flow channels, which were then pressurized to 245 MPa and are instantaneously injected from small nozzles at high speed. This allowed each fluid stream to collide in the chamber where they combined. The machine automatically permits repeated ultra high-pressure counter-collision treatments and our samples received one to ten collisions (passes).

Characterization

Scanning electron microscopy

Morphologies of the untreated cellulose as well as the freeze-dried cellulose, before and after the counter-collision treatment, were observed using a JSM-T330A (JEOL) scanning electron microscope. The samples were mounted on brass stubs and were coated with a thin layer of gold using an ion sputter coater. The freeze-drying treatment was applied for the aqueous cellulose suspensions as well as those obtained through solvent exchange described hereafter. Four freeze-drying methods were performed.

Regular freeze drying 1. Aqueous slurry of the cellulosic sample was placed in a glass bottle and placed in a conventional freeze drier (EYELA FD-550; Tokyo Rikakikai), cooled at −30°C and subjected to freeze drying.

Regular freeze drying 2. Aqueous slurry of the cellulosic sample was placed in a glass bottle and was frozen by immersing the bottle in liquid nitrogen. The sample was then subjected to conventional freeze drying.

Solvent exchange freeze drying. Imbibed water of cellulose aqueous slurry was exchanged to methanol and then to benzene or *t*-butyl alcohol, with several changes of each solvent. The gel containing benzene or *t*-butyl alcohol was subjected to regular freeze drying 2 as described above.

Rapid freeze drying. Imbibed water of cellulose aqueous slurry was exchanged to methanol and then to *t*-butyl alcohol, with several changes of each solvent. Several drops of the *t*-butyl alcohol-containing gel were put directly into liquid nitrogen stored within a dewar vessel and then subjected to the conventional freeze drying as above.

2.2. Results and discussion

At first, a planetary ball mill, a bead mill, and an ultra high-pressure homogenizer were all tried to be used to convert cellulose suspension into an appropriate nanopulverized state. The ultra high-pressure homogenizer was found to be the most suitable, because it can offer increased efficiency and flexibility over the others. In particular, the ultra high-pressure homogenizer offered the most straightforward means and ease of handling in separating the sample from the treating medium (balls, beads). The problems due to contamination of the treated sample with impurities derived from the medium can be avoided, and treatment of larger amounts of samples is possible within a shorter period of time. In this study, Star-Burst HJP-25080 was used as the ultra-high pressure homogenizer.

The additional feature of this ultra high-pressure homogenizer treatment is that the treated cellulose fiber can be homogeneously dispersed in water for a prolonged period. [5, 7] It is considered that cellulose fibers can be pulverized heavily to enlarge its specific surface area to give a much more highly hydrated state than that achieved with untreated cellulosic fiber. To visualize micropulverized states of the counter collision-pretreated cellulose, the morphological structure was studied by SEM as a function of drying method.

Figure 1 shows a SEM photograph of the counter-collision treated cellulose dried by the regular freeze drying 1 method in comparison with that of untreated cellulose. The Star Burst treated cellulose (Fig. 1b) is observed to have surface morphological features different from the untreated cellulose (Fig. 1a). The former has a fibrous surface when compared with the latter, but much of the surface area is severely coagulated to form aggregate structures. It is widely accepted that the relatively slow freezing process causes the formation of ice crystals and squeezes out cellulose microfibrils to form the aggregate structure. The growth of ice crystal on freezing is considered to be prevented by rapid cooling [10]. To confirm the descriptive validity, an SEM photograph was taken of the counter collision-treated cellulose dried by regular freeze drying 2 method (Fig.2).

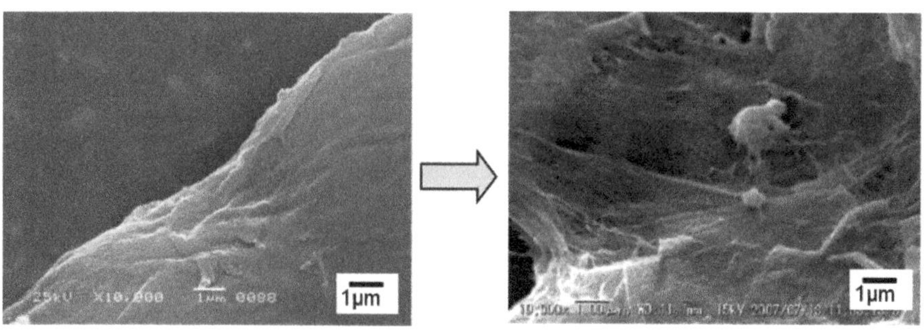

Figure 1. a, b. Scanning electron microscopy (SEM) photographs of cellulose fibers before (a) and after (b) Star Burst treatment (10 passes) followed by the regular freeze drying 1. KC flock W-400G was used as the cellulose sample.

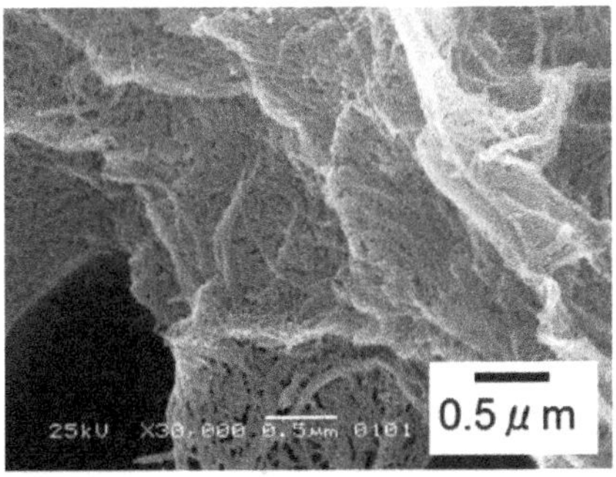

Figure 2. SEM photograph of cellulose fibers after the Star Burst treatment (10 passes) followed by the regular freeze drying 2. KC flock W-400G was used as the cellulose sample.

From Fig. 2, the Star Burst treated cellulose is observed to have surface morphological features more significantly different from that for the untreated cellulose (Fig. 1a). It loses the severe aggregate structure of Fig. 1b, but the whole surface is still regularly coagulated to form a laminate layer of thin leaves. Each thin leaf appears to be lamellae composed of cellulose microfibrils.

It is also known that preservation of cellulose fibrillar morphology is better achieved by solvent-exchange freeze drying using methanol and benzene or *t*-butyl alcohol [10]. Thus, it was planned that an additional solvent-exchange experiment was conducted. And, in addition to it, in order to accelerate the freezing process and to hinder ice crystal formation, rapid freeze drying was employed. To this end, drops of the exchanged slurry were directly put into liquid nitrogen stored in a dewar vessel after the solvent exchanges of water—methanol—*t*-butyl alcohol, and then subjected to the rapid freeze drying.

The resultant SEM photograph is shown in Figs. 3 and 4. When compared with the image in Fig. 2, it is clear that each thin leaf of cellulose fibrillar lamellae exists separately from others in these figures, noting that thin lamellae (leaves) are stuck to each other and produce laminated materials in the image in Fig. 2. This ease of coalescence among the thin leaves is caused by the easy formation of hydrogen bonding among the hydroxyl groups present on the different surfaces. That is, the thin leaves of cellulose fibrils have high density of –OH groups on the surface, which have a tendency to form hydrogen bonds between hydroxyl groups of adjacent molecules belonging to another thin leaf of cellulose fibrils.

Although the most desirable freeze-drying conditions for this experiment were employed, fibrous microfibrillated cellulose could not be directly obtained. As the ways of dealing with this problem, authors experimented with three investigations.

The first investigation was charging additional agitations for converting the thin leaves of cellulose fibrils into nanofibers. It was started from magnetic stirring during the solvent-exchange process before the freeze-drying. That is, cellulose slurry was pulverized by counter collision at a pressure of 245 MPa, after which the pulverized product was solvent-exchanged followed by freeze-drying. In the experiment, benzene was used as the final solvent of the solvent-exchange process and mechanical agitation was used during the solvent-exchange process. The SEM photograph of the corresponding freeze-dried product is shown in Fig. 5.

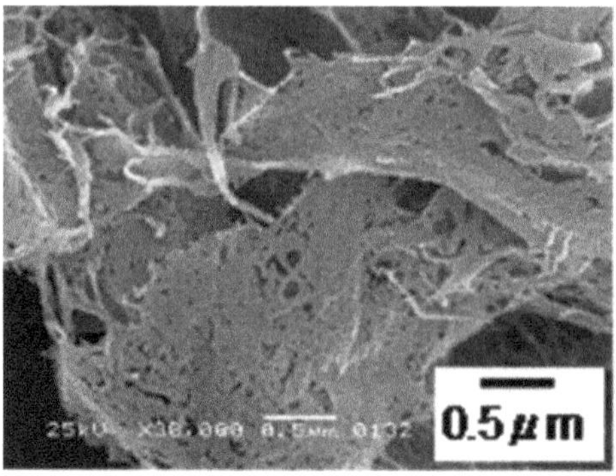

Figure 3. SEM photograph of cellulose fibers after Star Burst treatment (10 passes) followed by solvent exchange and rapid freeze drying.

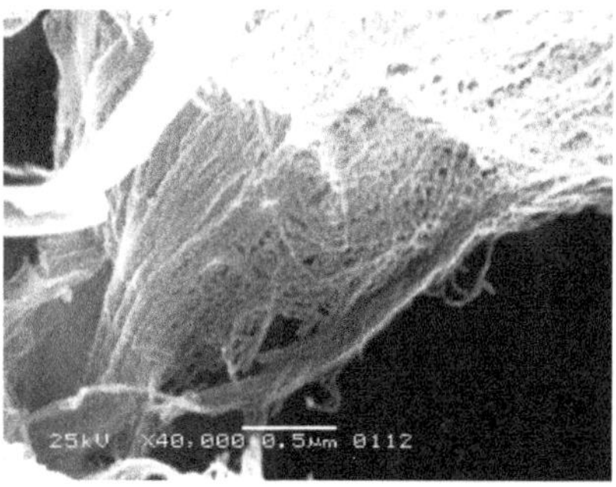

No. 112, ×40,000

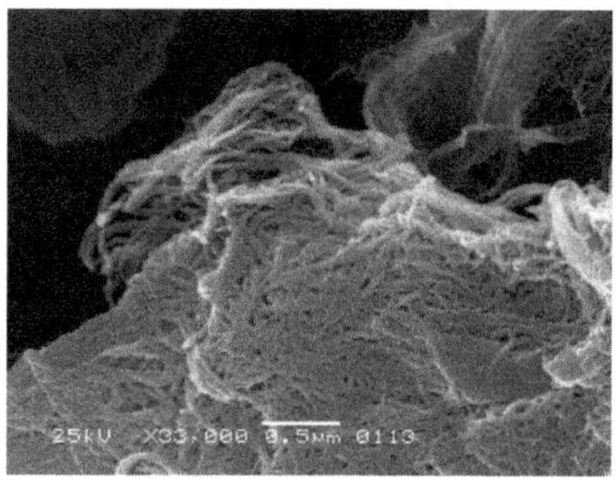

No. 113 ×33,000

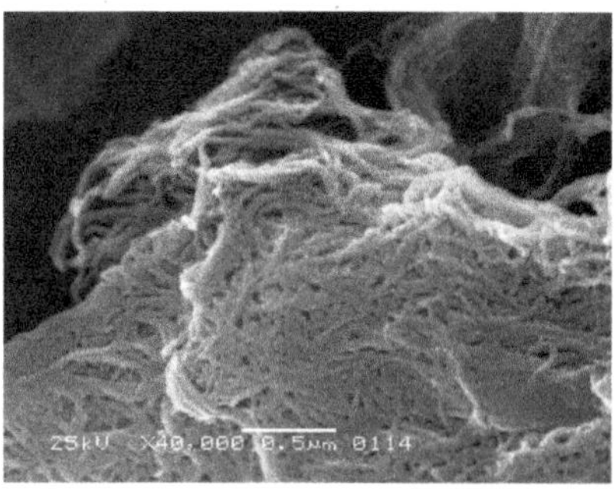

No. 114, ×40,000

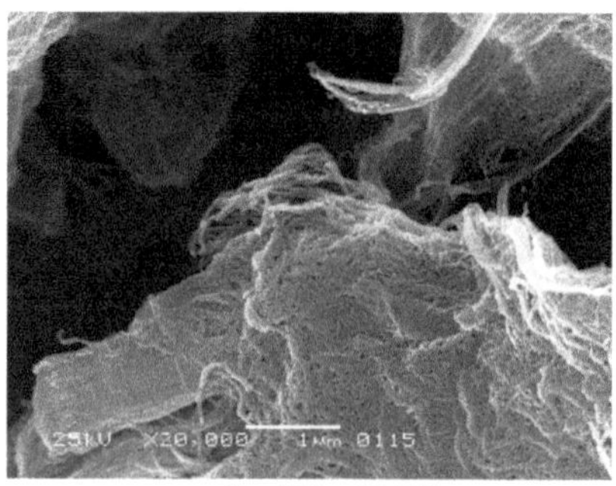

No. 115 ×20,000

Figure 4. Other examples of SEM photographs of cellulose fibers after Star Burst treatment (10 passes) followed by solvent exchange and rapid freeze drying.

Figure 5 shows that a large number of cellulose fibers are yielded on or near the surface but lamellae-like structures are observed behind. The cellulose fibers can be recognized as a nanofiber when taking the thickness (ca. 38nm) of the gold coating into consideration. Thus, it may be argued that although the untreated cellulose fibers with diameters of 8−20μm were pulverized to diameters of several nanometers, the formation of nanofibers was limited and incomplete (Fig. 5). This is despite the fact that the cellulose received the ultra high-pressure homogenizing treatment and stirrer agitation during the solvent-exchange process. When comparing the results of Figs. 2 and 5, however, it can be said that the magnetic-stirrer agitation having continued two days resulted in an insufficient formation of cellulose nanofiber.

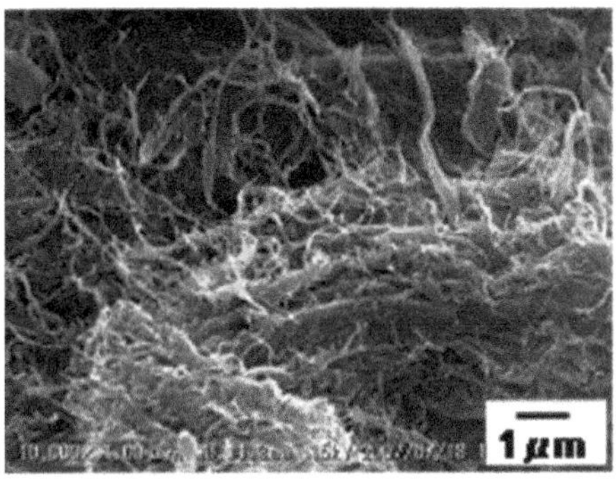

Figure 5. SEM photograph of cellulose fibers after Star Burst treatment (10 passes) followed by solvent-exchange and regular freeze drying 2.

Then, the degree of mechanical stimulations used for post-processing of the counter-collision were tried to be strengthened by use of T. K. ROBOMIX® (PRIMIX Co.), or by an ultrasonic processor (Sharp Co.; UT-205HS). An aggressive stirring with the rate of 15,000 rpm for 60 min was used by the former, and a treatment for 60 min at 40 kHz was taken for the latter.

After these treatments, both aqueous slurries were rapidly freeze dryed and SEM photograph were taken. As shown in Fig. 6, developments of CNF were clearly advanced, but even after these post-treatments, thin leaf structures of cellulose lamellae are still partially remaining. Effect of the treatment with T. K. ROBOMIX became larger compared with that with the ultrasonic processor [11].

It is considered that strong agitation followed by adept kneading with twin-screw extruder, would break the thin leaf structure of cellulose fibrillar lamellae, yielding nanofibrous microfibrillated cellulose. To examine this point, several experiments were performed and results are shown in the third section of this article.

The second investigation was to use half-esterified cellulose of dibasic acid anhydride (e.g. maleic anhydride) for the ultra high-pressure homogenizer treatment. By introducing small amounts (1~5wt%) of the half- esters on the amorphous region and microfibril surface of cellulose, followed by the counter collosion treatment, electrostatic repulsion and steric hinderance effects rose up and a rather well formation of nano-fibers was caused as shown in Fig. 7. It is understandable from this figure that the degree of conversion to nano-fiber is advanced.

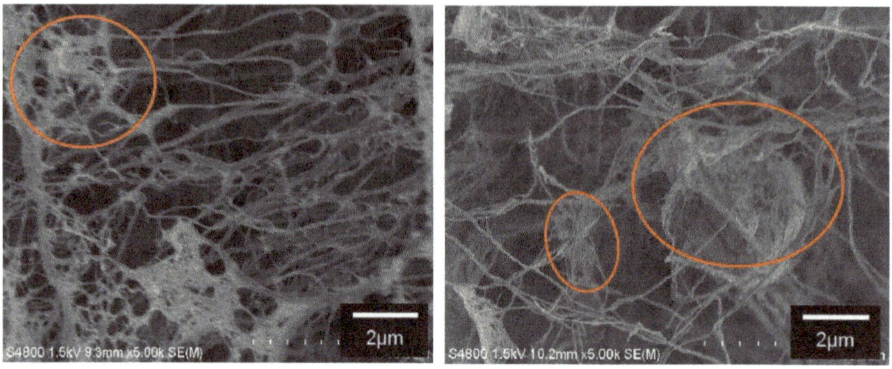

Figure 6. SEM photograph of KC flock W-400G after Star Burst treatment (10 passes) followed by an aggressive stirring with the rate of 15,000 rpm for 60 min (left), or by ultrasonic processor treatment (40 kHz) (right), and further followed by rapid freeze drying.

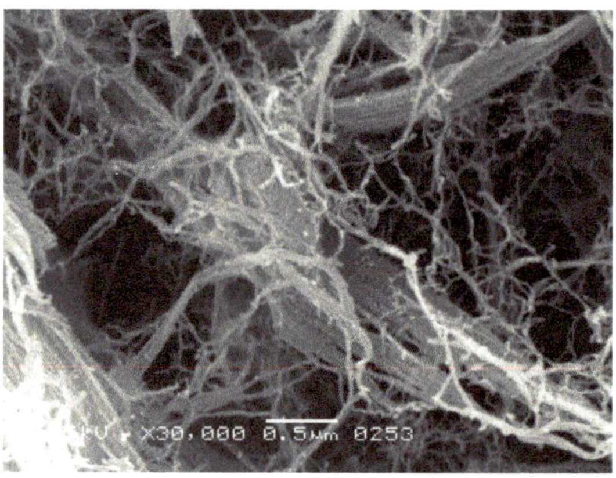

Figure 7. SEM photograph of half- esterified cellulose fibers after Star Burst treatment (10 passes) followed by rapid freeze drying.

The third investigation was to choose the species of cellulose. There are big differences in the nanofiber formation behavior between KC flock W-400G micronized cellulose powder and "Avicel", crystalline cellulose. Conversion to nano-fiber is easier for the latter when compared with the former. Even in non-treated state there are obvious differences between them as shown in Fig. 8. In Avicel's case, fibers are clearly observable and in KC flock W-400G's case, fibers are buried much more within the matrix binders.

On the other hand, however, when Avicel was used, nano-fiber formation was accomplished through 5 min of an aggressive post stirring with the rate of 15,000 rpm after the Star Burst treatment gave the tremendous effect in this cellulose species (Fig. 9). The results shown in Figs. 8 and 9, and also 6 reveals the big difference caused by selection of cellulose species.

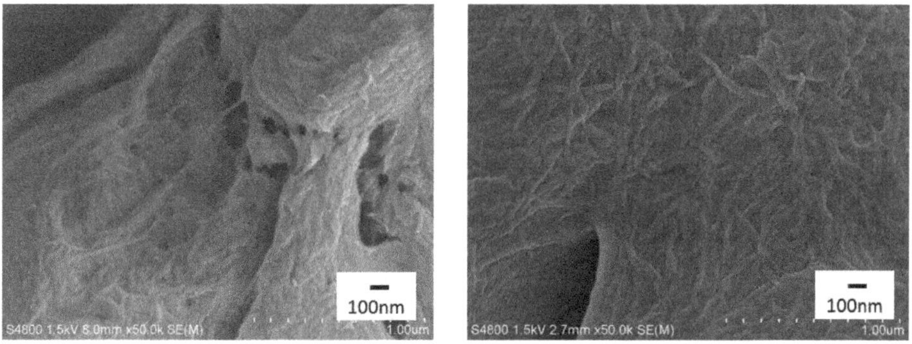

Figure 8. SEM photographs of the surfaces of untreated KC flock W-400G (right) and "Avicel"(left) powders.

Figure 9. SEM photograph of Avicel after Star Burst treatment (10 passes) followed by an aggressive stirring with T. K. ROBOMIX (stirring rate of 15,000 rpm for 5 min) and by rapid freeze drying.

Percentage amounts of nano-cellulose fiber composited to matrix resins have been an another problem. In general, investigators and scientists are tried to employ the values of 10% or more. At first, they used higher CNF concentration of 50~90 % in the corresponding composites.

On the other hand, there are arguments claiming that the smaller the particle, the smaller the clearance of the particle in particle/matrix resin composites, when the particle concentration is constant [12]. Nano-particles are those having near the smallest in dimension which cause their particle clearances the smallest. Being influenced by these arguments, authors have made choice of the concentration of CNF small, that is, less than 1 wt%. In this regard, 0.05 % Avicel nano-fiber aqueous slurry was prepared and its SEM images were taken after the rapid freeze drying (Fig. 10). The nano-fiber preparation was as follows: an aqueous mixture of Avicel fine powders having Avicel concentration of 0.05% was prepared, treated with Star Burst at 245 MPa counter collision (10 passes), followed by an aggressive stirring with T. K. ROBOMIX (stirring rate of 15,000 rpm for 5 min) and by rapid freeze drying. It is known from this figure that, though the slurry concentration is as low as 0.05 % , nano-fibers are indicating a closely-arranged morphology responsible for network structures [11]. That is, a crowded three dimensional network structure is shown.

It is known that CNF formation is much more attained when compared with Figs. 6, 7 and 10. This means again that choice of Avicel instead of KC flock W-400G makes the nano-fiber formation much easier.

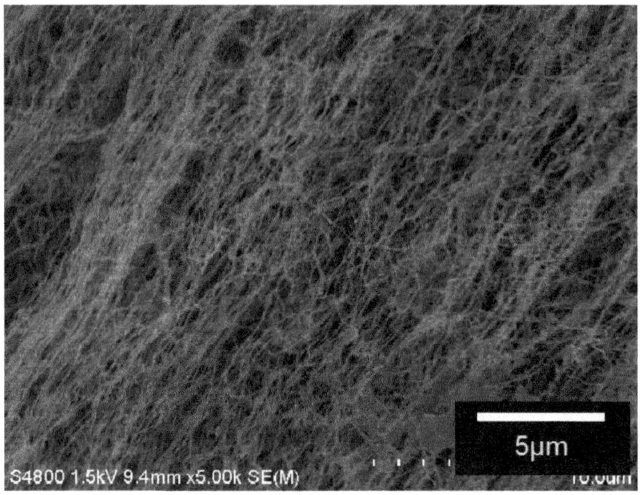

Figure 10. SEM photograph of the rapid freeze dryed 0.05% Avicel nano-fiber aqueous slurry. The sample was made from 0.05 wt % Avicel aqueous mixture after Star Burst treatment (10 passes) followed by an aggressive stirring with T. K. ROBOMIX (stirring rate of 15,000 rpm for 5 min) and by rapid freeze drying.

3. Usage of pulverized cellulosics prepared by ultra high-pressure water jet treatment in polymer nanocomposites

Using the resultant micropulverized cellulose emulsion prepared from KC flock W-400G, we attempted compositing with vinyl polymers through kneading and evaluated the effects of the counter-collision treatment. Use of KC flock W-400G has been known to result in less than perfect formation of CNF, however, adept kneading with twin-screw extruder was expected to break the thin leaf structure of cellulose fibrillar lamellae, yielding CNF, in the previous chapter. This point will be discussed in this chapter. This newly developed approach offers new opportunities for creating innovative products derived from biomass.

3.1. Experimental

3.1.1. Materials

Micronized cellulose powder (KC flock W-400G, average particle size: 24 μm) was supplied by Nippon Paper Chemicals. Deionized water was used throughout, including in suspension of sample powders. Agri-wood pellets, rice-type (BIOPOLY JOETSU Co.,Ltd.) including, Agri-wood(R)_1, Agri-wood(R)_2 (advanced second-generation) and Agri-wood(R)_3 (advanced third-generation), as well as their inflation-molded sheets were used. Both of Agri-wood(R)_2 and 3 are composed of old stockpiled rice (rice stored for almost 10 years since harvest), Wintec polyolefin WSX02 (polypropylene random copolymer; Japan Polypropylene), ultra high-pressure counter collision-pretreated cellulose aqueous slurry, and additives, in which the cellulose nanofiber concentration is around 0.05% (w/w). In the cases of Agri-wood(R)_2 and Agri-wood(R)_3 pellets, the ultra high-pressure counter collision-pretreated cellulosic slurry at around 0.05% nanofiber concentration was kneaded with Wintec polyolefin WSX02 (polypropylene random copolymer; Japan Polypropylene) and above mentioned old stockpiled rice in the presence of small amounts of additives, using a super high-torque twin-screw extruder (Supertex 77α, The Japan Steel Works), under a controlled moisture vapor release. In the case of Agri-wood(R)_1, the aqueous cellulose nanofiber slurry was not combined. Inflation-molded bags have been produced commercially from Agri-wood(R)_2 and 3 and widely used as municipal government-approved garbage bags. For our work, these garbage bag sheets were taken as test pieces. Recycled polyethylene (PE) was supplied by Ecos Factory.

3.1.2. Methods

Micropulverization of cellulose powders

The method for micropulverization of cellulose powders is the same that taken in the previous chapter.

Preparation of test pieces from inflation-molded bags of Agri-wood(R) pellets for tensile testing

Inflation molded bags were prepared using a blown film machine (Placo) from Agri-wood(R)_2 and Agri-wood(R)_3 pellets as well as first-generation Agri-wood(R) _1 pellets (not combined with cellulose nanofiber) as a control. Sample strips measuring 100×5×0.4 mm were cut from the molded sheets.

Preparation of test pieces of Agri-wood(R) pellets for tensile testing

Agri-wood(R)_1~(R)_3 pellets were molded into sheets at a prescribed temperature by hot pressing using a Toyo-Seiki 10 tons bench hot press. Each sample (approximately 3 g) was placed between polyethylene terephthalate (PET) sheets with a 0.4 mm-thick spacer. For molding, a gauge pressure ranging from 0 to 5 MPa was applied slowly over 3−5 min to allow air bubbles to dissipate. The pressure was subsequently raised to 15 MPa quickly and maintained there for 30 s. The samples were cooled for 10−15 min to room temperature by cold pressing under the same pressure. Sample strips measuring 80×5×0.4mm were cut from the molded sheets. Agri-wood(R)_1~(R)_3 pellets were directly subjected to the measurement of melt index (MI) [melt-flow rate (MFR)].

Compounding micropulverized cellulose with recycled PE

The ultra high-pressure counter collision-pretreated cellulosic slurry was kneaded with recycled PE at around 0.05 % nanofiber concentration using a super high-torque twin-screw extruder (TEX77α, The Japan Steel Works), under a controlled moisture vapor release. Compounded pellets were obtained from which films were prepared with a blown-film machine (Placo). Recycled PE with or without kneading by the Supertex 77α were used as the controls in the form of films.

Preparation of test pieces of micropulverized cellulose – recycled PE composites for tensile testing

The test pieces for tensile testing were prepared as described above. The compounded pellets of cellulosic slurry and recycled PE and their control were molded into specimens for the Charpy impact test by use of an Elject NEX110-12E (Nissei Plastic) high-performance injection molding machine.

Characterization

Scanning electron microscopy

The method for obtaining SEM photograph is the same that taken in the previous chapter. Fracture cross sections obtained by tensile elongation testing of the inflation-molded sheets prepared from Agri-wood(R)_2 and 3 were visualized by SEM.

Mesurements of mechanical properties of the Agri-wood(R), and recycled PE composite

An IM-401 Charpy-type Impact Tester (Tester Sangyo) was used for Charpy impact test, while a Universal Testing Machine AG10kNIS (Shimadzu) was used for tensile testing. Melt flow rate was measured by a Melt Flow Indexer IMC-1540 (Imoto Machinery).

Dynamic viscoelastic measurements were performed for the strips using a Rheogel-E4000 (UBM) under tension mode (frequency 10Hz, rate of temperature increase 3°C/min).

3.2. Results and discussion

Composite processing

Inflation films molded from pellets of Agri-wood(R)_2 and 3 were prepared, their tensile properties were measured, and fracture cross sections obtained were visualized by SEM. Figure 11 shows the SEM photographs for the fracture cross section of Agri-wood(R)_2 film. It is apparent from the figure that there is a scattering of fine starch of a few μm in diameter. The starch particles are almost completely exposed on the destroyed surface. This means that surface adhesion between the starch particle and the matrix resin, percolated by the microfibrillated cellulose fibers is poor. At the time of kneading, cellulose was added as Star Burst treated cellulose slurry (0.5% w/w) to give a concentration of ca. 0.05% in the pellet.

This kneading technique to prepare Agri-wood(R)_2 pellets was further advanced recently by modifying kneading conditions. Major improvement efforts have been focused on

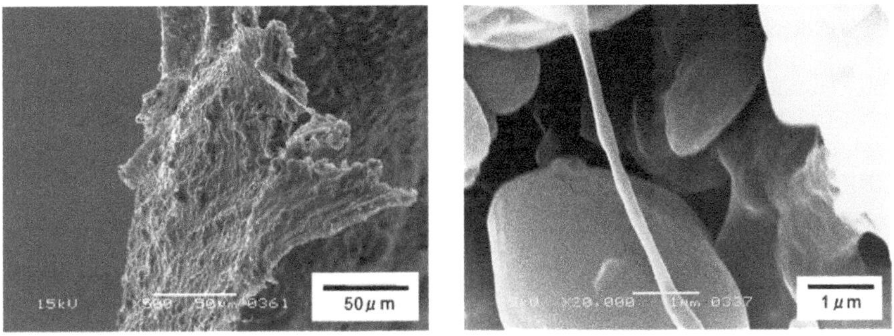

Figure 11. SEM photographs of fracture cross section of inflation film molded from Agri-wood(R)_2 [second-generation Agri-wood(R)] pellets obtained by tensile elongation test.

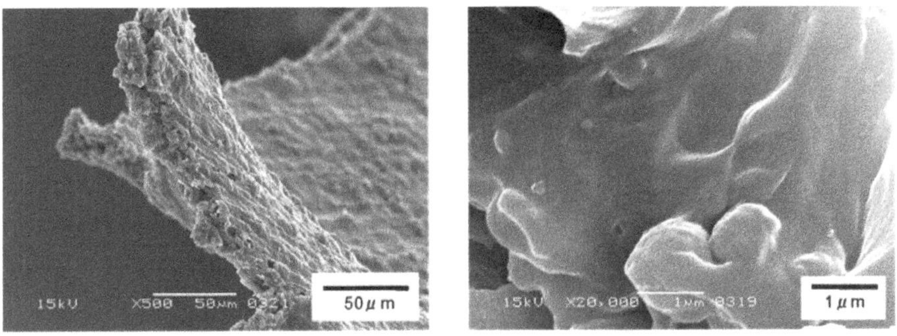

Figure 12. SEM photographs of fracture cross section of inflation film molded from Agri-wood(R)_3 [third-generation Agri-wood(R)] pellets obtained by tensile elongation test.

techniques of programmed evacuation of water vapor from the vents at pressurized, normal, and reduced pressures, as well as adopting appropriate paddle constructions [13, 14]. Such advancements were actually reflected on the SEM photographs of the fracture cross section of the film molded from Agri-wood(R)_3 pellets (Fig. 12). The starch particles are almost completely covered by the matrix resin, showing that surface adhesion between the starch particle and the matrix resin, percolated by the microfibrillated cellulose fibers is enhanced and morphological characteristics among these three components became more intimately unified. That is, cellulose nanofibers and starch particles (smaller size than those in Fig. 11) are homogeneously dispersed and combined with the matrix resin. Four photographs corresponding to and confirming Fig. 12 are shown in Fig. 13. The starch particles are also almost completely covered by the matrix resin, and penetrations of microfibrillated cellulose fibers through starch grains as well as matrix resin can be observed. That is, the mechanical interlocking caused by cellulose nanofibers are occurring, though the amounts are quite small. This argument is at least supported by the MFR values shown in Table 1. When comparing the MFR value for sample No. 2 with that for sample No. 3, the latter value is smaller than the former. The melt flowability of the matrix resin, Wintec polyolefin WSX02, is obviously much higher than that of rice starch and the microfibrillated cellulose fibers. It is thus considered that sample No. 2 composite is less blended than the advanced type and contains partly polyolefin rich portions in the composite due to the nonadvanced kneading technique. Therefore, the part of free polyolefin that is not well compatibilized with the rice starch and the microfibrillated cellulose fibers preferably passes through the melt flow indexer die head providing a higher MFR value.

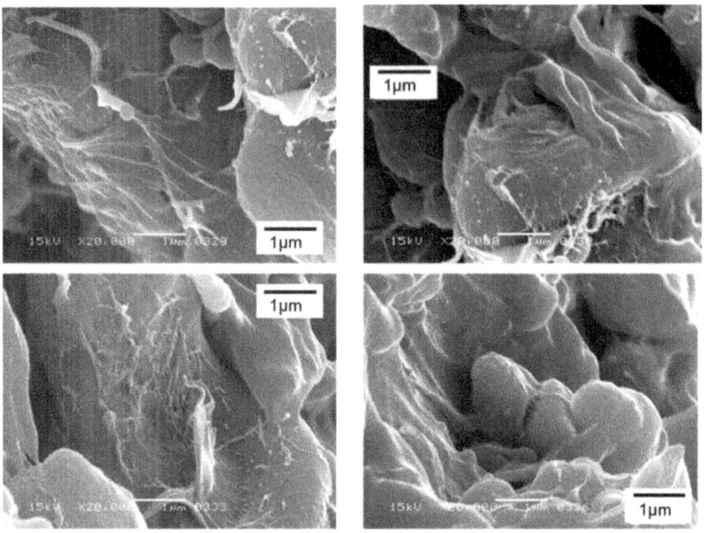

Figure 13. SEM photographs of fracture cross section of inflation film molded from Agri-wood(R)_3 [third-generation Agri-wood(R)] pellets obtained by tensile elongation test as the other examples for Fig. 12.

These observations strongly indicated that cellulose can exist more dispersedly within the nanocomposite after enhanced twin-screw kneading, in which 35%−70% amounts of rice was kneaded with 30%−65% of Wintec polyolefin WSX02 in the presence of small amount of Star Burst treated cellulose slurry. This argument is further enhanced by the results obtained from measuring the solid state properties of the molded test pieces. As shown in Table 1, all the tensile properties were enhanced in the order of the extent of nanofiber reinforcement. Especially, the value of tensile elongation at breakage can reflect the enhancement of compatibilization of the composite components of starch grain, matrix polyolefin, and microfibrillated cellulose fibers.

Concerning these findings, followings are found in the literatures. It can be stated as the first place, that cellulose nanofibers have a high density of hydroxyl groups, facilitating their agglomeration, and reducing interaction with surrounding matrix. Agglomeration is the formation of groups of cellulose fiber due to the hydrogen bonds between each of them. Overcoming strong hydrogen bond requires high energy. The high pressure and high energy in the presence of added water, afforded by the action of a twin-screw extruder were imparted to the cellulose fibers to defibrillate fibrils intensively in achieving acceptable dispersion levels [15].

No.	Sample	Tensile strenth (MPa)	Young's modulus (GPa)	Tensile elongation at break (%)	MFR[a] (g/10min)
1	Agri-wood(R)_1[b] (Control)	27.3	1.88	2.60	10.5
2	Agri-wood(R)_2[c]	30.6	2.04	3.14	9.96
3	Agri-wood(R)_3[d]	31.1	2.23	6.94	7.23

[a] 190℃, 10.58 kg

[b] First generation Agri-wood(R); no added cellulose nanofiber; polyolefin / rice = 30 / 70 (w/w)

[c] Second generation Agri-wood(R); polyolefin / rice / cellulose nanofiber = 30 / 70 / 0.05 (w/w)

[d] Third generation Agri-wood(R); polyolefin / rice / cellulose nanofiber = 30 / 70 / 0.05 (w/w)

Table 1. Mechanical properties of a series of Agri-wood(R) samples

The inherent polar and hydrophilic nature of the nanofiber and the non-polar characteristic of most thermoplastics result in difficulties in compounding the filler and the matrix. In view of these, there is a possibility that the presence of starch in the kneading system play a prominent role in achieving acceptable dispersion levels. It is possible that the starch molecule interacts with cellulose nanofibers, covers them and by that improves the dispersion of the cellulose nanofiber. Actually, Oksman *et.al.*, studied a manufacturing process of cellulose whisker / polylactic acid (PLA) nanocomposites and found by TEM observation that dispersed composite structures could be obtained when water-soluble polyethylene glycol (PVA) was included as an additive [16].

Figure 14 shows the storage modulus as a function of temperature for neat polyorefine (Wintec WSX02)(×), the first generation Agri-wood without cellulose nanofiber and the two kinds of Agri-wood/Cellulose nanofiber composites. The latter three are corresponding to Agri-wood(R)_1~3 in Table 1.

The curve corresponding to the pure matrix (x) shows typical of thermoplastic behavior. The neat polyorefine (Wintec WSX02) is a random copolymer of PP (with small amounts of PE) having a melting temperature of 125 °C. Thus, the starting temperature (room temperature) for the measurement is very close to Tg of the polymer, from which a rapid decrease in the elastic tensile modulus is occurring. Then, a rapid decrease in the elastic tensile modulus, by more than 3 decades, is observed, corresponding to the glass-rubber transition, after which it seems to go directly into flow region but the experimental setup fails to measure it.

The storage modulus in Fig. 14 shows that the three of the Agri-woods have higher moduli on the entire temperature ranges compared to that of the neat polyorefine (Wintec WSX02), which is caused by the starch filler effect. The corresponding data have shown large effect of starch fillers, which is further reinforced by the small coexistence of cellulose nanofibers, in order to obtain composites with good mechanical properties and increased stability at higher temperatures. The contents of cellulose nanofiber are remarkably small (ca.0.05%), its role can hardly be seen in the figure (Fig. 14).

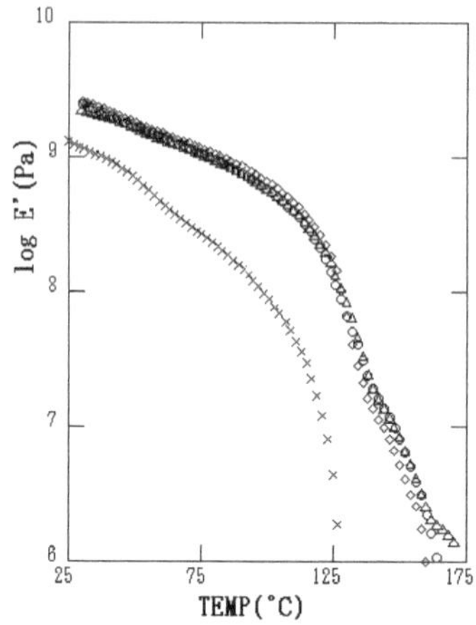

◇Agri-wood (R) 1st Generation (Control)(no added cellulose nanofiber; polyolefin/rice=30/70(w/w)); △: Agri-wood (R) 2nd Generation (polyolefin/rice/cellulose nanofiber=30/70/0.05(w/w)); ○Agri-wood (R) 3rd Generation (polyolefin/rice/cellulose nanofiber=30/70/0.05(w/w))

Figure 14. Temperature dependencies of storage modulus (E′) of neat polyorefine (Wintec WSX02) (×) and the Agri-wood/Cellulose nanofiber composites.

The influence of nanofiber existence for enhancing physical properties can be observed in the case of physical blending of microfibrillated cellulose fibers with thermoplastic polymer. An example is shown in Table 2. In this case, recycled PE was used instead of Wintec polyolefin WSX02 and kneaded with a small amount of Star Burst treated cellulose slurry. The resulting physical properties of the molded specimen were significantly improved when compared with the corresponding control, especially for impact strength and tensile elongation at breakage. This showed that nanofibrous microfibrillated cellulose was well dispersed within the molded specimen.

No.	Sample	Charpy impact strength / kJm^{-2}	Tensile strenth / MPa	Tensile elongation at break / %	MFR[1] / (g/10min)
1	Control 1 (Untreated recycled PE)	29	17	72	17
2	Control 2 (Kneaded recycled PE)	28	14	64	11
3	Cellulose nanofiber composite[2]	46	15	300	10

1) 190°C, 10.58 kg

2) Recycled PE / Cellulose nanofiber = 100 / 0.05 (w/w)

Table 2. Mechanical properties of the recycled PE / cellulose nanofiber composite

4. Usage of pulverized cellulosics prepared by ultra high-pressure water jet treatment in development of separator sheets equipped within the lithium ion battery (A feasibility study)

As shown in Table 1 in the previous chapter, the tensile properties were enhanced in the order of the additional or advantageous effect of nanofiber reinforcement. And also in Table 2, it was shown that when recycled PE was kneaded with a small amount of aqueous Star Burst treated cellulose slurry, the resulting molded specimens indicated improved physical properties when compared with the corresponding control. It is known, at the same time, that the presence of small amounts of cellulose nanofiber effects an enhancement of the heat resistance of the matrix resin [14]. These data encouraged us to apply these CNF reinforced HDPE to separator sheets equipped within the lithium ion battery.

An utilization of CNF reinforced polyolefin for the separator sheets equipped within the lithium ion battery [17] has been experimenting as a feasibility study. By use of the CNF from KC flock W-400G, half-esterified KC flock W-400G and half-esterified Avicel, the heat resistance and strength properties of the modified polyolefin sheets have been enhanced, which enables the sheets to be used for this purpose. Concerning data obtained up to now will be shown and discussed.

4.1. Experimental

4.1.1. *Materials*

Micronized cellulose powder (KC flock W-400G, average particle size: 24 μm) was supplied by Nippon Paper Chemicals. "Avicel", Crystalline Cellulose was from Asahi Kasei Co. Deionized water was used throughout, including in suspension of sample powders. Succinic anhydride was from nacalai tesque high-density polyethylene (HDPE)("HI-ZEH(HDPE) 7000F") was from Mitui Hi-Zex Co. Ltd.

4.1.2. *Methods*

Half esterification of cellulose with succinic anhydride

Half esterification of cellulose powders (KC flock W-400G and Avicel) with succinic anhydride was conducted by use of 500 mL Dispersion Mixer (MS Moriyama Co.). 100 weight parts of cellulose fine powders were mixed and reacted with 5 weight parts of succinic anhydride without solvent at 140°C for 40 min.

Micropulverization of the half esterufied cellulose powders

0.5 and 5% aqueous slurrys of cellulose or the prepared succinic anhydride half esterified cellulose fine powders were converted to nano-fiber by use of Star Burst mini (Sugino Machine Ltd.). The method for micropulverization of cellulose powders is the same that token in the previous chapter.

Compounding micropulverized cellulose (CNF) with high-density polyethylene (HDPE)("HI-ZEH (HDPE) 7000F")

The ultra high-pressure counter collision-pretreated cellulosic or half-esterified cellulosic slurry(5% and 0.5%) was kneaded with high-density polyethylene (HDPE ("HI-ZEH (HDPE) 7000F") at around 0.5 or 0.05 % nanofiber concentration with or without maleic anhydride modified PP (3%) and lubricant (BYK-P4101(1%)), using a super high-torque twin-screw extruder (Intermeshed co-rotation twin screw extruder (screw diameter= 48 mm, L/D=60), Technovel.Co.), under a controlled moisture vapor release. The compounding was done in the plant of Shiraishi Biomass Co. Ltd., located in Kyoto prefecture. Compounded pellets were obtained as shown in Table 3, from which films were prepared as described later.

Pellet No.	Composition	Cellulose	Additive	CNFcomposition amount	Slurry input amount	Slurry concentration	Star Burst treating frequency
①	HDPE			0	0		
②	HDPE			0	10phr	Water only	
③	HDPE+half-esterified CNF	KC flock W400G		0.1phr*	20phr	0.5wt%	10 passes
④	HDPE+half-esterified CNF+maleic anhydride modified PP	KC flock W400G	Maleic anhydride modified PP 3%	0.1phr	20phr	0.5wt%	↑
⑤	HDPE+half-esterified CNF	KC flock W400G		0.5phr	10phr	5wt%	↑
⑥	HDPE+half-esterified CNF	Avicel		0.1phr	20phr	0.5wt%	↑
⑦	HDPE+half-esterified CNF+lubricant	KC flock W400G	BYK-P4101 1%	0.1phr	20phr	0.5wt%	↑
⑧	HDPE+half-esterified CNF+lubricant	KC flock W400G	BYK-P4101 1%	0.5phr	10phr	5wt%	↑

*phr : per hundred resin

Table 3. CNF composited HDPE pellets

Preparation of biaxially-drawn porous film made by a phase separation method

Each compounded pellets of micropulverized non-treated or half-esterified cellulose with HDPE ("HI-ZEH (HDPE) 7000F") (Table 3) was mixed with liquid paraffin at a weight rate of 6:4, first in a beaker for the swelling of the pellets with liquid paraffin for 30 min, and then by use of a small size mixer at 180°C for 30 min, as shown in Figs. 15 and 16.

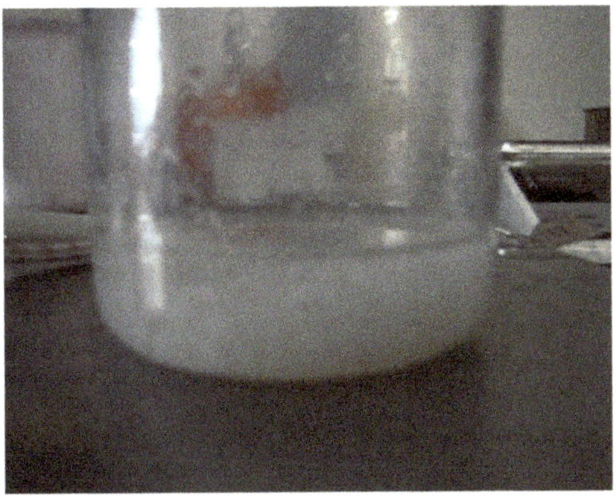

Figure 15. Swelling of the CNF compounded HDPE pellets with liquid paraffin.

Figure 16. The small size mixer used for kneading the pellets with liquid paraffin.

The kneaded products were molded with a metal mold (100mmΦ x 1mm thickness) at 180℃ under the pressure of 20 MPa. From obtained sheets, square shape sheets with a dimension of 80mm x 80mm were punched out, which were biaxially-stretched simultaneously in 5 x 5 multiples through the use of a table tenter (Fig.17). The stretching temperature was 110℃ and the rate of stretching was 8000 mm/min.

Then, from the prepared biaxially-drawn sheets, liquid paraffin was extracted with methylene chloride, obtaining biaxially-drawn porous films. These operations were done in the plant of The Japan steel works, Ltd., located in Hiroshima prefecture. Then, from the

SEM observations and the measurements of air permeability on the biaxially-drawn porous films, the potentiality of the obtained sheets for the application to the separator sheets equipped within the lithium ion battery was discussed.

Figure 17. The table tenter used for biaxially-stretching the sheets.

Characterization

Scanning electron microscopy

SEM observations were done after extracting liquid paraffin through methylene chloride, trying to obtain biaxially-drawn porous films. SEM photographs 18 and 19 were obtained by FE-SEM (Thermal field emission scanning electron microscope, JSM-7000F) (JEOL), after platinum deposition of ca. 3 nm thickness through the ion sputtering device, Super Fine Coater : ESC-101 (ELIONIX). The FE-SEM photographs were shown at the magnification ratio of 20,000.

Mesurement of air permeability

Air permeability measured by the Gurley densometer IM38-03-00 (air permeability measuring device; Testing Machines Inc.).

4.2. Results and discussion

Composite processing

As described in the experimental section, six kinds of CNF/higher molecular weight HDPE pellets have been prepared (Table 3). Also, as described in the experimental section, an array of biaxially-drawn porous films were finally prepared.

SEM observations of prepaired biaxially-drawn porous films

The characterful SEM imeges obtained for biaxially-drawn porous films are shown in Figs. 18 and 19, those film of which are prepared from pellet ④ and ⑥ in Table 3, respectively. Fibrous substances in several μm length are dispersed and unlevel pore like patterns are existing. The latter are presumed to be derived from shrinkage of the pores, caused by residual stress, arising from biaxially-stretching or liquid paraffin extraction. In this experiment, the second biaxially-stretching at certain temperature was not performed after the liquid paraffin extraction, which is usually conducted in the correspondent industrial plant. Thus, the shrinkage of the pores, caused by residual stress, would rise up and pore would be obstructed partly. From these observations, it can be said that the samples ④ and ⑥ have evidences of coming with good to excellent pores and are pointed out the possibility of expressing characteristics as separator sheets.

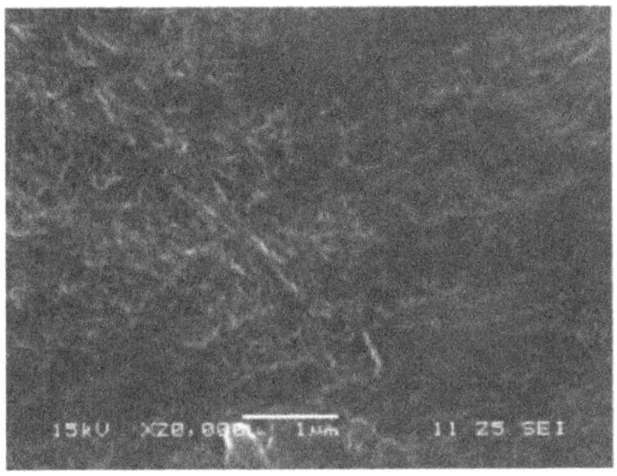

Figure 18. SEM photographs of the surfaces of biaxially-drawn porous films prepared from pellet ④ in Table 3. (20,000 magnifications)

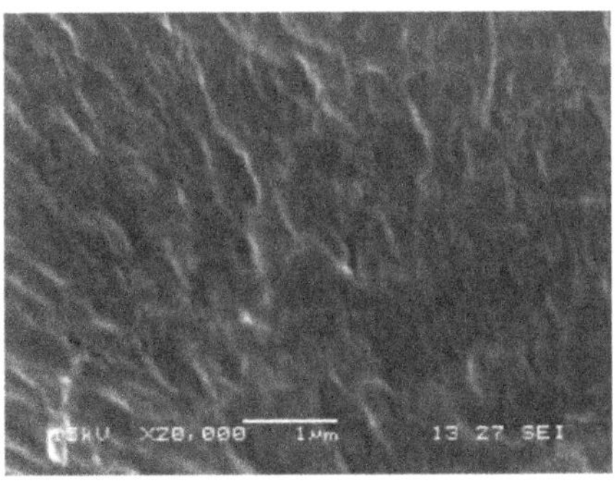

Figure 19. SEM photographs of the surfaces of biaxially-drawn porous films prepared from pellet ⑥ in Table 3. (20,000 magnifications)

Air permeability of prepaired biaxially-drawn porous films

Pore volumes and air permeabilities of the biaxially-drawn porous films prepaired from raw material pellets ③~⑧ are shown in Table 4. Pore volumes were derived from thickness and weight of 50 mm square sample sheets. As the density of samples, the density of used HDPE (HI-ZEH (HDPE) 7000F), that is, 0.95g/cm^3 , was adopted.

Air permeabilities were calculated from the obtained Gurley values, using the formulas shown in the footnote of Table 4.

Sheets prepared from raw material pellets ③ and ④ reviel Gurley values (times required 100ml air passing through the sheet) less than 1000, however, others have the values more than 3000, reflecting the small pore existence rate. Especially, the sheet prepared from raw material pellet ④ exhibits the air permeability value almost equal to that obtained for the standard sample, showing the existence of pass-through pores. The previous SEM observation also pointed out that the sheet prepared from raw material pellet ④ gave the result showing the existence of pass-through pores.

From these data, it can be said that the second biaxially-stretching at certain temperature should be performed after the liquid paraffin extraction. In addition, there is an argument in our research group that as the matrix resin, more appropriate high molecular weight HDPE should be adopted. Under these considerations and others, the related developing studies are continuing.

Pellet No.	Average thickness (μm)	Pore volume (%)	Gurley value (s/100cc)	Air permeability 1) (μm/(Pa·s)	Remarks
③	25	21.4	356	0.38	
④	13	33.8	1000	0.14	Accuracy of thickness : good
⑤	27	9.34	>3000	0.05>	
⑥	44	27.2	> 3000	0.05>	Unevenness of thickness : large (high heat stability)2)
⑦	18	18.1	>3000	0.05>	
⑧	24	11.39	>3000	0.05>	(high heat stability)2)
Standard sample	—	—	280	0.34	

1) Air permeability : p=135.3/t(time to 100ml permeable), t: Gurley value(s/100cc)

2) Heat stability: Stability during SEM observation.

Table 4. Pore volumes and air permeabilities of prepaired biaxially-drawn porous films

5. Conclusions

Cellulose fine powders were found to be effectively pulverized and emulsified by an ultra high-pressure counter collision treatment in an aqueous suspension state. The emulsion was visually stable and did not show any phase separation even after a few years of storage. The fiber diameter of cellulose was reduced from an average of 24 μm to several nanometers by the described treatment. The different effects of different freeze-drying methods were visualized by SEM observations. The coalescence of microfibrillar structures was observed easily when the freeze-drying conditions were inadequate. While the degrees of the microfibrillation done by use of homogenizers including the Star-burst were unsatisfactory for use as fillers for the preparation of polymer nanocomposites, the situation was found to become acceptable one by subsequent after treatment such as using proper kneading techniques. Presence of rice starch or hydrophilic polymers in the CNF and polyolefin kneading system has a big role in improving the dispersion of the CNF in the composite. It is possible that the starch molecule interacts with CNF, covers them and by that improves the dispersion of the CNF. To use half-esterified cellulose of dibasic acid anhydride (e.g. maleic anhydride) enhanced the formation of CNF. By introducing small amounts (1~5wt%) of the half- esters on the microfibril surface of cellulose, followed by the counter collosion treatment, electrostatic repulsion and steric hinderance effects rose up and a rather well formation of nano-fibers was caused. To choose the species of cellulose is one of the factors obtaining a high quality CNF. There are big differences in KC flock W-400G micronized cellulose powder and "Avicel", crystalline cellulose, for example. Conversion to nano-fiber is easier for Avicel when compared with KC flock W-400G. Percentage amounts of nano-cellulose fiber combined to matrix resins have been an another key factor. Usually, the values of 10% or more are preferably employed. However, there are arguments claiming that the smaller the particle, the smaller the clearance of the particle in particle/matrix resin composites, when the particle concentration is constant. Nano-particles are those having near the smallest in dimension which cause their particle clearances the smallest. Being influenced by these arguments, authors have made choice of the concentration of CNF less than 1 wt%. It is known that the pysical properties were enhanced by this nanofiber reinforcement and that the presence of small amounts of CNF effects an enhancement of the heat resistance of the matrix resin. As an application study of these CNF reinforced HDPE, preparations of separator sheets for lithium ion battery have been proceeded. From SEM observations, it is known that fibrous substances in several μm length were found to dispers and pores for lithium ion transference could be formed in the biaxially-drawn films prepaired from CNF/HDPE pellets. Further studies are required for this application. Especially, the second biaxially-stretching trials for the separator sheets after the liquid paraffin extraction are required, and HDPE with more adequately high molecular weight should be used. Under these situations and considerations, the developing studies are continuing.

Author details

Mariko Yoshioka* and Yoshiyuki Nishio
Div of Forest & Biomaterials Sci., Graduate School of Agriculture, Kyoto University

Satoru Nakamura, Yoshiyuki Kushizaki, Ryo Ishiguro and Toshiki Kabutomori
The Japan Steel Works, Ltd.

Takeo Imanishi and Nobuo Shiraishi
Shiraishi Biomass Co., Ltd.

6. References

[1] Endo T, Kitagawa R, Kabeya H, Hirotsu T (1998) Solid-phase compounding for cellulose and polyethylene glycol. In: Abstract of the 48th Annual Meeting of the Japan Wood Research Society, Shizuoka, p. 367

* Corresponding Author

[2] Endo T, Hirotsu T, Hosokawa J (1999) Novel microparticle aggregate of natural polymer and method for manufacturing the same. Japan Patent 2979135 (Applied on March 4, 1997)
[3] Endo T, Kitagawa R, Zhang F, Hirotsu T, Hosokawa J (1999) Mechano-chemical Preparation of Novel Cellulose-Poly(ethylene glycol) Composite. Chemistry Letters: 1155-1156
[4] Endo T (2000) Mechano-chemical Preparation of Novel Cellulose Composite. Cellulose Communication 7(2):63-66
[5] Kondo T, Morita M, Hayakawa K, Onda Y (2004) The wet pulverization method for polysaccharide. Japan Patent Publication Unexamined 2005-270891 (Applied on March 26, 2004)
[6] Kondo T (2005) Nano-pulverization of native cellulose fibers by counter collision in water. Cellulose Communication 12 (4):189-192
[7] Yoshioka M, Ohno T (2005) Liquid composition including dispersed biomass and method for manufacturing the same and products from the same. Japan Patent Publication Unexamined 2006-289164 (Applied on April 6, 2005)
[8] Herrick FW, Casebier RL, Hamilton JK, Sandberg KR (1983) Microfibrillated Cellulose: Morphology and Accessibility. J Applied Polymer Science: Applied Polymer Symposium 37:797-813
[9] Turbak AF, Snyder FW, Sandberg KR (1983) Microfibrillated Cellulose: A New Cellulose Product. J Applied Polymer Science: Applied Polymer Symposium 37: 815-827
[10] Jin H, Nishiyama Y, Wada M, Kuga S (2004) Nanofibrillar cellulose aerogels. Colloids Surfaces A Physicochem Eng Aspects 240:63-67 ; Alince B (1975) Porosity of swollen solvent-exchanged cellulose and its collapse during final liquid removal. Colloid & Polymer Sci., 253: 720-729
[11] Kitagawa Y (2012) Elucidation of the effect of mixed solvent composition on woody biomass / phenol liquefaction and that of conditions for compositing liquefied products with cellulose nanofiber. Thesis for master's degree 2011, Kyoto University.
[12] Chujo K (2000) "Nanocomposite World", Kogyo Tyosa-kai, Tokyo, K Books 157 20-26
[13] Ohno T, Shiraishi N. Devices and methods for manufacturing polymer composites, Japan Patent 4660528 (2011.1.7)
[14] Yoshioka M, Sakaguchi, K., Ohno, T., Nishio, Y., Shiraishi, N. Fabrication of pulverized cellulosics by ultra high-pressure water jet treatment and usage in polymer nano-composites and graft copolymerization, Journal of Wood Science, 55 (5), 335-343 (2009)
[15] Wang B, Sain M (2007) Dispersion of soybean stock-based nanofiber in a plastic matrix, Polymer International 56:538-546
[16] Oksman K, Mathew A.P, Bondeson D, Kvien I (2006) Manufacturing process of cellulose whiskers / polylactic acid nanocomposites, Composites Science and Technology 66:2776-2784
[17] Yoshioka M, Report on Japan Science and Technology Agency(JST) A-STEP "Adaptable and Seamless Technology Transfer Program through Target-driven R&D" Feasibility study (FS) stage, "Studies on application of polyolefin composited with cellulose nanofiber for the separator sheets equipped within the lithium ion battery", March 31 (2011)

Permissions

The contributors of this book come from diverse backgrounds, making this book a truly international effort. This book will bring forth new frontiers with its revolutionizing research information and detailed analysis of the nascent developments around the world.

We would like to thank Theo van de Ven and Louis Godbout, for lending their expertise to make the book truly unique. They have played a crucial role in the development of this book. Without their invaluable contribution this book wouldn't have been possible. They have made vital efforts to compile up to date information on the varied aspects of this subject to make this book a valuable addition to the collection of many professionals and students.

This book was conceptualized with the vision of imparting up-to-date information and advanced data in this field. To ensure the same, a matchless editorial board was set up. Every individual on the board went through rigorous rounds of assessment to prove their worth. After which they invested a large part of their time researching and compiling the most relevant data for our readers. Conferences and sessions were held from time to time between the editorial board and the contributing authors to present the data in the most comprehensible form. The editorial team has worked tirelessly to provide valuable and valid information to help people across the globe.

Every chapter published in this book has been scrutinized by our experts. Their significance has been extensively debated. The topics covered herein carry significant findings which will fuel the growth of the discipline. They may even be implemented as practical applications or may be referred to as a beginning point for another development. Chapters in this book were first published by InTech; hereby published with permission under the Creative Commons Attribution License or equivalent.

The editorial board has been involved in producing this book since its inception. They have spent rigorous hours researching and exploring the diverse topics which have resulted in the successful publishing of this book. They have passed on their knowledge of decades through this book. To expedite this challenging task, the publisher supported the team at every step. A small team of assistant editors was also appointed to further simplify the editing procedure and attain best results for the readers.

Our editorial team has been hand-picked from every corner of the world. Their multi-ethnicity adds dynamic inputs to the discussions which result in innovative outcomes. These outcomes are then further discussed with the researchers and contributors who give their valuable feedback and opinion regarding the same. The feedback is then collaborated with the researches and they are edited in a comprehensive manner to aid the understanding of the subject.

Apart from the editorial board, the designing team has also invested a significant amount of their time in understanding the subject and creating the most relevant covers. They scrutinized every image to scout for the most suitable representation of the subject and create an appropriate cover for the book.

The publishing team has been involved in this book since its early stages. They were actively engaged in every process, be it collecting the data, connecting with the contributors or procuring relevant information. The team has been an ardent support to the editorial, designing and production team. Their endless efforts to recruit the best for this project, has resulted in the accomplishment of this book. They are a veteran in the field of academics and their pool of knowledge is as vast as their experience in printing. Their expertise and guidance has proved useful at every step. Their uncompromising quality standards have made this book an exceptional effort. Their encouragement from time to time has been an inspiration for everyone.

The publisher and the editorial board hope that this book will prove to be a valuable piece of knowledge for researchers, students, practitioners and scholars across the globe.

List of Contributors

Jose M. Moran-Mirabal
Department of Chemistry and Chemical Biology, McMaster University, Hamilton, ON, Canada

Matheus Poletto and Ademir J. Zattera
Laboratory of Polymers (LPOL), Center of Exact Sciences and Technology (CCET), University of Caxias do Sul (UCS), Caxias do Sul,RS, Brazil

Vinícios Pistor
Institute of Chemistry, Federal University of Rio Grande do Sul, UFRGS, Porto Alegre, RS, Brazil

Viren Chunilall
Council for Scientific and Industrial Research, Forestry and Forest Products (CSIR-FFP), Congella, Durban, South Africa

Tamara Bush
Department of Chemistry, University of KwaZulu-Natal, Durban, South Africa

Per Tomas Larsson
Innventia AB, Stockholm, Sweden
Wallenberg Wood Science Centre, KTH Royal Institute of Technology, Stockholm, Sweden

Polina Mikshina, Tatyana Chernova, Svetlana Chemikosova, Nadezhda Ibragimova, Natalia Mokshina and Tatyana Gorshkova
Kazan Institute of Biochemistry and Biophysics, Kazan Scientific Centre, Russian Academy of Sciences, Kazan, Russia

Tamer A. Tabet
School of International Tropical Forestry, University Malaysia Sabah, Kota Kinabalu, Sabah, Malaysia

Fauziah Abdul Aziz
Physics Department, Centre for Defence Foundation Studies, National Defence Universiti of Malaysia, Kem. Sg. Besi, Kuala Lumpur, Malaysia

Carina Olsson
Department of Chemical and Biological Engineering, Chalmers University of Technology, Gothenburg, Sweden

Gunnar Westman
Department of Chemical and Biological Engineering, Chalmers University of Technology, Gothenburg, Sweden

Wu Lan, Chuan-Fu Liu and Feng-Xia Yue
State Key Laboratory of Pulp and Paper Engineering, South China University of Technology, Guangzhou, PR China

Run-Cang Sun
State Key Laboratory of Pulp and Paper Engineering, South China University of Technology, Guangzhou, PR China
College of Material Science and Technology, Beijing Forestry University, Beijing, PR China

Alexandar Metodiev Zhivkov
Institute of Physical Chemistry, Bulgarian Academy of Sciences, Acad. G. Bonchev Str., Sofia, Bulgaria

Ferenc Kristály
University of Miskolc, Faculty of Earth Science and Engineering, Miskolc-Egyetemváros, Hungary

Annie King
Department of Animal Science, University of California, Davis, USA

Hiroyuki Kusano and Shin-ichi Kimura
Tottori Institute of Industrial Technology, Wakabadai, Tottori, Japan

Masahiko Kitagawa
Department of Information and Electronics, Graduate School of Engineering, Tottori University, Koyama, Tottori, Japan

Hiromi Kameya
National Food Research Institute, Tsukuba-shi, Ibaraki, Japan

Mitsuko Ukai
National Food Research Institute, Tsukuba-shi, Ibaraki, Japan
Hokkaido University of Education, Hakodate-shi, Hokkaid, Japan

Ludmila Golova, Igor Makarov, Ludmila Kuznetsova, Elena Plotnikova and Valery Kulichikhin
A.V. Topchiev Institute of Petrochemical Synthesis, Russian Academy of Sciences, Moscow, Russia

Mariko Yoshioka and Yoshiyuki Nishio
Div of Forest & Biomaterials Sci., Graduate School of Agriculture, Kyoto University

Satoru Nakamura, Yoshiyuki Kushizaki, Ryo Ishiguro and Toshiki Kabutomori
The Japan Steel Works, Ltd.

Takeo Imanishi and Nobuo Shiraishi
Shiraishi Biomass Co., Ltd.